Contact Loading and Local Effects in Thin-walled Plated and Shell Structures

International Union of Theoretical
and Applied Mechanics

V. Křupka · M. Drdácký (Eds.)

Contact Loading and Local Effects in Thin-walled Plated and Shell Structures

IUTAM Symposium Prague/Czechoslovakia
September 4 – 7, 1990

Springer-Verlag Berlin Heidelberg GmbH

Distributed by Springer-Verlag, Berlin, Heidelberg, NewYork, London, Paris,Tokyo,
Hong Kong, Barcelona, Budapest, throughout the world with the exception of
Albania, Bulgaria, China, Cuba, Czechoslovakia, Hungary, Mongolia, North Korea,
Poland, Rumania, U.S.S.R.,Vietnam, and Yugoslavia

Library of Congress Cataloging-in-Publication Data
Contact loading and local effects in thin-walled plated and shell structures:
ITUAM symposium, Prague, Czechslovakia, September 4–7, 1990 / V. Křupka, M. Drdácký (eds.).
p. cm.
Includes bibliographical references and index.
ISBN 978-3-662-02824-7   ISBN 978-3-662-02822-3 (eBook)
DOI 10.1007/978-3-662-02822-3
1. Thin-walled structure--Congresses. 2. Shells (Engineering)--Congresses.
I. Křupka, V. (Vlastimil), 1927-. II. Drdácký, M. (Miloš), 1945-.
III. International Union of Theoretical and Applied Mechanics.
TA660.T5C66   1991
624.1'776--dc20   91-2823

2161/3020 543210

PREFACE

Processes taking place in the contact of two bodies and having an influence on strength and service life of machinery or of the whole structure have been in the interest of physicists and engineers for a very long time. The first scientific base had been created by H. Hertz (see *Hertz, H.: Über die Berührung, fester elastischer Körper, J. reine und angewandte Mathematik, 1882, 92, pp. 156-171*) more than one hundred years ago. Since that time (proportionally to the need of industry) countless numbers of other studies have originated. But practice always has been an inspiration for the solution being more general and profound.

At the beginning only the problem of the contact of two solid elastic bodies was solved. It was followed with the problems of local load by forces on an elastic half-plane and half-space (see *Boussinesq, J.: Application des potentials, à l'étude de l'équilibre et du mouvement des solides élastiques, Paris 1885*). The integration then has enlarged the point contact into a surface one so as to correspond to geometry, rigidity or physical properties of material in the place of contact. At present days even the effect of different frictions, origin and propagation of cracks are being solved. Not only mathematics and computers but even modern experimental methods are involved in the solution of problems.

The use of thin-wall structures has been spread within the last fifty years. The procedure of their solution has been developed similarly as for the contact problems of solid bodies. The solution of the thin-wall structures, however, has its own specialities. While in the case of solid bodies the stability has only exceptionally and effect to the overall behaviour of the bodies in contact then in case of thin-wall structures it is hardly possible to leave it out. In this branch we meet more often with non-linear problems both from physical and geometrical points of view. It makes the solution of the problems more interesting but even more complicated.

Our symposium cannot solve all the problems. It its the first of this kind and that is why it will do if it attracts the attention to some problems and initiates further research and experiments. But it introduces undoubtedly even contributions that are so elaborated that they can be directly applied for dimensioning of structures and included into Standards and Recommendations for designers.

It is true that our symposium is the first of its kind but I am convinced that not the last one. We lay the foundations for a branch needed for practice where, for the time being, the practice often uses primitive methods that can result and even actually sometimes have resulted in failures with designing. For this reason let us wish to proceed quickly forward within the frame of our mutual cooperation. And this is the sense of our symposium.

Financial support for the Symposium was generously provided by the *International Union of Theoretical and Applied Mechanics (IUTAM)*. Further, it was sponsored by the *Institute of Theoretical and Applied Mechanics of the Czechoslovak Academy of Sciences* in Prague.

The Scientific Committee was composed as followed

H. BUFLER (FRG)	J. HULT (Sweden)
P.J. DOWLING (UK)	V. KŘUPKA (Czechoslovakia) chariman
F.D. FISCHER (Austria)	R. MAQUOI (Belgium)
E.I. GRIGOLIUK (USSR)	C.R. STEELE (USA)
N. HAJDIN (Yugoslavia)	M. ŠKALOUD (Czechoslovakia)

The local organisation was possible by the participation of the staff members of the Institute of Theoretical and Applied Mechanics of the Czechoslovak Academy of Sciences in Prague

M. DRDÁCKÝ (chairman)	M. ZÖRNEROVÁ
J. KRATĚNA	P. VOKROJ
M. MICKA	

The most important work in the preparation, during the Symposium and afterwards was done by the Symposium secretaries Mrs. H. HADAČOVÁ and Miss D. KOPECKÁ.

Prague, October 1990

Vlastimil KŘUPKA
Chairman of the Symposium
Scientific Committee

VI

CONTENTS

PART I. - SHELL STRUCTURES   1

S. Al Athel   Localized buckling of elastic shells
M.S. El Naschie   in the light of modern nonlinear
                     dynamics   3

K. Brandes   Contact force distribution between
                     thin-walled tubes and saddle supports   11

H.D. Bui   Splitting of three-dimensional solution
S. Taheri   into local and global effects   20

V. Křupka   Plastic squeeze of circular shell due
                     to saddle or lug   28

A. Muc   On the contact of cylindrical shells
                     with an elastic or rigid foundation   34

Lars Å. Samuelson   Effect of local loads on the stability
                     of shells subjected to uniform pressure
                     distribution   42

J.G. Teng   A study of buckling in column-supported
J.M. Rotter   cylinders   52

A.S. Tooth   The support of cylindrical vessels on
F.A. Motashar   rigid and flexible saddles - an improved
                     analysis   62

P. Vokroj   Expert like system for cylindrical shell
                     support   70

DISCUSSIONS   75

M.T. Alimzhanov   Load carrying capacity of shells with
                     regard to cumulation of damages   77

A. Muc   Effects of fibre orientations on
                     failure modes of composite shells
                     subjected to unilateral constraints   83

B.L. Pelekh   Contact problem in the theory of
N.N. Shcherbina   anisotropic cylindrical shells   91
D.D. Matieshyn

J.M. Aribert   Modelling and experimental investigation
A. Lachal   of plastic resistance and local buckling
M. Moheissen   of H or I steel sections submitted to
                    concentrated or partially distributed
                    loading   101

M. Drdácký   Non-stiffened steel webs with flanges
                    under patch loading   111

Bo Edlund   Buckling analysis of trapezoidally
R. Luo   corrugated web under patch loading
                    using spline finite strip method   119

M. Elgaaly   Behavior of webs under eccentric
                    compressive edge loads   128

T. Höglund   Local buckling of steel bridge girder
                    webs during launching   135

A. Kolesov   Elastic and plastic states of metallic
B. Lampsi   structures under bending with allowance
V. Gusev   for local stresses   140

I. Kutmanová   Ultimate load behaviour of longitudinally
M. Škaloud   stiffened steel webs subject to partial
K. Januš   edge loading
O. Löwitová   148

I. Kutmanová   "Breathing" of longitudinally stiffened
M. Škaloud   steel webs subject to repeated partial
K. Januš   edge loading   158

C.M. Menken   Buckling of thin-walled beams under
G.M. van Erp   concentrated transverse loading   165

J. Raoul   Tests of buckling of panels subjected
I. Schaller   to in-plane patch loading
J.-N. Theillout   173

S. Shimizu   Behaviour of stiffened web plates
S. Horii   subjected to the patch load
S. Yoshida   184

I. Spinassas   Parametric study on plate girders
J. Raoul   subjected to patch loading
M. Virlogeux   192

R. Maquoi   Load-introduction resistance of column
J.P. Jaspart   webs in strong axis beam-to-column
                    joints   204

J. Studnička   Behaviour of thin-walled cold-formed
                      wide profiles under concentrated
                      loading   212

A. Venkatesh   A finite element formulation for the
J. Jirousek   analysis of local effects   218

DISCUSSIONS   229

M. Drdácký   Early prague tests on welded plate
                      girder webs under partial edge
                      loadings   231

M. Drdácký   Comments on constructions of empirical
                      formulas   235

S. Krenk   Models of thin-walled beam connections   240
L. Damkilde

PART III. - POSTERS   253

R. Bogacz   On stability of crushed columns under
Sz. Imiełowski   non-conservative contact loading   255

P. Brož   Interaction of structures   262

V.Z. Gristchak   The influence of edge effects on the
A.N. Pisanko   stress concentration around a hole
                      in a thin spherical shell   267

M. Hýča   Local effects in thin-walled beams
                      subject to bending and shear   271

E. Inan   The nonlocal theory of thermoelastic
                      plates and surface problems   277

I. Ivanova   Cylindrical shells under dynamic
                      axial impact   281

V. Kartopolsev   Stability of double-steel beams
                      elements beyond the elasticity limit   285

G. Mateescu   Local buckling effects on the supports
V. Gioncu   of continuous trapezoidal profiled
                      steel sheets   288

J. Náprstek   Nonstationary vibrations of rectan-
O. Fischer   gular plate excited by concentrated
                      force with linearly variable frequency   292

IX

B.L. Pelekh   Generalized nonlinear theory of shells
M.V. Marchuk   and plates applicable for contact
                       loads problems   296

J. Szlendak   Influence of the weld size on the
                       strength of rectangular hollow
                       section joints   300

D.D. Zakharov   Vibrations of composite plate
I.V. Simonov   containing circle delamination
                       on boundary of layers   304

X

PART I.

SHELL STRUCTURES

LOCALIZED BUCKLING OF ELASTIC SHELLS IN THE LIGHT OF MODERN NONLINEAR DYNAMICS

M.S EL NASCHIE

Sibley School of Mech. & Aero. Eng.
Cornell University
112 Upson Hall
Ithaca, N.Y. 14850, U.S.A.

S. AL ATHEL

KACST, Riyadh, Saudi Arabia.

Abstract

A critical review of the classical postulates of shell buckling theory is undertaken in the light of recent results of soliton, chaos and fractals research.

Introduction

As documented by everyday experience and countless experimental observations, many fundamental phenomena in physics, chemistry and biology are often confined to a relatively small sub-space. These localization problems have triggered a great deal of theoretical and experimental work starting with the F.P.U. model and culminating in the discovery of two of the most important subjects in nonlinear science, namely soliton and chaos [1-4]. Shell buckling also shows features of these two fundamental aspects of modern nonlinear dynamics. Buckling waves are predominantly localized and appear to spread over the shell surface in a disordered manner which furthermore seems to be extremely sensitive to certain initial conditions [5-13].

Recent research has revealed that this likeness is not coincidental. Localized buckling does indeed represent a form of homoclinic soliton in the stringent mathematical sense [5-10]. They are a spacial counterpart to Poincare's homoclinic orbits only with the fixed point laying at infinity in space. Consequently, these potentially chaotic buckling waves of permanent form will inevitably be perturbed into spacial chaos due to the unavoidable spacial imperfection even if this imperfection would have been a completely periodic and deterministic harmonic waves[5-8]. This has all been shown to be true using numerous computer simulations. As so often happens however, the results seemed to pose at least as many new questions as those they answered. We know for instance that imperfection is not deterministic, but stocastic in nature and we know that soliton owes its existence to the delicate balance between dispersion and very small nonlinear terms where as all shell equations are only an approximation, to mention only two obvious points [14].

In the present work we continue previous efforts and give a critical review of shell buckling results and equations and their consistency with the insights gained from the new results of nonlinear dynamics.

Basic ideas

The main idea upon which the work is based is that localized buckling
waves represent a separatrix loop emerging from a hyperbolic saddle point.
It is therefore most essential to begin by defining exactly what we mean
with this separatrix [5].

First let us consider the classical Euler elastica. Below the buckling
load a well known ball analogy is that of a ball at the bottom of a
concave circular valley. This equilibrium is usually described as stable,
implying that any dynamical disturbance will eventually decay with the
passing of time. However what most text books omit to mention is that for
this to be possible, friction is necessary. In the absence of friction the
motion of the ball will not be that corresponding to a spiral attractor in
phase space. It would correspond to a simple undamped harmonic oscillation that means a centre. The amplitude of oscillation of such motion will
depends on the initial conditions. Clearly this is not a stable situation
in the ordinary sense referred to as asymptotic stability [5].

Next let us consider the situation when the critical load is just exceeded. The analogy here is that of a ball at the top of a circular hill. By
contrast to the stable situation friction is completely of no consequence.
For a perfectly round ball and a perfectly circular hill the slightest
disturbance will lead to loss of stability regardless of the magnitude of
friction [5]. Such a point is referred to as a saddle. In effect a saddle
is as unstable as a node or a repeller corresponding to unstable
oscillation. Both are hyperbolic singular points. Sometimes they are both
referred to as saddles. This is very misleading however because there is a
very important difference between them. A repeller is a dynamically
unstable point which can in principle be stabilized by introducing
sufficient friction to the system. It can be reduced to a centure then
reversed to a stable focus by varying the damping. By contrast the
unstable character of the saddle can not be destroyed or changed by
damping. For this reason, relative to the saddle, the centres may be
regarded as stable. This might seem as a pathological and atypical
situation. Nevertheless there is a wide range of technically very
important problems whose stability behaviour is adequately described by
such a theory. It is also a thriving theory and usually referred to as
stability theory of elastostatics. Within the assumptions of this theory,
statical and dynamical calculation alike leads to essentially the same
results [5,11-13].

We know of course that even in structural engineering this is not entirely
true and that there are situations where we have to consider the dynamics
in order to find the correct stability behaviour. Structural engineers
encountered this for the first time in connection with the so called
configuration dependent loading or as more commonly referred to,
structures loaded by follower forces [5]. The well known Beck problem and
the Leipholz problem are examples of this kind which buckles by flutter.
That means not due to a saddle like in the case of the Euler strut, but
due to a repeller. These type of hyperbolic points are affected by damping
as we have mention earlier.

In its simplest form damping is a velocity dependent quantity. That is why dynamics must be involved in investigating such problems. But where does damping come from in a structure made of perfectly elastic material without any damping such as the Beck and the Leipholz strut. Instability in both cases is observed as we simply increase the loading. The answer becomes of course trivial when we look at the differential equations. In the Leipholz problem for instance a first derivative of the deflection appears there multiplied with the loading parameter. Mathematically this is indistinguishable from Newton's model of linear damping.

The critical point and the associated separatrix which we will be discussing here is neither the first one associated with the Euler strut or what is referred to in the mathematical literature as pitchfork bifurcation (remember that our investigations are purely statical). Nor is it the second type of a hyperbolic point associated with the Beck and Leipholz problem and referred two in the mathematical literature as Hopf bifurcation. It is thus not associated with the buckling point. Rather, it is a way of viewing the complex situation, arising after buckling has taken place, in terms of a generalization of the concept of criticality and separatrix. It is in fact an alternative explanation to that of localization based on nonlinear mode coupling [12].

Next we discuss Koiter's theory of initial postbuckling [5,11,12]. Koiter, like von Karman has seen the key to understanding shell buckling in the nonlinear effects. He realized that the incorporation of imperfection in an exact asymptotic analysis around the critical point might explain the reduction in the actual buckling loads found experimentally. Now these are all secured results which were confirmed experimentally by J. Roorda. However the experimental verification for shells was not as near to perfect as it was for the simple structures of Roorda. This was convincingly attributed however to the degree of complications associated with the higher dimensionality of shell structures and not to any qualitatively new elements playing a role. The discrepancies were considered more or less mere technical difficulties. Before the new insights and possibilities opened by deterministic chaos and soliton it was indeed virtually impossible to think of anything else. On the other hand from the theory of nonlinear dynamics and chaos we know by now exactly that nonlinearity itself generates a kind of a pseudo random behaviour which is indistinguishable from real randomness if there is indeed anything like real randomness in nature [5-8]. This random behaviour we must stress is not a consequence of the large degrees of freedom or the physical complexity of the system and it can appear in the simplest of objects with deceptively simple nonlinearity such as the logistic discrete map of R. May, or the Roessler attractor [5]. In mechanics the simplest well known system which exhibits such behaviour is the planar pendulum and the analogous problem of the infinitely long elastica [6]. Consequently we think we have sufficient reason to argue that at least a part of the randomness of the buckling configuration is due to self generated "imperfection" in addition to the undoubtedly present usual imperfection. It is our main intention here to give a very simple model for the mechanics which lead to the generation of such "internal imperfection".

Shell equations and the new mechanics

There is however another important point which might also be controversial and for which we have here more questions than answers. To start any kind of realistic analysis we need of course the governing equations. Now in driving the "generally accepted" shell equations there are a considerably more simplifications and neglections of small terms necessary than usually called for in structural mechanics. The linear differential equations are of course, in the meantime well established. On the other hand the nonlinear shell equations involve considerably more restrictive assumptions and simplifications. We know again from the study of chaos in Hamiltonian systems, the role of irrational numbers in preserving quasi periodic orbits from destruction. They play here a role similar to that played by damping in the stability of equilibrium points of dissipative systems. We know also that the destruction of orbits with strongly irrational winding numbers is an indication of transition from quasim periodicity to chaos. If all these findings would have their contraparts in the statical spacial world, then it might be that the "generally accepted" shell equations could fail completely to describe chaotic behaviour. Terms whose smallness and irrelevance have been rigorously established within traditional mechanics might be essential within the new mechanics of chaos. According to the KAM theorem for instance [5], there will be infinitely many cases where shell equations apply. However, there will also be infinitely many cases where it does not apply. In the present work we show that the inclusion of a very minor term of higher order can be at least in principle decisive for understanding localization and chaos in the simple model given here. To exaggerate the point we may say that small terms in shell equations might have a Lorenzian butterfly effect on the results [4].

The nonlinear partial differential equation of localized buckling waves

Consider the following partial differential equation

$$\propto W'''' + \sigma W'' + C_1 W - C_2 W^2 - C_3 W^3 + S \ddot{W} = 0$$

setting $S = C_2 = 0$, one finds the equation of a buckled strut on a non-linear elastic foundation. On the other hand for $\sigma = C_2 = 0$ the equation of vibration of a strut on an elastic foundation is found. The third and for the present discussion most important interpretation is that of the axsymetrical buckling of a cylindrical shell with a logarithmic radial strain definition when setting $C_3 = 0$. Consequently we may regard W as the radial displacement, σ as the axial pressure, $C_1 = E\delta/r^2$, $C = E\delta/r^3$, S is the inertia, r is the radius and t is the thickness of the cylinder, $(') = d()/dx$, $(\cdot) = d()/dt$.

Now we seek a perturbative reduction of the order of this equation using the method of strained coordinates. The simplest reduction is to use one single slow space $x_1 = \varepsilon x$ where the loading increment $\varepsilon = (\sigma^c - \sigma)^{\frac{1}{2}}$ is used as a perturbation parameter and with prior knowledge that we are dealing with a symmetric unstable bifurcation. That way and omitting the details of the calculations one finds an O.D.E. for the complex amplitude of deflection A

$$\sigma^c \overset{**}{A} - \frac{1}{2} n_c^2 A + \frac{1}{2} \gamma A |A|^2 = 0 \qquad , (\overset{+}{)} = \frac{d()}{d x_1}$$

6

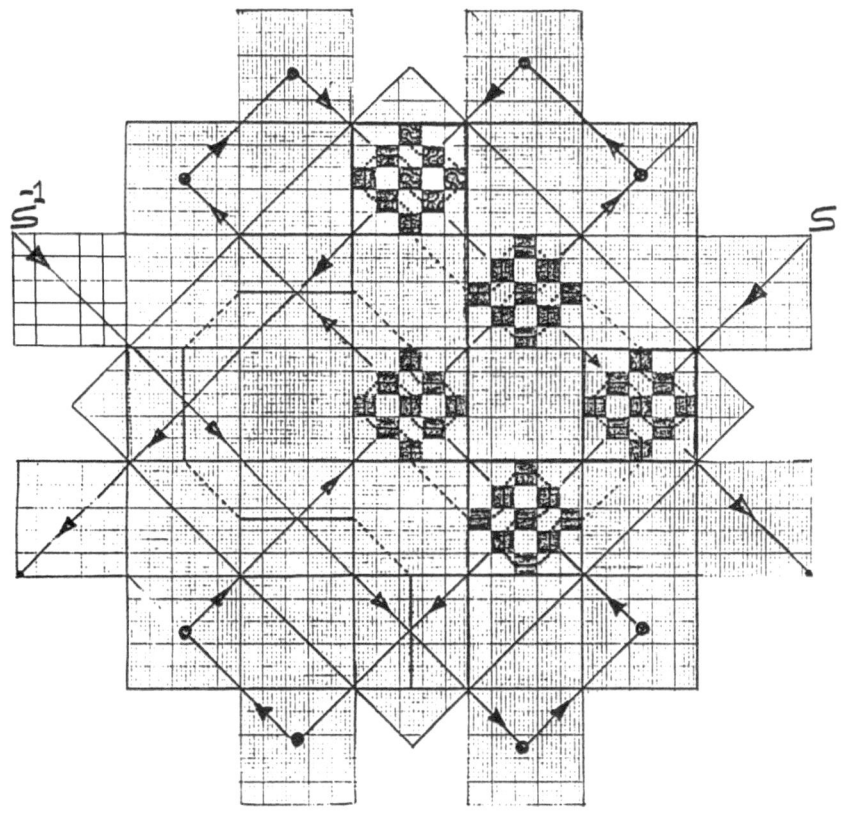

_ FIG. 1 -

Piano dynamics map representing stretching, pressing, twisting
and folding. The Hausdorf dimension of this contracting map tends
toward an exact integer $d_c = 2$.

where σ^c is the well known classical critical pressure and n_c is the associated critical wave number and $\gamma = 38/9r^4$.

Our original fourth order P.D.E. in the deflection W has thus been reduced to a second order O.D.E. in the amplitude A. In this drastic reduction the time evolution is lost. In fact the solution which may be regarded as that of a stationary nonlinear Schroedinger equation, does not differ at all from the one obtained from purely statical consideration.

Returning to our complex equation and seeking a solution we start by moving from the complex A_c to the real A_R. This is easily achieved by a simple scalling of the result using $A_R = 2 A_c$ and finding that γ is now simply $\gamma_R = \frac{1}{4} \gamma = 19/18 \, r^4$. Setting for simplicity $\alpha = c_1 = c_2 = c_3 = r = 1$ one finds the real form of the amplitude equation

$$4 \overset{\rightarrow\rightarrow}{A} - A + 19/18 \, A^3 = 0$$

This way is much easier than the some what cumbersome and unnecessary use of trigonometric functions. Now this equation although nonlinear, is elementary integrable for a localized solution. It can be shown using the inverse scattering transformation that this localized solution has the properties of a soliton, that means a localized wave of permanent form

$$A = 6/\sqrt{19} \, \text{sech} \, 19/12 \, X_1$$

which is a "sech" localized solution. Before going any further two important observations may now be made. First the spacial separatrix interpretation of the soliton shows that this solution must be homoclinic, that is to say any perturbation must lead to severe dependency on initial conditions and consequently chaotic distribution of this envelope soliton as shown in Ref. [16]. Second, the same conclusion may be made by a close look at the last amplitude equation which represents mathematically a Duffing equation with negative linear spring. Consequently a sinusoidal perturbation of this equation will lead in the appropriate range of parameters to chaotic behaviour. This chaotic behaviour, well known from dynamics, will manifest itself in our infinitely long cylinder spacially as a randomly spaced sequence of envelope soliton. The envelopes are of course only mathematically present and only behave theoretically like a soliton wave enveloping the deformation wave. A comprehensive treatment of localization and some highly interesting results may be found in a paper by C. Lange and A. Newell [15].

Imperfection, piano dynamics and strange nonchaotic behaviour

We have already mentioned that even determanistic periodic imperfection can produce chaos and that imperfection is of course noisy. However, recent research has revealed that noise can eliminate chaos and in the presence of dissipation, could lead to strange nonchaotic behaviour [16]. This recently discovered phenomena can in turn be understood in terms of cantor set-like objects with positive Lebesgue measure. It would correspond to a horse shoe map which is twisted as well as stretched and compressed. The immediate consequence of the piano-like dynamics is that the Hausdorf dimension is $d_c = 2$. The mechanics underlying this phenomena is shown in Fig. (1).

Estimation of the load carrying capacity of elastic shells

The idea of obtaining an estimation of the lower stability limit based on the theory of geometrical bending was outlined on several previous occasions [9,10]. Using this theory and a damped mode of deformation

$$W/a = \bar{e}^k (\cos kx + \sin kx)$$

where a and k are constants, a lower stability limit $\sigma^c \approx 0.322\ E\delta/r$ was found. It is therefore of interest to repeat the same analysis based on a sech soliton solution. Thus we may use the following modulated sech function

$$W = a \cos kx \operatorname{sech} kx$$

That way one finds

$$\sigma^c = 0.271\ E\delta/r$$

Although this value is less than half of the classical buckling load $\sigma = 0.6\ Et/r$ it is still not satisfactory. Nevertheless considering that a horizontal tangent at x=0 as a geometrical compatibility condition imposed on such a simple Ritz function is very restrictive, this is not a bad result. It could probably be improved essentially by relaxing the tangent condition.

Conclusion

In analogy to dynamical chaos, spacial chaos may be generated by the introduction of spacial deterministic fluctuation that is to say imperfection, to an analogus spacial homoclinic orbit. The homoclinic soliton loops of the Euler elastica is by now a well known example. In the case of a shell-like structure we hope to have shown here that it is reasonable to assume that at least a chaos related phenomena is in principle not only experimentally observable but also mathematically possible. The possibility of nonchaotic strange behaviour was also considered. Finally a simple method for estimating the load carrying capacity of a cylindrical shell is given.

References

1. Fermi, E., Pasta J. and Ulam S.: Studies in nonlinear problems, I. Los Alamos report LA. (1955) pp. 1940.
2. Tabor, M.: Chaos and integrability in nonlinear dynamics. Wiley, New York (1989).
3. Zabusky, N. and Kruskal, M.: Interaction of solitons in a collision-less plasma and the recurrence of initial states. Phys. Rev. lett. 15, (1965) pp. 240.
4. Lorenz, E.: Deterministic nonperiodic flow. J. atmos. sci. 20, (1963) pp. 130.
5. El Naschie, M.S.: Stress, stability and chaos in structural engineering. McGraw Hill, London (1990).
6. El Naschie, M.S. and Al Athel, S.: On the connection between statical and dynamical chaos. Z. fuer Naturforschung A44 No. 7, (1989) pp.645-650.

7. El Naschie, M.S., Al Athel, S. and Walker A.C.: Localized buckling as statical homoclinic soliton and spacial complexity. In proceeding of IUTAM symposium: Nonlinear dynamics in engineering system. Editor: W. Schiehlen. Springer, Berlin (1990).

8. El Naschie, M.S.: On certain homoclinic soliton in elastic stability. J. of the phys. Soc. of Jap. 58 No.12. December (1989) pp. 4310-4321.

9. El Naschie, M.S.: High speed deformation of shells. In proceeding of IUTAM symposium: Hight velocity deformation of solids. Editor: K. Kawata. Springer, New York (1977) pp.363-376.

10. El Naschie, M.S.: Local postbuckling of compressed cylindrical shells. Proc. Instn. Civ. Engrs. part 2, No.59 (1975) pp.523-525 and No.61 (1976) pp.483-488.

11. Koiter, W.T.: Elastic stability and postbuckling behaviour in nonlinear problems. Editor: R.E. Langer.University of Wisconsin press, Madison (1963) pp.257-275.

12. Sewell, M.J.: Maximum and minimum principles. Cambridge University press. Cambridge (1987).

13. Koiter, W.T.: Omzien in Verwondering maar niet in wrok. In: Trends in solid Mechanics. Editor: J.F. Besseling and A.M.A. van der Heijden. Delft University press, Delft (1979) pp.237-246.

14. Drazin, P. and Johnson, R.S.: Soliton. Cambridge University press, Cambridge (1989).

15. Lange, C.G. and Newell A.C.: The postbuckling problem for thin elastic shells. SIAM J. Appl. Math. 21 No.4. Dec. (1971) pp.605-629.

16. El Naschie, M.S. and Kapitaniak, T.: Soliton chaos models for mechanical and biological elastic chains. Physics Letters A, vol. 147 No. 5,6, 16 July (1990) pp. 275-281.

CONTACT FORCE DISTRIBUTION BETWEEN THIN-WALLED TUBES AND SADDLE SUPPORTS

K. BRANDES

Bundesanstalt für Materialforschung und -prüfung (BAM)
Unter den Eichen 87
D-1000 Berlin 45
F.R. Germany

Summary

Hybrid analytical-experimental techniques imply the fruitful
coupling of experiments with the analytical-numerical treatment
of a physical problem.
A special hybrid technique is applied to solve the question of
how is the distribution of contact forces between a saddle support and a thin walled tube on this support.
The tube lies on a saddle support the radius of which is slightly larger than the outer radius of the tube. Consequently, the
region of contact varies with the total load.
To find the force distribution, tests have been performed and
strains at several points of the tube wall have been measured.
For the evaluation, a hybrid method was applied.

1. Introduction

Very often, tubes and vessels are supported on saddle supports
and the local stresses are substantial for the behaviour and the
reliability of these thin-walled structures. The radius of the
saddle support is somewhat larger than the outer radius of the
tube or vessel. Thus, the region of contact between the tube's
wall and the saddle support, which can modelled as line support,
varies with the total support force, fig. 1.

This problem has been solved analytically [1] [2]. Additionally,
tests have been performed to confirm the analytical solution.
The distribution of the contact forces is qualitatively given in
fig. 1. For one of the tests, the method is described and results are presented.

2. Test

A steel tube of 508 mm outer diameter and 10 mm thick wall has
been tested, supported as a simple beam at the two ends, 3,992 m
span, and loaded centrally. The force was transferred by a
nearly rigid gauge to the tube (fig. 2). The gauge, made from
20 mm thick steel sheet, had a diameter of 514 mm.

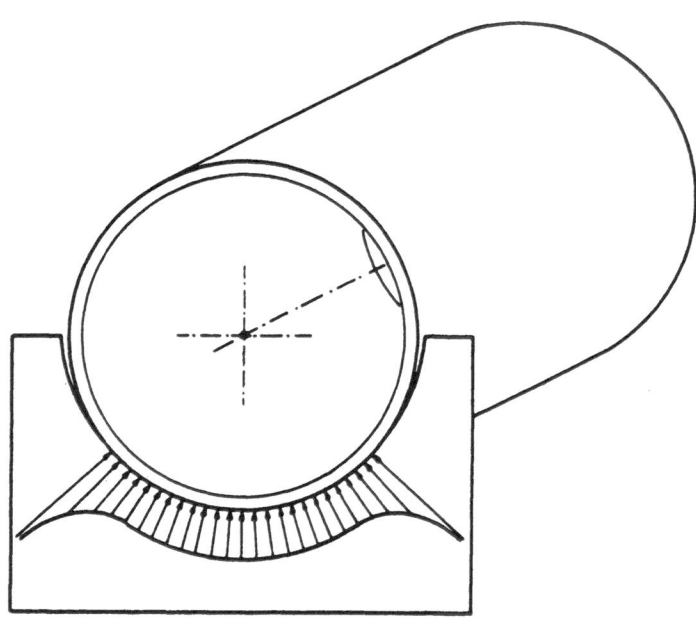

Fig. 1.
Support of a
circular
cylindrical
tube on a
saddle sup-
port the
radius of
which is
somewhat
larger than
the outer
radius of
the tube.

The strains were measured at 51 points on the outside and inside
surfaces of the wall with strain gauges, at 17 points with rosettes (3 strain gauges of 120° angle between) (fig. 3).

For several values of the load, measurements have been carried
out and the distribution of the internal forces of the shell was
evaluated. Only for measuring points with rosettes, the bending
moments and the membrane forces could be evaluated referring to
the analytical functions of the boundary value problem.

Fig. 2.
Test arrangement

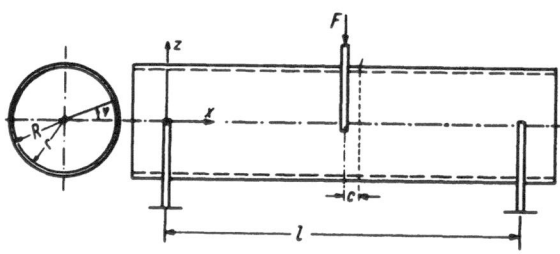

A typical stress distribution in a cross-section c = 6 cm dis-
tant from the center (see fig. 2) is presented in fig. 4. It can
be recognized that the stress distribution is dominated by the
membrane forces in the tube wall which can be calculated applying simple beam theory. Besides this, the bending moments in the
wall can not be neglected.

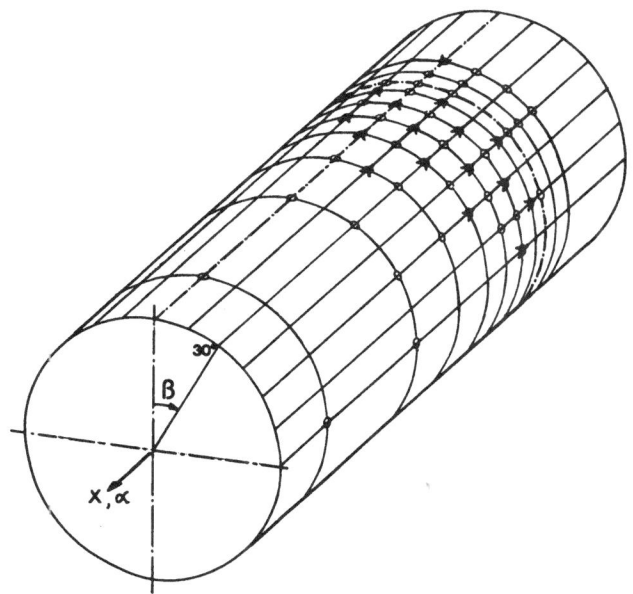

Fig. 3.
Location of
the measur-
ing points
on the tube
wall. (The
drawing is
180° turned
compared
with fig.
1.) The
rosettes are
characterized
by full
circles.

The stress distribution presented in fig. 4 was calculated by
applying constraint equations to the regression formulation to
ensure that the general equations of equilibrium for the beam
(fig. 2) are satisfied [3].

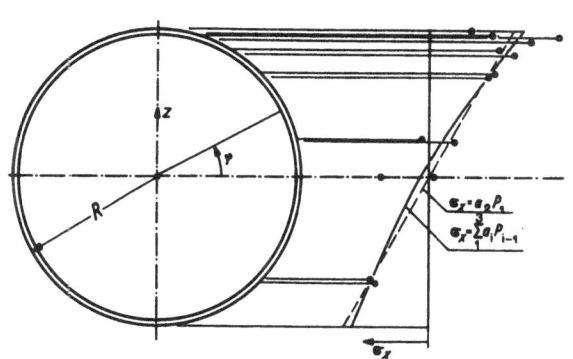

Figure 4.
Stress distribution at
the inside and outside
surface of the tube
wall for F = 294 kN.

13

3. Evaluation of the contact force distribution

In a first step, the boundary value problem of the thin elastic
circular cylindrical shell has to be solved. Here the solution
in terms of a stress function is applied as established by
Wlassow [4], especially of the form as given in [5].
One of the boundaries is defined for the central cross section
of the tube, α = 0. The transverse force there is (for definitions see fig. 5)

$$\overline{Q}_1(\alpha =0) = Q_1 - \frac{1}{R} \frac{\partial M_{12}}{\partial \beta}\bigg|_{\alpha =0} = \sum_n p_n \cos n\beta, \qquad (1)$$

a Fourier Series in β-direction. The correlated function of the
radial displacement w for α = 0 is

$$w(\alpha =0) = \sum_n w_n \cos n\beta$$

$$= \sum_n p_n H_n \cos n\beta \qquad (2)$$

where the H_n imply the solution of the shell equations [5].

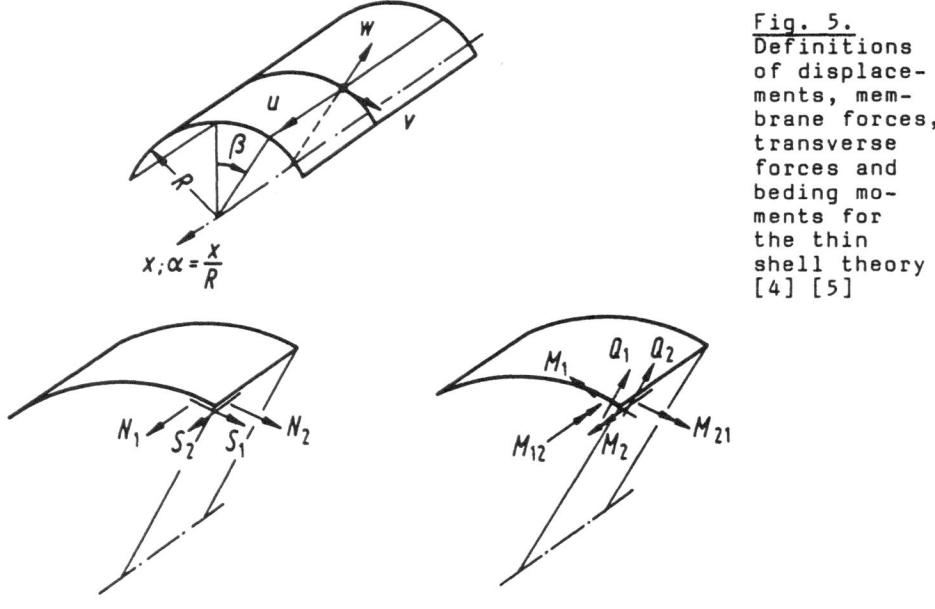

Fig. 5.
Definitions
of displace-
ments, mem-
brane forces,
transverse
forces and
beding mo-
ments for
the thin
shell theory
[4] [5]

As a first attempt to come to a numerical solution, the coefficients p_n can be evaluated by calculating the strains at the
measuring points as functions of p_n and apply a classic regression analysis. This solution is insuffient because it turned out

that the distribution of contact forces is such that also in the region of no contact forces occur.

The result was improved slightly after setting up several additional dummy measuring points in the region of no contact because in this region no measurement data were available. The strains in these points were set to be as large as calculated underlying the simple beam theory for the tube. Nevertheless, the results were improved only slightly.

However, there is a way, to formulate this mixed boundary value problem as a stress boundary value problem with a shape function for $\overline{Q}_1$ in the region of contact, $-\beta_0 \leq \beta \leq \beta_0$, and to require that the contact forces vanish in the region $\beta_0 < \beta < -\beta_0$ as suggested in fig. 6: In the region of contact, $-\beta_0 \leq \beta \leq \beta_0$, a simple polynomial is used

$$\overline{Q}_1(\alpha = 0) = A_0 + A_2\beta^2 + A_4\beta^4 + \ldots = \sum_{j=0,2\ldots} A_j\beta^j. \qquad (3)$$

In the region without contact, $\overline{Q}_1$ vanishes. The Fourier series expansion of (3) yields

$$\overline{Q}_1(\alpha = 0) = \sum_n p_n(A_j) \cos n\beta \qquad (4)$$

$$\overline{Q}_1 = \sum_j \left\{ A_j \sum_n K(\beta_0, j, n) \cos n\beta \right\} \qquad (5)$$

The parameters A_j are determined by means of a hybrid experimental-analytical method referring to the measured values of strain at the measuring points (α_k, β_1). This procedure is a generalized indirect regression method including linear constraint equations in terms of the unknown parameters A_j, [2] [6] [7].

Let be $\varepsilon_{kl}^{(A_j)}$ the strain at (α_k, β_1) related to $A_j = 1$ and $\varepsilon_{kl}^{(meas)}$ the measured strains there, than the set of linear equations

$$\begin{bmatrix} \varepsilon_{11}^{(A_0)} & \varepsilon_{11}^{(A_2)} & \cdots\cdots \\ \varepsilon_{12}^{(A_0)} & \varepsilon_{12}^{(A_2)} & \cdots\cdots \\ \vdots & \vdots & \\ \vdots & \vdots & \varepsilon_{kl}^{(A_j)} \cdots \\ \vdots & \vdots & \cdots\cdots\cdots \end{bmatrix} \left\{ \begin{matrix} A_0 \\ A_2 \\ \vdots \\ \vdots \\ \vdots \end{matrix} \right\} - \left\{ \begin{matrix} \varepsilon_{11}^{(meas)} \\ \varepsilon_{12}^{(meas)} \\ \vdots \\ \vdots \\ \vdots \end{matrix} \right\} = \left\{ \begin{matrix} r_{11} \\ r_{12} \\ \vdots \\ \vdots \\ \vdots \end{matrix} \right\} \qquad (6)$$

can be stated. The strains $\varepsilon_{kl}^{(A_j)}$ at the points (α_k, β_1) are calculated underlying the bending theory of thin shells as given in [5]. The elements r_{kl} are the errors or residuals at the measuring points. In matrix terms, equ. (6) is formulated as

$$[E] \{A\} - \{E^{(meas)}\} = \{r\} \qquad (7)$$

The equations (6) resp. (7) are the observational or error equations as stated in the theory of observations [7]. The least squares solution

15

$$\sum r_{kl}^2 = \min \qquad (8)$$

generates the Gauss-Transformed, the normal equation in the theorie of observations,

$$[E]^T[E]\{A\} - [E]^T\{E^{(\text{meas})}\} = 0, \qquad (9)$$

$[E]^T$ being the transposed of $[E]$. The (one) constraint equation is derived from the overall equilibrium that the applied force $\overline{F}$ equals the force as is given by the parameters A_j,

$$\overline{F} + \sum_j A_j\, a_j = \overline{F} + \{a\}^T\{A\} = 0 \qquad (10)$$

Following the method of variational principle the constraint equation can be added to equ. (8)(Lagrangrian multiplier λ),

$$\sum [r_{kl}^2] + \lambda[\overline{F} + \{a\}^T\{A\}] = \min, \qquad (11)$$

the so called Lagrangian [7]. Than, the Minimum of equ. (11) is derived from

$$\begin{bmatrix} [E]^T[E] & \vdots & \{a\}^T \\ \hline \{a\} & \vdots & \end{bmatrix} \left\{ \begin{array}{c} A \\ \hline \lambda \end{array} \right\} = \left\{ \begin{array}{c} [E]^T\{E^{(\text{meas})}\} \\ \hline \overline{F} \end{array} \right\} = \{0\} \qquad (12)$$

In particular, the constraint equation (12) is [5]

$$\overline{F} + 2R \sin \beta_0\, A_0 + 2R[(\beta_0^2-2)\sin \beta_0 + 2\beta_0 \cos \beta_0]\, A_2 + \ldots = 0$$

and, in the example, $R = 24,9$ cm, $\beta_0 = 74°$, $F = 366$ kN:

$$366 + 45,7\, A_0 + 19,6\, A_2 = 0.$$

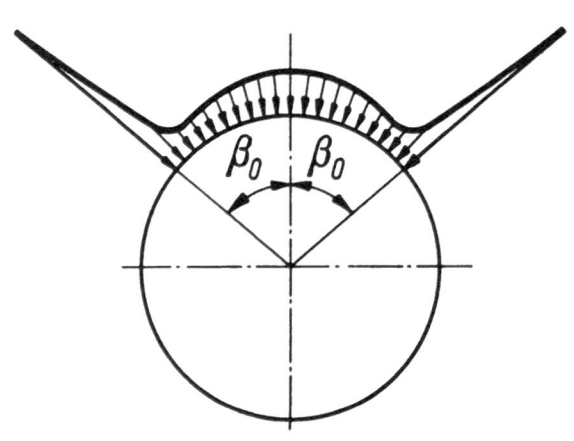

Fig. 6. Contact force distribution between saddle and tube wall for $\alpha = 0$ (central cross section of the tube as shown in fig. 2), equ. (1).

16

4. Results

Now, equ. (12), for only two parameters, A_0 and A_2 is read

$$
\left[\begin{array}{cc|c} & & 45,7 \\ [E]^T[E] & & \\ & & 19,6 \\ \hline 45,7 & 19,6 & 0 \end{array}\right]
\left\{\begin{array}{c} A_0 \\ A_2 \\ \hline \lambda \end{array}\right\}
-
\left\{\begin{array}{c} [E]^T\{E^{(meas)}\} \\ \hline 366 \end{array}\right\}
= \{0\}
\qquad (13)
$$

The result is (including standard deviation)

$$A_0 = -7,7 \ kN/cm \pm 3,1$$

$$A_2 = -0,8 \ kN/cm \pm 0,5$$

As a consequence of the large scatter of the measuring data, the standard deviation of the determined values of A_0 and A_j is rather large. Thus, the result is not quite satisfactory. However, the mean values are roughly in accordance with numerically evaluated ones [5].

The distribution of the membrane force N_2 in circumferential direction is plotted in fig. 7, conveying an impression of the scatter.

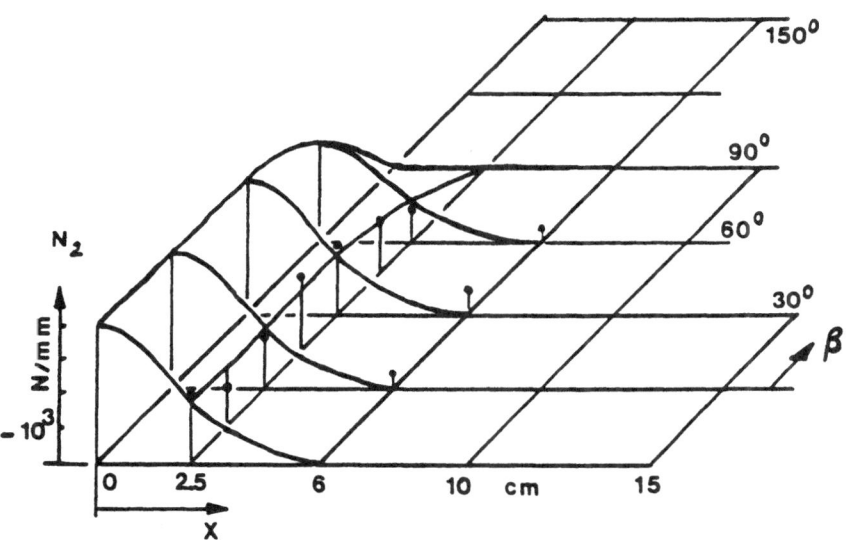

Fig. 7. Distribution of the membrane force N_1 for a load of F = 366 kN (see fig. 2 and fig. 5)

Some of the results in terms of shell internal forces are presented in the figures 8 and 9. Only the membrane force N_1 does not tend to zero with α because of the inhomogenous part representing simple beam theory load transfer.

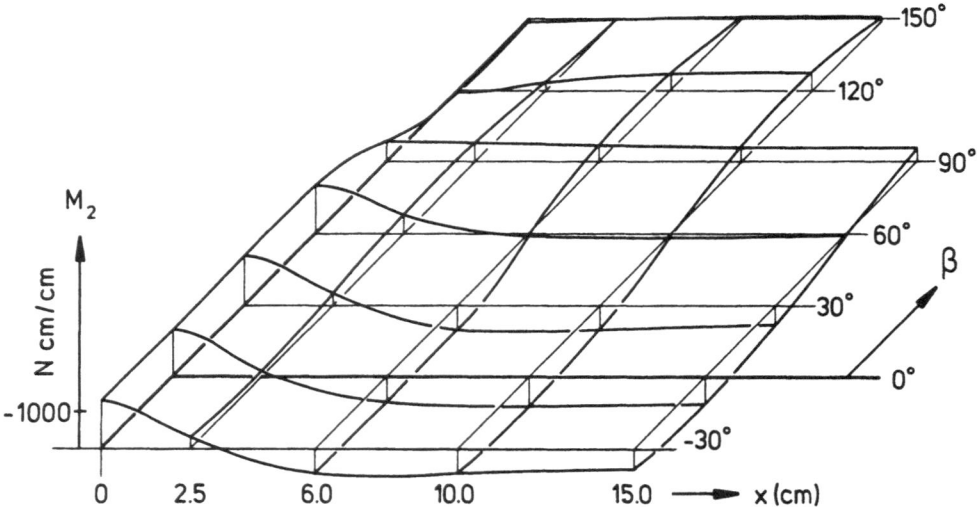

Fig. 8. Shell bending moment M_2 for a total load of 366 kN.

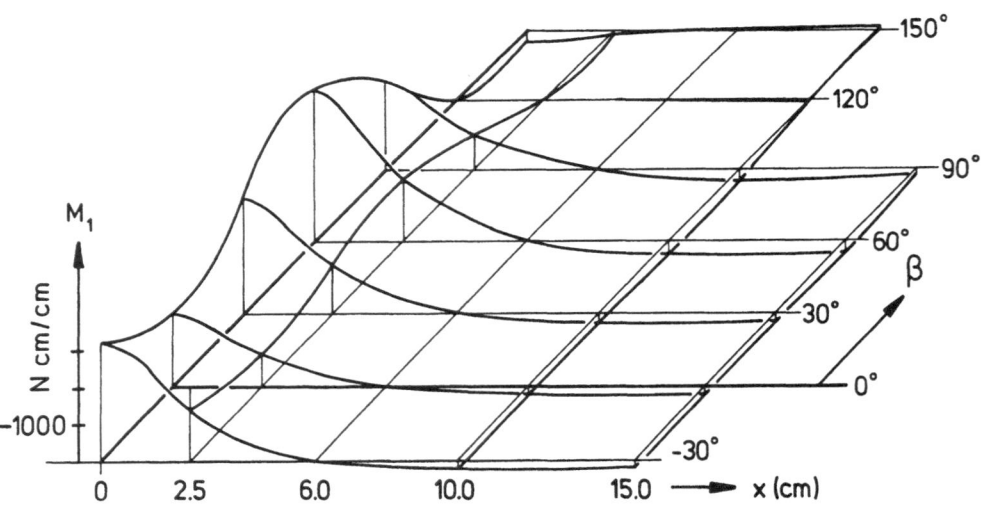

Fig. 9. Shell bending moment M_1 for a total load of 366 kN.

5. Conclusion

The presented method can be generalized: A mixed boundary value problem is transposed to a stress boundary value problem satisfying the boundary conditions in some parts, and than a hybrid experimental-analytical method should be applied.

5. References

1. R.J. Del Gaizo. Wirkung der Änderung der Kreisform am Sattelhorn des Sattellagers für liegende Behälter. Bautechnik 63, 351-357 (1986).

2. K. Brandes. Application of experimental-analytical hybrid techniques to civil engineering problems (in German). VDI-Berichte No. 73, 367-381 (1989).

3. K. Brandes. Zur Regression von Versuchsergebnissen bei Berücksichtigung von Nebenbedingungen. Materialprüfung 20, 265-269 (1978).

4. W.Z. Wlassow. Allgemeine Schalentheorie und ihre Anwendung in der Technik. Akademie-Verlag, Berlin (1958).

5. K. Brandes. Die Lagerung des Kreiszylinderrohres auf einem starren Linienlager. Stahlbau 40, 298-310 (1971).

6. K. Brandes. Solving engineering mechanics problems with experimental-analytical hybrid techniques. 9th International Conference on Experimental Mechanics, Copenhagen, 20.-24. August 1990, Proceedings.

7. Encyclopedia of Statistical Sciences, Vol. 4. John Wiley & Sons, New York, 1988.

SPLITTING OF THREE-DIMENSIONAL SOLUTION INTO LOCAL AND GLOBAL EFFECTS

H.D. BUI[1] and S. TAHERI

Electricité de France, 92141 Clamart, France
([1]) & Laboratoire de Mécanique des Solides, Ecole Polytechnique, 91128 Palaiseau, France.

Abstract.
A high stress level is observed at the free surface near the interface of a bimaterial subjected to thermomechanical loads. It is known that logarithmic singularities appear when the elastic properties are discontinuous. There are two another kinds of discontinuity due to thermal and plastic behaviours. Analytical relations about the discontinuities of stresses, thermal strain and plastic strains are proposed. For a thin tube, the three-dimensional solution is decomposed into a global effect (Love Kirchhoff solution) and a local effect independent of the mean curvature of the shell (plane strain solution) . The extension of local effect is of order of the thickness and is very sensitive to the spatial derivatives of the thermal load. The results have been extended to thick tubes and also to other type of structures. A shear shell solution is also used to approximate the local effect. The maximal error on the stress is derived. Finally the above analysis have been applied to rolled tube-sheet joints.

1. Introduction

Discontinuous thermal properties or more generally discontinuous thermal strains are the well-known causes of microcracking at the surface of structures, particulary under cyclic loads. The initiation of the edge cracks can be explained by the high value of the stress component parallel to the free surface. This stress component is discontinuous at the crossing point between the free surface and the interface, resulting in a very high stress variation under cyclic thermal load. Discontinuous plastic strains which result from different material properties (yield stress and strain-hardening) are also the causes of high stresses at the crossing point. There is a jump relation between the plastic strain and some stress component which is the analogue of the well-known jump relation on the thermal stress.
The interface between dissimilar elastic materials has been widely analysed in the literature,[1] to [5]. In this paper, we are interested in the stress singularity near the crossing point A, between the free surface (x_2 =0) and a surface of discontinuity (x_1 =0) of thermal strain $[\alpha T] \neq 0$ or plastic strain $[\epsilon^p] \neq 0$, by assuming continuous elastic behavior. This situation typically corresponds to welding of steel : the basic metal and the weld are generally designed so that they have the same elastic constants, but due to heat and phase transformations the mechanical properties of the weld zone are at lower strength than that of the basic metal. We found a new type of singularities, weaker than classical unbounded ones. Instead of logarithmic singularities or negative power behaviors, etc.. we found that the principal stress parallel to the free surface is continuous inside the body, but discontinuous on the free surface and its gradient is unbounded, [6], [7], [8].
In [9], we proposed a simplified method for estimating local effects in cylindrical shells under thermomechanical loads, which consists of splitting the three dimensional solution into local and global effects, using the decomposition of the three dimensional solution into odd and even functions of the through thickness variable

$x=r-r_m$ .The local effect having a characteristic length of the order of the shell thickness can be estimated by a plane strain solution. The global effect is practically given by the Love-Kirchhoff (L-K) linear shell theory. In this paper, we extend the method of [9] to more general geometries of shells. Both analytical and numerical analysis using the Finite Element Method will be used. In the case of thermal loading , a shear shell solution will also be used to approximate the local effect.

The decomposition method is also applied to the problem of expanding a tube in a circular hole of a plate, in order to determine the axial stress on the internal radius of the tube.

2. Singularity analysis.

■ Tube under axisymmetric thermomechanical loading : The elastic solution of a long tube submitted to a thermal loading T(z) and internal and external pressures P_{int} ,P_{ext} can be obtained by the use of Love's displacement function $\Phi(r,z)$ [6].

$$2\mu\, u_r(r,z) = - (\partial^2/\partial r\partial z)\Phi(r,z)$$
$$(1) \quad 2\mu\, u_z(r,z) = 2(1-\nu)\Delta_r\, \Phi(r,z), \quad \Delta_r = \partial^2/\partial z^2 + \partial^2/\partial r^2 + \partial/r\partial r$$

The Love function satisfies the equilibrium equation (E : Young modulus, ν : Poisson coefficient) $\Delta_r\Delta_r\Phi(r,z) = E\alpha T'(z)/(1-\nu)(1-2\nu)$. These equations have been used to analyse the decomposition of 3D axisymmetric solution in local and global effects, [9]. For instance, we are interested in the expression of singularities obtained in the case of a thermal shock.

■ Tube under a thermal shock : In order to study analytically the stress field near the interface z=0 of a discontinuous thermal strain $[\epsilon^T]\neq0$, we consider a long circular tube (coordinates r, z). The discontinuity of ϵ^T may be caused by either the discontinuity of thermal coefficient $[\alpha]\neq0$ or the discontinuity of the temperature $[T]\neq0$ or both $[\alpha T]\neq0$. Without loss of generality, we consider the second case, and we assume that the temperature is prescribed as follows $T'(z) = \Delta T/\eta\sqrt{2\pi}\ \exp(-z^2/2\eta^2)$,

where ΔT is the temperature jump. The choice of this particular temperature field is very convenient for the purpose of Fourier's transform $T(z) \rightarrow$

$T^*(s) = i\Delta T/s\sqrt{2\pi}\ \exp(-s^2\eta^2/2)$ which leads to analytical solution. This analytical solution has been used to check the finite element calculation [7]. The thermal shock experiment corresponds to the limiting value $\eta\rightarrow0$. In the latter case, the analytical solution derived in [6] for the case $p_e=p_i=0$ shows that at $x_2=r=r_e$ (or $x_2=r_i$) the Fourier transform of the axial principal stress σ^*_{11} admits asymptotically, for the first term of its Laurent series, the expression $\sigma^*_{11}(s,x_2=r_e) \simeq iE\alpha[T]/(1-\nu)(2\pi s) + ..$ This relation in Fourier transform shows that the physical stress σ_{11} is discontinuous at $x_1=0$, $x_2=r_e$, e.g. at the interface point A of the free surface; the discontinuity relation for the general case $[\alpha T]\neq0$ at the point A is given by

$$(2) \quad [\sigma_{11}] = E[\alpha T]/(1-\nu)$$

The same method has been used as in the case of plastic shock [7], [8]. Let us recall some results of the thermal and plastic shock problems, by assuming that the shock is located at the interface $x_1=0$.

■ The Thorn singularity : Fig.1 shows the 3D representation of the field $\sigma_{11}(x_1,x_2)$ in the case of a thermal shock inside the tube. The stress σ_{11} is antisymmetrical with respect to the interface $x_1=0$. For large x_1, we recognize the wave form of the L-K shell solution with the characteristic length $\gamma=(hr_m)^{1/2}$ .Near the point $x_1=0$, we observe a very localized and discontinuous stress field σ_{11}, see [6]; however along the interface $x_1=0$, the stress σ_{11} inside the tube is continuous. Suggestively we called this localized stress the "Thorn Singularity". We can see that this singularity is characterized by the following properties:-boundedness of the stress -discontinuity at the free surface point A -continuity inside the domain -unbounded gradient. It is worth

21

noticing that a finite element calculation with a coarse mesh, without the knowledge of the Thorn singularity, cannot exhibit the local effect. The presence of a localized high stress near the interface and the free surface point is a general feature of thermal shock loading, even for arbitrary geometry of structures, and also in the case of plastic shock as shown in the next sections.

■ Tube under pressure shock : For a infinite half plane under a constant pressure p applied on $-a < x_1 < a$,$x_2 = 0$, we know that $[\sigma_{22}] = [\sigma_{11}] = p$. We have compared this analytical value with numerical results obtained on the free surface of a long rectangular beam of finite thickness h under two opposite shocks Fig.2. Obviously, there is no effect of the thickness on the singularity. On a tube we apply pressure shocks on the external and internal skins (discontinuity of the shocks at z=0). On the Fig.2. we show σ_{11} on the external skin, we may remark that the discontinuity is usually negligible with respect to the maximum stress (which may be obtained by a L-K solution), while that is not the case for a thermal shock or a plastic shock.

■ Plastic shock : We study the residual stress field in a composite solid, consisting of two materials whose mechanical properties are only different by their plastic behavior. We wish to study the asymptotic behavior of the solution near the point A of the interface intersecting normally the free surface. We consider the following problem : Assume that the plastic strain ϵ^p is piecewise constant in each material domain, or more simply $\epsilon^p = \gamma H(x_1)$, where H is the Heaviside step function, H=0 for $x_1 \langle 0$, H=1 for $x_1 \rangle 0$. The domain Ω in consideration is the 3D rectangular bar. We wish to analyse the residual stress σ_{11} at $x_1 = x_2 = x_3 = 0$. Without loss of generality, we assume that $\gamma = (\gamma_{11}, \gamma_{12}, \gamma_{22}, \gamma_{33})$ is a constant tensor and $\gamma_{ii} = 0$. The equations are :

(3) $\sigma = L\epsilon(u) - 2\mu\epsilon^p$, div $\sigma = 0$ in Ω, $\sigma.n = 0$ on $\partial\Omega$.

Let $s = L\epsilon(u)$, $\gamma = \gamma^1 + \gamma^2$, $\gamma^1 = (0, \gamma_{12}, 0, 0)$, $\gamma^2 = (\gamma_{11}, 0, \gamma_{22}, \gamma_{33})$. We search a solution u^1, for $\epsilon^p = \gamma^1 H(x_1)$, such that :

(4) $L\epsilon(u^1) = 2\mu\gamma^1 H(x_1)$ or $\sigma \equiv 0$.

Since the right hand side of (4) is a piecewise constant function of x_1, $\epsilon(u^1)$ is a piecewise constant shear strain, thus the above equation is integrable. This means that the plastic deformation $\gamma_1 H(x_1)$ is compatible , in the sense of Kröner. In other words, no residual stress arises from the compatible plastic strain γ^1. However, there is a residual stress arising from the incompatibility of the plastic strain γ^2. This can be seen by the following physical arguments [7],[8]. Let the solid be separated into two parts along the interface $x_1 = 0$. This operation releases the residual stress and changes the areas of the common interface in different manner, the right bar (II) in Fig.3. undergoes the homogeneous deformation γ_{11}, γ_{22}, γ_{33}. Let the deformation of the right part of the beam come back to zero by using uniform surface forces, while the left part is unstressed. Then we stick the two parts together and apply opposite surface forces on the whole structure. These operations are equivalent to the initial problem and, using discontinuous pressure load result, lead to :

(5) $[\sigma_{11}(A)] = 2\mu[\epsilon^p_{22}]$

The above relation has been derived in the simple example of rectangular bar with a piecewise constant plastic deformation. As a matter of fact, this relation is valid for the interface in a rectangular beam and for the general case where the plastic strain is discontinuous at the point A, but not necessary a piecewise constant field. In this case, the plastic strain can always be decomposed into a piecewise constant field and a continuous field throughout the solid where the continuous component does not produce any discontinuity of the stress. This argument may be extended to more complex geometry with, possibly, a different ratio $[\sigma_{11}]/[\epsilon^p_{22}]$ [to be published].

Fig.3. shows the 2D finite element solution for a bimetallic beam, with the same elastic properties but with different plastic characteristics. The plastic deformation ϵ^p_{22} versus x_1 curve has been shifted vertically to show that the discontinuity relation (5) is satisfied with precision. We find also that the residual stress σ_{11} in the beam presents the same thorn singularity as shown in Fig.1.

Let us give some values for steel. For a discontinuous temperature of 100 C° with $\alpha = 1.2 \ 10^{-5}$, E=200000 MPa, ν=0.3, the discontinuity of thermal stress is $[\sigma]$=340 MPa . For a discontinuity of α equal to 10^{-6} (the interface weld-metal case) and a service temperature of 550 C°, the discontinuity of σ_{11} is 160 MPa . For a plastic deformation discontinuity of order .001, the stress discontinuity is about 150 MPa . These values show how important is the localized stress at interface, which we call the thorn singularity. Relations (2) and (5) give us the explanation of some phenomena such as the formation of surfaces microcracks in thermal stripping, as observed particularily in cyclic loading, or the initiation of cracks at the weld region.

3. Decomposition of three-dimensional solution in local and global effects.

■ Thin tube : In the previous sections we noticed that the discontinuity of σ_{11} due to a thermal shock is the same in both plane strain and axisymmetric cases. Moreover, it is easy to show that the values of stresses at $x_1=\pm0$ on the external and internal skins for plane strain is the same as in axisymmetrical case, equal to $E[\alpha T]/2(1-\nu)$ (in fact we know that in plane strain $[\sigma_{11}]= E[\alpha T]/(1-\nu)$ and σ_{11} is symmetrical with respect to $x_1 = 0$). It has to be noted that σ_{11} obtained by Love-Kirchhoff is equal to zero at $x_1 = 0$ and thus at this point the 3D solution is exactly the sum of a plane strain solution and a Love Kirchhoff solution. This has been generalized as following. It has been shown [9] that, for a thin tube (r_m/h>15) under a thermomechanical axial loading, in the case of small perturbation, the 3D solution can be decomposed into a local effect and a global effect. The local effect is independent of the mean radius and may be approximated by a plane strain solution, while the global one may be approximated by the classical LK linear theory. To show this, a decomposition of functions into odd and even functions of the through thickness variable $x_2=r-r_m$ is used through Love previous equations [9]. The local effect has an extension of an order of the thickness , and is extremely sensitive to the derivative of the temperature profile T"(z) which appears in the Beltrami equations. Fig.4. shows the axial stress on the external skin, for two temperatures fields [9], which are sligthly different in a small region just by their second derivatives T"(z). The comparison has been made between three-dimensional solutions 3D1, 3D2 and Love-Kirchhoff solutions LK1, LK2. It has to be noted that it is very difficult to obtain a good approximation of T"(z) using a T(z) experimentale data. That gives a particular importance to the limiting case of a thermal shock for which we have an analytical value of the maximal stress.

■ Thick Tube : The above mentioned analysis has been extended to the case of thick tubes [9] . It seems that for tube where r_m/h > 4 the local effect may be considered as independent of the mean radius with an error of less than 5 percent on the maximal stress Fig.5. This may be a criterion for the separation of thick shell and three dimensional structure. In fact we may write:
$$\sigma_{zz} = \sigma_{zz}{}^e + \sigma_{zz}{}^0 = \{\alpha_0(z) + \alpha_2(z)x^2 + ..\} + \{\alpha_1(z)x + \alpha_3(z)x^3 ..\}$$
$$\sigma_{rz} = \sigma_{rz}{}^0 + \sigma_{rz}{}^e = \{\beta_1(z)x + \beta_3(z)x^3 + ..\} + \{\beta_0(z) + \beta_2(z)x^2 ..\}$$
where e and o stand for even and odd parts of functions of $x=r-r_m$. The numerical results for the case of a thermal loading (axisymmetric and constant through the thickness) on a long tube show that:
- For a thin tube, in the case where a local effect has some importance, we have $\sigma_{rz}{}^e \simeq 0$ and $\sigma_{rr}{}^0 \simeq 0$; $\sigma_{zz}{}^0$ is given by the L-K solution which is linear in x; $\sigma_{zz}{}^e$, $\sigma_{rz}{}^0$, $\sigma_{rr}{}^e$ are given by a plane strain solution with a polynomial representation in the x function which can be of high order [10].
- For a thick tube ,$\sigma_{zz}{}^e$,$\sigma_{rz}{}^0$,$\sigma_{rr}{}^e$ are represented by the same plane strain solution. A cubical term appears in global part $\sigma_{zz}{}^0$ [10].
- In the case of small local effect in thin or thick tube, the three-dimensional solution is given by L-K solution [9].

■ A shell theory with shearing effect for thin tube : We found that the L-K solution can not describe the local effect correctly. A question may be raised about the capability of a shear shell theory to give a better description of the local effect. The answer is no, as we will see below through an analysis where we consider a thin tube under previous thermal loading. The first terms which we may add to the L-K solution to have a better solution for σ_{zz} is: $\sigma_{zz}^e = A_0(z) + A_2(z).x^2$ .This is an even term, so to obtain A_0 and A_2 we use plane strain equilibrium equations and Beltrami equations which give a cubical shear effect. Neglecting the pinching effect σ^{rr}, we finally obtain the values of A_0 and A_2 [9]. Fig.6. shows the comparison between this local effect and three-dimensional local effect obtained by the difference between 3D solution and K-L solution. The difference between the two results becomes important when we approach the thermal shock. An analytical result [9] shows that in this case the maximal stress (at $x_1=0$, $x_2=0$) is $2E\alpha\Delta T/(2-\nu) \simeq 1.17E\alpha\Delta T$, while the three-dimensional value is $E\alpha\Delta T/2(1-\nu) \simeq 0.71E\alpha\Delta T$, hence an error of 70 percent (the maximal axial stress obtained by K-L solution is $0.29E\alpha\Delta T$). We also show in [9] that, for the thermal shock, the local effect is described by the function $\exp(-\alpha z/h)$, while it is well known that the L-K solution is of the form $\sin(\beta z)\exp(-\beta z/(hr_m)^{1/2}$,($\alpha$ and β are positive constants).

■ Other type of results : To show that some of the previous results are independent of particular shape of structure, we calculate by the finite element method the membrane stresses on ring, sphere for a thermal shock . Fig.7 shows the independence of the local efect with respect to the curvature. In case of a thermal shock. Using the notion of local effect, it is qualitatively shown that, generally, a classical linear theory is not valid at a clamped edge, because of the existence of a local effect [9]. Numerical results reported in [8] shows the decomposition into local and global effect and the parity observed in these effects for a bimaterial tube under traction. Fig. 8. shows comparison between solutions for a bimaterial : 1) a 2D plane strain solution, 2) an axisymmetric solution for a tube, 3) a L-K solution for the tube. Quite the same local effect is found in solutions 1) and (2). Outside the region of local effect, the same global behavior is observed in solutions 2) and 3).

■ Application to a contact problem in plasticity : The above analysis has been applied to the rolled tube-sheet joints problem, in order to explain the feature of the residual stresses in the transition zone. Fig.9. shows the calculated axial stress on the internal skin of the rolled tube. One may remark global and local effects with their characteristic lengthes, which have been related to local solution and global L-K solution [11]. Experimental studies of the local stress are difficult. The extension of local effect is too small for direct observation by strain gages etc. The same difficulty exists for the case of thermal shock. However, the evidence of local effect is clear from the observation of surface microcracks (thermal striping). In fact, in cylic loadings of tubes under thermal shock [12], in the rolled tube where the plastic deformation may be discontinuous, and in the interface of weld-metal plastic deformation, the presence of microcracks indicates the high tensile stress level at the crossing point between the interface and the free surface. The formation of cracks along the interface release the tensile stress. Observations of the irreversible step on the free surface which have been reported in the literature give the best evidence of the jump of the plastic strain ϵ^P_{22} corresponding to the jump of the stress σ_{11}.

4. References

1. F. Erdogan and V. Biricikoglu. Int. J. Eng. Sci. 11, p.745 (1973).
2. F. Erdogan and G.D. Gupta. Int. J. Fract. 11, p.13 (1975).
3. R.I. Zwiers, T.C.T. Ting and R.L. Spilker. J. Appl.Mech. 49, p.561 (1982).
4. M. Comninou and J. D. Dundurs. J. Appl. Mech. 46, p. 849 (1979).
5. J. R. Willis. J. Mech. Phys. Solids 19, 1971, p.353 (1971).
6. H.D. Bui. Choc thermique sur le tube générateur de vapeurs. Rapport interne LMS-3, Ecole Polytechnique, (1981).
7. H. D. Bui and S. Taheri. Singularité Epine dans les bimatériaux en thermo-élasto-plasticité. C. R. Acad. Sci. Paris. 309, II, p.1527 (1989).
8. H. D. Bui and S. Taheri. Stress singularity and discontinuites on the free surface of a thermo-elastoplastic bimaterial body. Euromech-Colloquium 225 Padeborn. (1990).
9. S. Taheri. Three dimensional local effect and shell theories. Int. J. Ves. and Piping. 36, p.225 (1989).
10. S. Taheri. Effet local et théorie des coques. Rapport interne Electricité de France. (1987).
11. F. Voldoire. Calcul des contraintes dans les tubes de générateurs de vapeur. Rapport interne, Electricité de France. (1990).
12. M. Cousin, J.F Julien, H. Lauer, F. Vouilloux and J. Casier. Rochet thermique: Experimentation. SMIRT 6, L7/6 (1981).

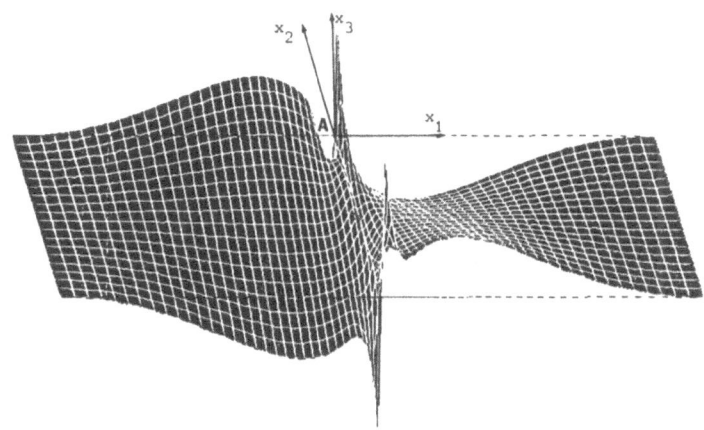

Fig. 1. The Thorn singularity of the stress σ_{11} parallel to the free surface, [7].

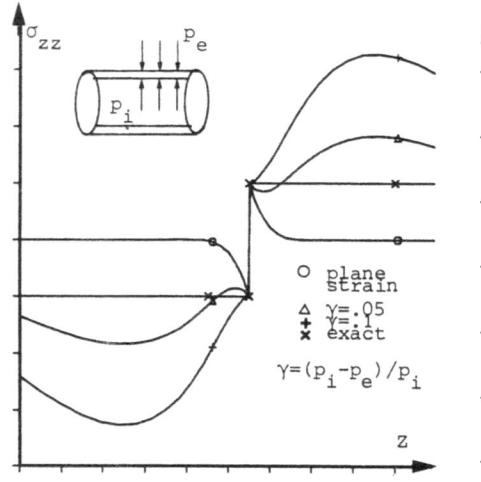

Fig. 2. Stress σ_{11} due to presure shock.

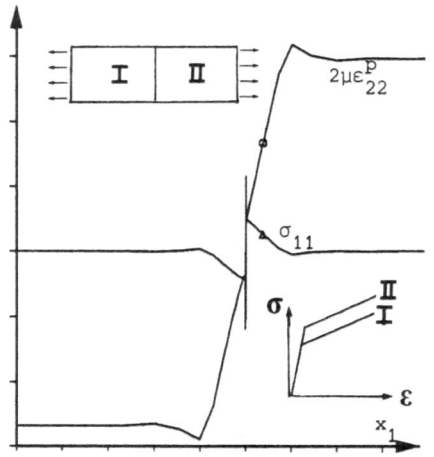

Fig. 3. Comparison between σ_{11} and $2\mu\epsilon_{22}^p$

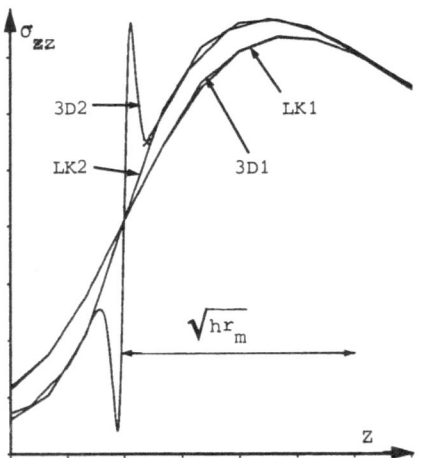

Fig. 4. Comparison between 3D and LK
solutions for 2 temperatures.

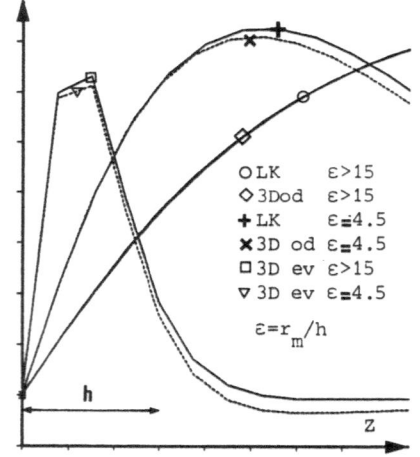

Fig. 5. Even and odd parts of σ_{zz};
LK Solutions.

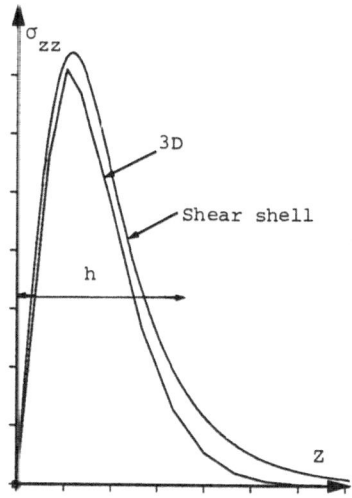

Fig. 6. Local effect : shear shell and 3D solutions.

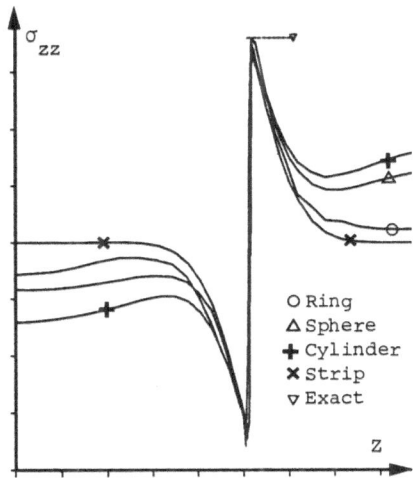

Fig. 7. Curvature independent local effect

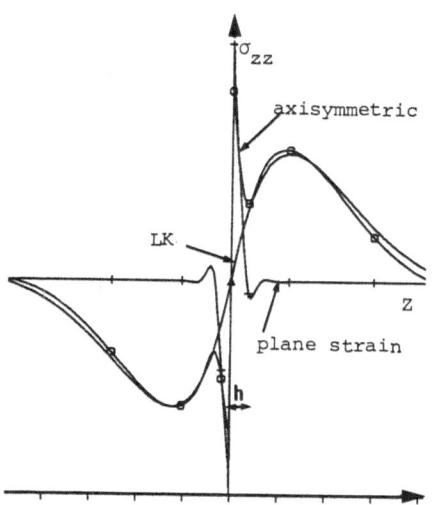

Fig. 8. Comparison of solutions :
   i)   2D plane strain
  ii) Axisymmetric 3D
 iii) LK solutions

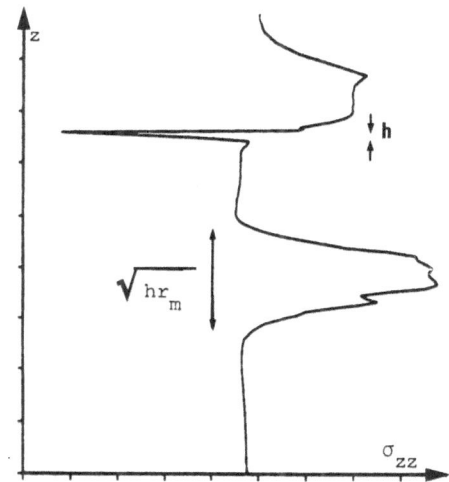

Fig. 9. Residual stress on internal skin of a rolled tube in a tube-sheet [12].

PLASTIC SQUEEZE OF CIRCULAR SHELL DUE TO SADDLE OR LUG

V. KŘUPKA

VÍTKOVICE - Institute of Applied Mechanics,
611 00 Brno 11; Box 32
Czechoslovakia

Abstract

The plastic collapse of cylindrical shell, with closed circular cross-section, supported by rigid saddle supports or lug is examined both experimentally and in a theoretical manner. Simple analytical relations are found and compared with the different tests.

1. Introduction

The solution of saddle support or lug being in contact with a vessel or tank shell has tended for more than 20 years to the solution of stresses, considering the elastic state only. The highest stress originates along the circumference of the saddle or lug, effects particurally the origin and growth of fatigue cracks under cycling load. A series of studies deal with a suitable arrangement of rigidity in contact. The shape of a saddle can substantially effect the value of stresses. The location of a wear plate is particularly of an importance for the reduction of high stress concentration. However such arrangements are in the case of the statical loading of a less importance for the increase of total ultimate load-carrying capacity. The aim of the submitted paper is to prove such a statement.

2. Solution of saddle supports

Let us give first the experimental results of saddles not welded to the vessel (shown in Fig. 1). The wall thickness $t = 5$ mm, mean radius $r = 807,5$ mm, yield point $R_y = 240$ MPa. Two saddles were tested with different width of wear plate, $2b = 120$ mm and $2\theta = 90°$ (including the overhanging end of the wear plate).

In the first test the saddle load was applied at the middle of a tested vessel being $7\ 500$ mm in length. It was found by testing that the ultimate load-carrying capacity is equal to 248 kN for the wide saddle $(2b = 900$ $mm)$ and 234 kN and 268 kN for the narrow saddle $(2b = 120$ $mm)$. It is seen that the ultimate load-carrying capacity was practically the same for all the saddles independent of the width $"b"$.

Table 1 shows the results. The second column gives the load in the saddle produced in the horn by the circumferential stress reaching its yield point. It is seen that to reach the yield point at the horn of a wide saddle we need nearly three-times greater force than in the case of a narrow saddle. However the experiment proved that the ultimate load-carrying

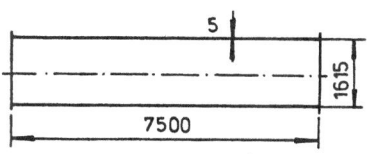

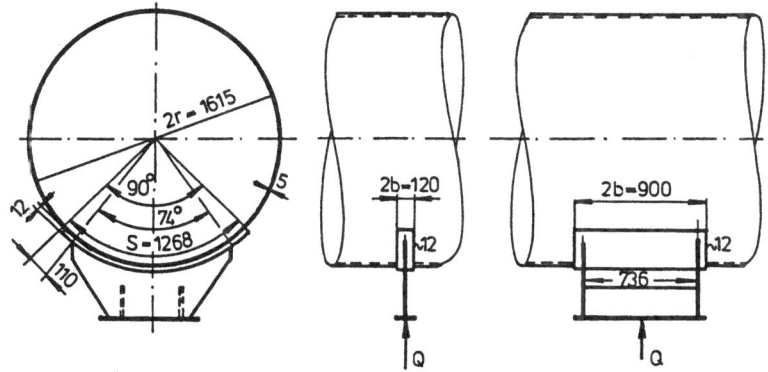

Fig. 1.

capacity is the same for both the saddles (see the third and fourth column). The table gives firstly the calculated values and the last column gives the ultimate load-carrying capacity measured in experiments. The measured load carrying capacity coincides in both cases perfectly with the following teoretical solution (see [1], [2]).

$$Q = 2R_y \cdot s \cdot t \cdot \sqrt{\frac{t}{r}} = 2 \cdot 240 \cdot 1268 \cdot 5\sqrt{\frac{5}{807,5}} = 239\ 466\ N \qquad (1)$$

where $s = 1\ 268\ mm$ is the legth of the saddle in contact with the shell.

The mentioned experiment has distinctly proved that *the different stress concentration values at the saddle horn have no influence on the total plastic carrying capacity at the place of the saddle.*

Table 1

SADDLE SUPPORT	Width	Theory		Experiment
	2b [mm]	Q_s [kN]	Q_{lim} [kN]	Limit carrying capacity [kN]
	1	2	3	4
WIDE	900	75,3	240	248
NARROW	120	24,6	240	234 268

In order to prove this result even in other cases the saddle was placed near to the reinforced end at different distances $a = 750, 680, 250$ mm. The experiment has given the values being approximately the same as with the load in the middle $Q = 257$ kN; 265 kN; 268 kN. That means that the place of application has no substantial influence on the value of limit load. Angle $2\theta = 2 . 45° = 90°$ has been got together with a wear plate. The rigidity of the wear plate was so high that it practically crushed into the shell. Other experiments followed after cutting off the overhanging part of the wear plate along the circumference. The angle decreased to $74°$ instead of $90°$. Experiments were repeated, as in the previous case, for different widths of the saddle and different places of application (being $a = 250$ mm away from the stiffened faces up to the middle $a = 3$ 750 mm). Limit load varried from 192 kN to 207 kN. Its theoretical value calculated by (1) is 197 kN. Even in this case the experimental solution clearly confirmed the theoretical considerations and relations given in [1] and [2].

Another prove has been got by an unprepared experiment with real underground wine storage tanks. Tanks were supported very near to their faces that should give a sufficinet reinforcement effect as expected. Due to unfavourable geological conditions one end of the tank was overloaded by soil which resulted in a plastic squeeze near to the stiffened face (see Fig. 2 and 3). Even in this case the limit load carrying capacity correspondes to eq. (1).

Fig. 2.

On the other hand it is necessary to underline that the equation (1) will not valid for wider angles θ since by saddles welded to the vessel with increasing angle $\theta > 60°$ the influence of shearing forces (being neglected with the derivation of eq. (1)) starts to predominate at the place of contact

Fig. 3.

(see results of experiments in [3]). Our results (form the analytical solution (1)) will be on the safe side because in the case of welded saddles the shearing forces increase the load carrying capacity.

3. Solution of lugs

These conclusions can encourage to apply eq. (1) even for the calculation of the effect to lugs (see Fig. 4).

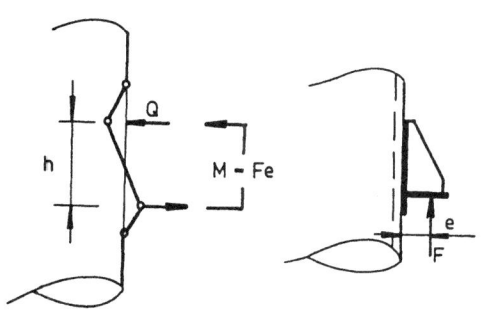

The basic consideration is shown in Fig. 4. We suppose that the lug will crush inside and outside in the upper and lower parts in accordance with the same principle as in the case of eq. (1) (see [1], [2]). The influence of the shearing forces is also neglected. Then

$$M_{lim} = Q_{lim} \cdot h$$

Fig. 4.

31

where eq. (1) was used to calculate Q_{lim}.

If we suppose a perfectly rigid lug with respect to the shell rigidity then we can write as given in Fig. 4.

$$F \cdot e = Q \cdot h$$

and from there

$$F = Q \cdot \frac{h}{e} = 2R_y \cdot s \cdot h \cdot t^{1,5} \cdot e^{-1} \cdot r^{-0,5} \qquad (2)$$

where eq. (1) can be substituted for Q.

Under this consideration we neglect the effect of shear force F being produced at the contact of the lug with the shell. This effect can be neglected if $e \gg h$ With thin shells and short distance e , however, it can come into effect and should be solved since this tangential force originates not only local stress concentration but could result even in buckling. To solve the problem theoretically is a rather complicated matter. Approximate solution similar to [4] is possible. To get an imagination, at least, we have made a series of tests as tabulated in Tab. 2. It also compares the measured and calculated results with the help of an approximate formula (2).

Table 2

	e [mm]	800	500	300	250	200	100	75	50
Experiment*	$F_{buckling}$ [N]	65	78	130	165	195	383	495	670
Experiment*	$F_{squeeze}$ [N]	57	66	107	130	155	298	380	490
Theory	F_{lim} [N]	41,8	66,9	111,5	133,8	167,2	334,4	445,9	668,9

* The own weight of the arm is included in·the results.

A shear flow at the place of lug contact with a shell is a cause of buckling and thus even a temporary reduction of carrying capacity. The buckling occurs due to shear or due to combination of shear and compression. After buckling the carrying capacity will again rise. Squeeze also usually limits the limit carrying capacity and results in overall exhaustion of carrying capacity.

Table 2 shows the comparison of measured and calculated values. Measured values with the first loss of stability due to buckling are in the first row, beginning of crushing in the second one. The base was of the following dimensions: $s = 54$ mm, $h = 51$ mm, $r = 65$ mm, $t = 0,3$ mm, $R_y = 340$ MPa, arm e varies from $e = 50$ mm up to $e = 800$ mm.

If $e < h$ then the shear force comes remarkably into effect and thus also the buckling resulting from it. The drop

32

of carrying capacity due to buckling is a considerable one. It is of interest, however, that a postcritical recovery appears. Overall differences are, as a matter of fact, not great.

The problem remains rather complicated even with used modern calculating methods since separate deformation processes take place in the region being geometrically and physically nonlinear and mainly they are complicated by different kinds of bifurcations being sometimes unforeseen. Thus the introduced solution is only the first step necessary for the first estimation of dimensions [5] and for the understanding of physical principle of solution. It must be followed by the application of computer finite methods. Without this first step, however, a researcher of our branch could be compared with a blind man whom is not even shown the direction of the way in an unknown countryside. If he is not correctly directed he surely will lose his way. Even a research worker will reel through a labyrinth of bifurcations without any success in spite of the fact that he has up-to-date computers at his disposal. On the other hand, when we choose a correct and phenomenological complex approach we can get with the help of computers considerably more reliable and at the same time more general results.

4. References

[1] V.Křupka. Buckling and Limit Carrying Capacity of Saddle Loaded Shells, in ECCS Colloquium on Stability, Ghent 1987, 617-622.

[2] V.Křupka. Saddle Supported Unstiffened Horizontal Vessels, in Acta Technica ČSAV No. 4, 1988, 472-492.

[3] A.S.Tooth and N.Jones. Plastic Collapse Loads of Cylindrical Pressure Vessels Supported by Rigid Saddles, Journal of Pressure Strain Analysis, Vol. 17, No. 3, 1982, 187-198

[4] A.Samuelson. Buckling of Cylindrical Shells Under Axial Compression and Subjected to Localized Load, in Colloquium Euromech 200 Postbuckling of Elastic Structures, Matrafüred 1985.

[5] E.G.Heinz. Erweiterte Anwendung von DIN 28083 Teil 2 zur Ermittlung zulässiger Pratzenmomente, in Konstruktion 32, (1980), 140-147.

ON THE CONTACT OF CYLINDRICAL SHELLS WITH AN ELASTIC OR RIGID FOUNDATION

A. MUC

Technical University of Cracow,
Institute of Mechanics and Machine Design,
ul. Warszawska 24,
31-155 Kraków,
Poland

Abstract

In the present paper we formulate unilateral friction boundary problem with the help of shell theories (geometrically linear or nonlinear) having various number of parameters (three, five or six) in order to study their influence on shell prebuckling deformations. Some problems connected with buckling analysis are also investigated herein. The theoretical considerations are illustrated on the example of a cylindrical shell loaded locally by external pressure and restrained by a rigid or elastic outer wall.

1. Introduction

Contact problems for shells constitute a particular class of problems to be frequently encountered in engineering practice. They are connected with a special class of boundary conditions (called unilateral boundary conditions) where it is not known a priori which part of boundary works itself loose from the support and which part remains in contact with it. This type of boundary conditions involves a completely different (than classical) mathematical formulation of the problem in the form of variational inequalities because unilateral boundary conditions can be expressed strictly in the form of distributions only. A broader discussion of these problems can be found in [1-4]. It has also its reflection in the formulation of the appriopriate numerical procedures used in the solution of contact problems (see eg. [5-8]). For instance in the finite element analysis a special class of boundary elements is introduced in order to take into account the considered type of boundary conditions. It should be emphasized that analytical or semi-analytical methods of solution are also applied here. However, their use is strictly limited to the analysis of geometrically linear problems only. The examples of these considerations are presented in [9-12].

34

It has been proved (see eg. [13,14]) that a shell theory which relaxes the Love-Kirchhoff hypothesis should be used in contact problems. It involves automatically the change in the form of contact reactions as well as in the localisation of the contact regions. These problems are discussed for instance in [15,16]. It is necessary to mention also about the influence of geometrical nonlinearities on both contact pressures and areas. In this case two opposite phenomena are possible: shrinkage or extensions of contact zones in comparison with results obtained with the use of geometrically linear shell theories. Dundurs [17] introduces here to the description of those effects the conception of the receding and advancing contact.

2. Formulation of the problem

In contact problems two types of boundary conditions occur. The first determined on the shell boundaries $\partial\Omega$ is prescribed in the equality form (bilateral conditions),whereas the second type formulated on the contact domain S (uknown in advance) is described in the inequality form (unilateral boundary conditions) and takes the following fashion:

$$S_N + g(v_3)1(v_3) = 0 \qquad \text{for } x_\alpha \in S \qquad (1)$$

S_N denotes the normal reaction of the foundation and 1 is the Heaviside distribution. The form of the reaction is described by the function (or operator) $g(v_3)$ of normal shell deflections v_3 which is assumed to be a convex one. For the rigid foundation it is modelled by the operator $S_o \dfrac{d}{dv_3}$ (S_o is an uknown constant) and for the elastic one by kv_3 (k is a positive constant). The above relation is supplemented by the friction conditions on S :

if $|\vec{S}_T| < \mu S_N$ then $\vec{v}_T = 0$,

if $|\vec{S}_T| = \mu S_N$ then $\vec{v}_T \neq 0$ and $\vec{v}_T = -\lambda\vec{S}_T$, $\lambda > 0$ (2)

$\vec{S}_T$ ($\vec{v}_T$) is a vector of tangential forces (displacements,rsp.)to the shell midsurface (on the plane x_1, x_2)and μ denotes Coulomb's friction coefficient. Additionally, we assume the following division of the components of the vector $\vec{S}_T$ in the domain S i.e., as sliding of the shell is possible

$$\frac{S_{T1}}{S_{T2}} = \frac{v_1}{v_2} \qquad (3)$$

One of the aim of the present paper is to compare the effects of different approximations used in shell theories. Therefore, we formulate the functional of total potential energy in general form, i.e. for six-parametrical shell theory. In this case the displacements $u_\alpha(x_1,x_2,x_3)$ of an arbitrary point of shells can be approximated in the following way

$$u_\alpha(x_1,x_2,x_3) = v_\alpha(x_1,x_2) + x_3\gamma_\alpha(x_1,x_2), \quad \alpha = 1,2,3 \qquad (4)$$

where v_α denotes displacements of the shell midsurface and γ_α are angles of rotation of the normal to the midsurface. The functional of total potential energy describing the nonlinear elastic deformation of cylindrical shells loaded by an external normal pressure p and subjected to the unilateral constraints (1),(2) takes the following form

$$J = J_1 + J_2 \qquad (5)$$

where,

$$J_1 = \frac{1}{2} \int_\Omega [N_{\alpha\beta}\varepsilon_{\alpha\beta} + M_{\alpha\beta}\varkappa_{\alpha\beta} + p(2 + e_{ii})v_3 - pv_i\vartheta_i]d\Omega \qquad (6)$$

$$J_2 = \frac{1}{2} \int_\Omega [(2+e_{ii})(S_N v_3 + S_{Ti}v_i) - S_N\vartheta_i v_i + S_{Ti}(v_{j,i}v_j + v_3\vartheta_i)]d\Omega \qquad (7)$$

$$e_{ii} = v_{i,i} - c_{ii}v_3, \quad \vartheta_i = v_{3,i} + c_{ij}v_j, \quad i,j = 1,2, \quad \alpha,\beta = 1,2,3.$$

$N_{\alpha\beta}$, $M_{\alpha\beta}$ mean stress resultants and stress couples, respectively, but M_{33} is identically equal to zero due to the assumed approximation of displacements (4). The above quantities are related to direct $\varepsilon_{\alpha\beta}$ and bending $\varkappa_{\alpha\beta}$ strains by the classical relations obtained from the Hook law. Ω is the two-dimensional space occupied by the shell. J_2 represents here the influence of unilateral friction boundary conditions. The contact forces S_N (1) and the friction reactions $\check{S}_T$(2) are defined by the set of inequalities and they do not have their potentials in classical sense. From the pure mathematical point of view functional J may be treated as so-called "superpotential" which is, in some sense, a generalisation of classical potentials and has been widely used in the abstract mathematical theorems of unilateral boundary value problems. This new concept permits a compact formulation of unilateral boundary conditions.

The solution of the problem considered consists of finding the stationary points of the total energy J on the space of kinematically admissible displacements (ie. satisfying strain displacements relations, the unilateral boundary conditions on S and some bilateral conditions on $\partial\Omega$). Strictly speaking the proof of the existence and uniqueness of solutions of the problem described by the functional (5) and of its equivalence to the problem represented by the system of Eqs.(1)-(3), strain displacements relations and equilibrium equations base on the notions of convexity (for linearly elastic bodies) or polyconvexity (for nonlinearly elastic bodies). However, due to the existence of nondifferential term J_2-(7) in the functional J, the relations (1),(2) cannot be obtained by the simple classical differentiation J as well as the principle of the virtual work cannot be written in the equality form.

It is worth to mention also an interesting feature of contact problems. Unilateral constraints (normal - (1) or frictional - (2)) introduce nonlinearities in the form of generalised functions. In this way, even for geometrically linear analysis, the problem with unilateral constraints may be treated (from mathematical viewpoint) as the loss of stability, because the space of fundamental solutions V (without unilateral constraints) is contracted by the unilateral constraints to the space $V_1 \subset V$. A broader discussion of those problems can be found in [4].

3. Comparison of results for various shell theories

It is wellknown that for shells geometrical relations arise as an approximation of three-dimensional case. For shells subjected to rigid (normal or frictional) constraints (ie. as $g(v_3) = S_o \frac{d}{dv_3}$ in (1)) those effects are especially visible in the form of contact pressures. In general, in terms of normal deflections v_3 one can describe briefly the contact mechanism in the following way: as a shell sticks to the rigid foundation v_3 and $v_{3,i}$ are constant or equal to zero but at the edge of contact domains ∂S the derivatives $v_{3,i}$ change their values. The precise mathematical description of $v_{3,i}$ depends directly on the type of shell theories applied. Both for three- ($\gamma_\alpha = 0$ in (4)) and five- ($\gamma_3 = 0$ in (4)) parametrical shell theories the derivatives $v_{3,i}$ at points on ∂S have discontinuities in the class of classical functions and may be expressed as distributions. For shells basing on Love-Kirchhof's hypothesis (three-parametrical theory) both shearing forces N_{i3} and bending moments M_{ii}, $i=1,2$ are discontinuous on ∂S, whereas for five-parametrical shell theories shearing forces N_{i3} only. For six-parametrical shell theories (ie. as the compression of normal to the shell midsurface is taken into account) all stress $N_{\alpha\beta}$ and bending $M_{\alpha\beta}$ resultants are expressed by continuous functions in classical sense. Therefore, in the latter case contact pressures are described by continuous functions without any localised (in the form of Dirac's delta distributions) forces at the edges of contact domains ∂S. For three- and five-parametrical shell theories concentrated forces can satisfy entirely the system of differential equations obtained after minimisation of the functional J - (5).

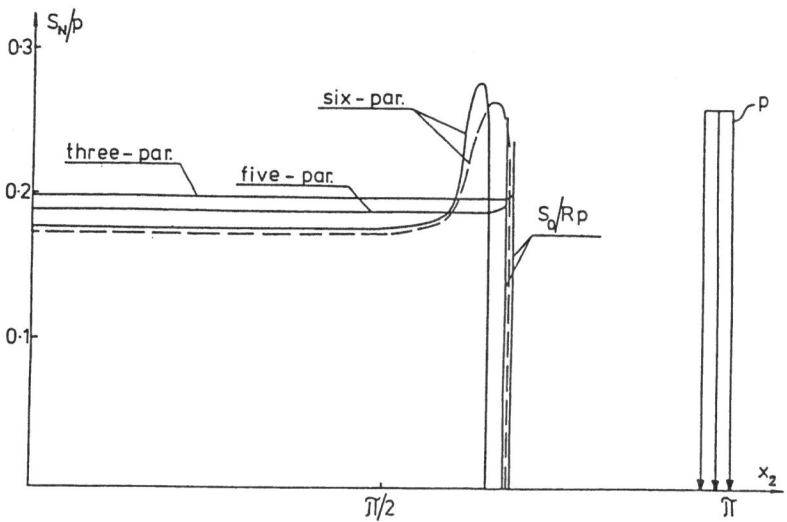

Fig.1 Influence of the type of shell theory on variations of contact reactions (---- linear, —— nonlinear).

37

In order to illustrate the above-mentioned problem let us consider the contact problem of an infinite cylindrical shell surrounded by a rigid wall and loaded locally by an external pressure p. The bilateral boundary conditions are taken herein as the symmetry conditions for displacements v_α and angles of rotations γ_α at $x_2 = 0$ and $x_2 = \pi$. the numerical procedure used in calculations is analogical to that presented in [3,18]. In general, it bases on the finite difference disretisation of the functional (5) and is associated with the Gauss - Seidel overrelaxation method.

Fig.1 is a plot of contact normal reactions S_N - (1) for various number of parameters describing shell deformations (see Eq.(4)). As it may be seen the increasing number of parameters reduces contact pressures inside the contact domain S and simultaneously results in the growth of contact forces at the edges of S. It is also connected with the shrinkage of contact regions. For three- and five- parametrical shell theories concentrated contact forces appear at ∂S, and for six-parametrical theories they are replaced by a continuous distribution of contact pressures over the whole region S including its edges. It is visible also that the application of geometrically linear shell theory extends contact regions but reduces maximal contact pressures (up to 4.5%) in the comparison with the case as geometrically nonlinear theory is used. Let us notice also that for three- and six- parametrical theories (geometrically nonlinear approach) the difference in the maximal contact pressures reaches 18%.

For shells resting on the elastic foundation the influence of shell theories is observed for very high values of the k parameter and from the practical point of view the foundation may be treated as the rigid one.

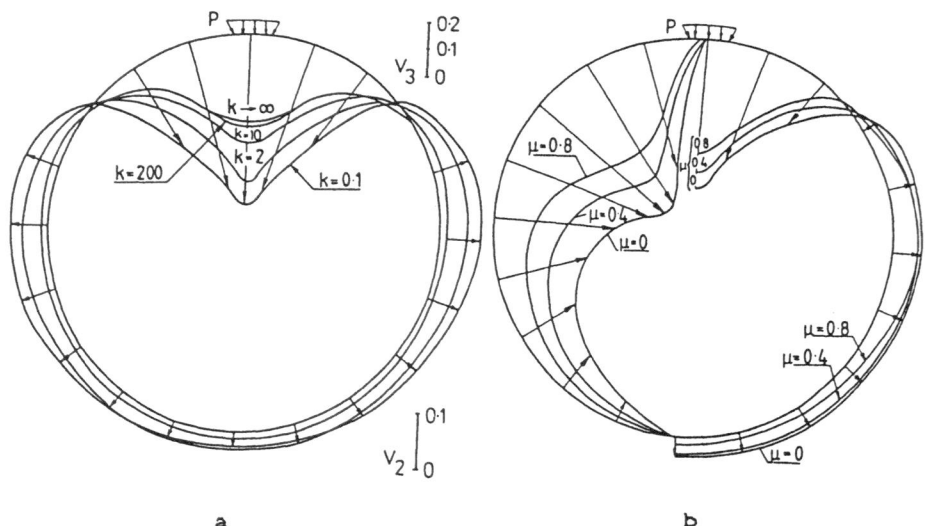

a   b

Fig.2 Influence of foundation rigidity and friction on distributions of normal and tangential displacements($2h/R$ = 0.02, $pR(1+\nu)/2Eh$ = 0.1).

4. Cylindrical shells restrained by elastic foundation

Let us consider now the numerical results for the case of Winkler's foundation , ie. as $g(v_3) = kv_3$ in (1). The second type of unilateral boundary conditions (the rigid foundation) can be treated as the limit state of elasic one (ie. for k → ∞) The influence of the foundation rigidity k on the distributions of the normal displacements v_3 is shown in Fig. 2a. In this case the contact area S is a simply connected domain and decrease with the growing value of k. Friction extends the contact areas – Fig. 2b, however, it reduces the normal v_3 as well as the tan-gential displacements v_2. Additio-nally, it is seen that in the con-tact domain S the relation(2b) is always valid (sliding of the shell is only possible – $v_2 \neq 0$).

More illuminating are the results for the case of calculations of displacements fields under the whole shell with free edges $\partial\Omega$ (however, the symmetry with respect to $x_1 = 0$ and $x_2 = 0, x_2 = \pi$ planes is assumed). The effects of the lengthwise of the external pres-sure p on the locations of con-tact domains S are illustrated by Fig. 3. The extension of the pres-sure penetration in the circumfe-

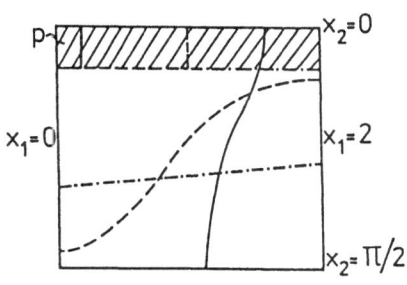

Fig. 3 Variations of contact domains with the lengthwise of external pressure (2h/R= 0.02, L/R = 2, pR(1+ν)/2Eh = 0.1, k(1+ν)/2Eh = 0.01).

rential direction x_2 results in the extension of contact regions; however, one cannot obtain the limit case of the circumferential symmetry due to the influence of boundary conditions on $\partial\Omega$.

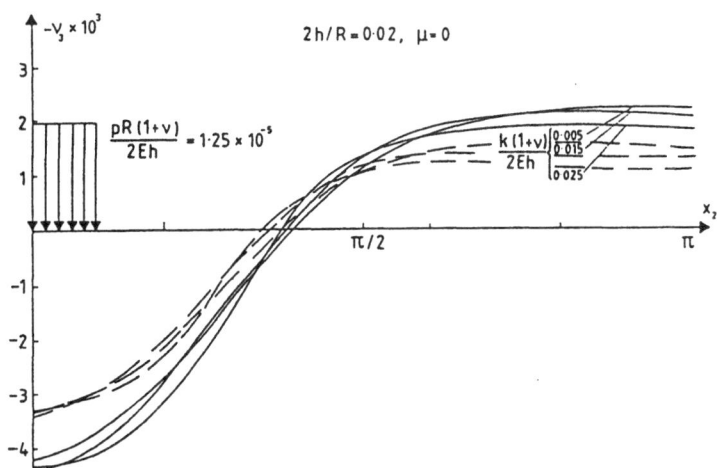

Fig. 4 Comparison of results for geometrically linear and nonlinear shell theory (----- linear, ——— nonlinear).

39

The comparison of the results obtained on the base of large and
small displacement five-parametrical shell theory is presented
in Fig. 4. Geometrical nonlinearities do not change the
character of contact areas. They cause only shrinking of
contact domains (so-called receding contact according to the
nomenclature introduced by Dundurs [17]) and the growth of the
normal deflections v_3. More mumerical results is discussed in
[18].

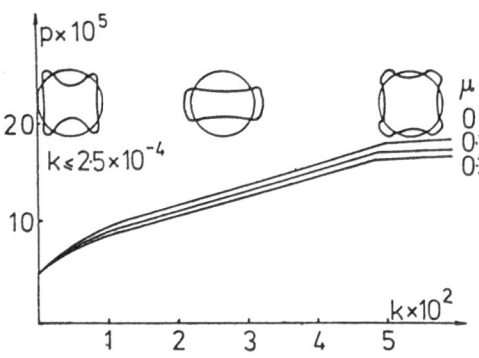

Fig. 5 Variations of buckling loads vs the foundation rigidity k

Taking into accunt of geometrical nonlinearities allows
to study buckling problems
of unilaterally constrained
shells. We apply the similar
approach as discussed in[4],
ie. we investigate the sign
of the second functional derivative in the Frechet sense of the functional J -(5).
The analysis of this problem
has been carried out on the
example of the infinite cylindrical shell placed in
the elastic medium. For the
case considered the limit
curves as well as the buckling modes are plotted in

Fig. 5. The limit curves are almost linear for greater values of
k and change their slope with respect to to the variations of
buckling modes. Friction reduces critical load and it is
associated with the simultaneous shift of the positions of new
instability forms. This result is very controversial because
according to the common opinion in the static approach friction
should enlarge buckling loads. The possible explanation of such
a phenomenon is presented in [19].

References

1. I. Ekeland and R. Temam, Convex analysis and variational problems, North-Holland, American & Amsterdam Elsevier, New York
   (1986).
2. P. D. Panagiotopoulos, Inequality problems in mechanics and
   applications, Birkhäuser, Basel (1985).
3. A. Muc, Theoretical and numerical aspects in contact problems
   for shells, ZAMP 35, 890-905 (1984).
4. W. Krzyś and A. Muc, Buckling and post-buckling behaviour of
   shells under unilateral constraints, Post-buckling of elastic structures, Proc. Euromech Coll. 200 (edts J. Szabo
   et al), Elsevier, Amsterdam, 171-184 (1986).
5. F. D. Fischer, Zur Lösung des Kontaktproblems elastischer
   Körper mit ausgedehnter Kontaktfläche durch qudratische
   Programmierung, Computing 13, 353-384 (1974).
6. R. Glowinski, J. L. Lions and R. Tremolieres, Analyse numerique
   des inequations variationnelles, Dunod, Paris (1976).
7. K. J. Bathe and A. Chaudhary, A solution method for planar
   and axisymmetric contact problems, Int. J. Num. Meth. Eng. 21,
   65-88 (1985).
8. E. Stein and P. Wriggers, Stability of rods with unilateral

constraints, a finite element solution,Comp.Str. 19,205-211 (1984).

9. J.T. Tielking and R.A. Schapery, A method for shell contact analysis, Comp.Meth.Appl.Mech.Eng. 26, 181-195 (1981).

10. E.I. Grigolyuk and V.M. Tolkatchev, Contact problems in plates and shell theory, Maschinostroyenie, Moscow (1980 in Russian).

11. B.L. Pielekh and M.A. Sukhorolski, Contact problems in elastic anisotropic shells theory, Naukova Dumka,Kiev (1980 in Russian).

12. V.I. Mossakovski, V.S. Gudramovich and J.M. Makeev, Contact reactions of elements of shell structures, Naukova Dumka, Kiev (1988 in Russian).

13. F.Essenburg and S.T. Gulati, On the contact of two axisymmetric plates, J.Appl.Mech. 33, 341-346 (1966).

14. K.Z. Galimov, Shell theory with transverse shear effects, Izd. Kaz. Univ., Kazan (1977 in Russian).

15. G.J. Popov and V.M. Tolkatchev, Contact problems of rigid bodies with thinwalled structures, Mekh. Tverd.T. , 192-206 (1980 in Russian).

16. V.M. Blokh, Choice of the model in contact between thin-walled bodies,Prikladnaya Mekh. 13,34-42(1977 in Russian).

17. J. Dundurs, Properties of the elastic bodies in contact, Proc. Symp. The mechanics of contact between deformable bodies, Delft UD, 54-65(1975).

18. A.Muc, Deformations in an infinite cylindrical shell restrained by elastic or rigid walls, Acta Mech. 73, 1-10(1988).

19. A.Muc, Influence of friction on critical loads for structures on elastic foundation, ZAMM 68, 51-53(1988).

EFFECT OF LOCAL LOADS ON THE STABILITY OF SHELLS SUBJECTED TO UNIFORM PRESSURE DISTRIBUTION

Lars Å Samuelson,

The Swedish Plant Inspectorate
P. O. Box 49306,
S-10028 Stockholm,
Sweden

Abstract

Rules for design against buckling of shells according to current code cover only complete shell segments subjected to uniform or nearly uniform load distributions. The effect of initial imperfections is recognized but, local disturbances such as holes, local external forces or nonuniform stiffness distributions are not yet considered. Such disturbances are common in practical applications and may have a drastic influence on the carrying capacity of the shell. The deflection of a shell caused by a small local load may be analyzed by use of linear shell theory. If it is assumed that its effect is equivalent to that of an initial imperfection, rules may be established for the analysis of the carrying capacity of the shell. This hypothesis was tested for cylindrical shells under axial compression through finite element analyses and tests on small scale plastic specimens and for spherical caps by computer analyses. It was found that the assumption is realistic and it yields, in its simplest form, slightly conservative results for cylinders but is nonconservative for spheres. Conversely, buckling tests on point loaded shells appear to give valuable information on the initial imperfection sensitivity and it may be possible to utilize such tests to develop more accurate guidelines on tolerances for manufacturing.

Introduction

The effect of initial imperfections on the stability of certain types of shells is well understood and current design methods give rules for determination of tolerance levels in order to ensure a safe estimate of the carrying capacity. On the other hand, finite element analysis of the carrying capacity of a shell with known initial imperfections and built in stresses is very complicated and is as a rule not feasible in practical cases.

Other types of disturbances, such as nonuniform load distributions at the supports and variable pressure distributions, are also known to lower the carrying capacity but have been studied to a considerably lesser degree. This is rather unfortunate because such disturbances are very common in practical applications but are left to the designer to consider. The present investigation considers a specific class of problems encountered in many fields involving stability of shell structures, namely the effect of local loads or local bending moments acting on the shell surface. Typical examples are pipe attachments, cantilever beams acting as supports for ladders or external equipment etc.

A point load acting perpendicular to a cylindrical or spherical shell surface causes deflections similar to an initial imperfection. In [1] the assumption was made that the two types of initial deflections are equivalent and a design criterion was developed where allowable local loads could be determined. A similar method was used in [2] for spherical shells and the results were adopted in [1].

There is, however, a distinct difference between an initial imperfection and a deflection caused by a local load: The imperfection distribution is of a stochastic nature whereas the local load yields a deformation which is deterministic. Thus it may be assumed that the stability behavior may be similar to that of a shell with a hole as was demonstrated in [3]. The main objective of the present investigation is, therefore, to check through experiments and finite element analyses whether the dependence of the carrying capacity can be determined as a simple function of the point load (or moment) and the geometric parameters of the shell. An analytical solution of the problem is obviously not possible to obtain due to the highly nonlinear nature of the governing equations. See for instance [4].

Current design practice

The design recommendations given in [5] may be taken as representative of current codes for design against buckling of a number of different shells subjected to uniform loads. For a cylinder under axial compression the design criteria may be written as follows:

The theoretical estimate of the buckling stress is given by the classical formula (or the bifurcation stress obtained in a linear finite element analysis)

$$\sigma_{cr} = \frac{1}{\sqrt{3(1-\nu^2)}} E \frac{t}{r} = 0.6 \, E \frac{t}{r} \quad \text{if } \nu = 0.3 \tag{1}$$

This value is known to provide a nonconservative estimate of the carrying capacity and has to be reduced by a factor α determined from experiments. This factor $\alpha = \alpha_0$ is given as follows in [5]:

$$\alpha_0 = \frac{0.70}{\sqrt{0.1+0.01 r/t}} \quad \text{provided } r/t > 212 \text{ and } w_r/l_r < 0.01 \tag{2}$$

If a tolerance limit of 0.02 is assumed, the reduction factor is to be taken as $\alpha = 0.5\alpha_0$. The carrying capacity which may be utilized for design purposes is then given as

$$\sigma_u = 0.75\alpha\sigma_{cr} \quad \text{if} \quad \alpha\sigma_{cr} < \sigma_y/2 \tag{3}$$

A slightly more optimistic reduction function was adopted in [1] as illustrated in Fig.1.

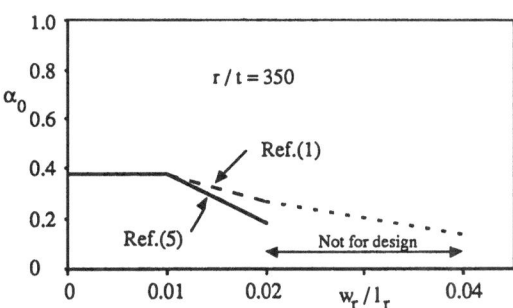

Fig. 1 Recommended reduction factors as functions of the imperfection magnitude for a cylindrical shell under axial compression.

A local load acting perpendicular to the shell surface causes a deflection which is similar to an initial imperfection. If it is assumed that the effect is equivalent, the deflection under the local load may be characterized by the same function of w_r/l_r as the initial imperfection, see Fig.2. This was done in the design recommendations given in [1]

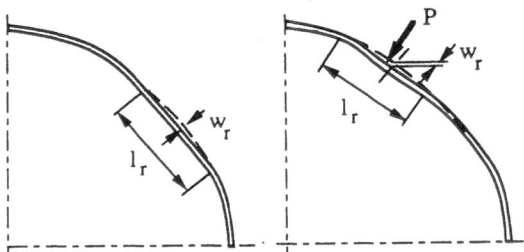

Fig. 2 Initial imperfections and local deflections caused by a small concentrated load P.

Experimental investigation

A series of buckling experiments was carried out with plastic cylinders under axial compression and subjected to varios kinds of disturbances. One of the advantages with plastic test specimens is that buckling occurs in the elastic range and the loading may be repeated a large number of times without any change in the carrying capacity. The influence of small disturbances may therefore be studied on the same test specimen which precludes the influence of other parameters important to the carrying capacity of the shell. The procedure has been used successfully in the study of shells with cutouts, see for instance [6] and [7].

Test specimens and loading system

The geometry of the two test specimens is shown in Fig. 3. The cylinders had the same diameter d = 140 mm, thickness t = 0.19 mm and length 100 mm.

The material properties were assumed to be given by the following values measured on plastic strips from the same sheet and given in [8]: E = 4700 MPa, $v = 0.38$. This implies a classical buckling limit of:

$$\sigma_{cr} = \frac{1}{\sqrt{3(1-v^2)}} E \frac{t}{r} = 7.96 \text{ MPa}$$

corresponding to a critical load of

$$F_{cr} = 2\pi rt \cdot 7.96 = 665 \text{ N}$$

According to the ECCS Recommendations for design of shells against buckling, [5], the lower limit of the carrying capacity is:

$$F_u = 0.75\alpha_0 F_{cr} = \frac{0.75 \cdot 0.70}{\sqrt{0.1+0.01r/t}} F_{cr} = 161 \text{ N}$$

Axial compression was applied in a standard testing machine whereas the load disturbances were produced by use of a lever system as indicated in Fig. 3.

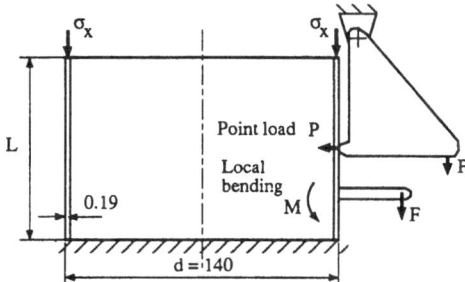

Fig. 3 Cylinder geometry and point load / moment lever arrangement

Test results

A number of test series were carried out where the type of local loading was varied. The test results are presented in Figs. 4a through 4f. Each series of experiments was started by testing the undisturbed cylinder thus defining the carrying capacity of the (imperfect) shell. Subsequently, a small disturbance was applied and the buckling test was repeated. The disturbance level was then increased stepwise and the degradation of the load carrying capacity was recorded. Fig 4a shows typical results in the point load case.

The imperfections of the test specimens were not measured, mainly since such measurements could not be easily carried out by means of regular test methods involving contacting deflection sensing equipment. It was assumed that the buckling tests would show if there was interaction between the effect of the local load and the initial imperfections. As a matter of fact, the results given below strongly support this assumption.

The effect of a local load on the carrying capacity differed depending on where along the circumference the load was applied. In some cases a small load caused a noticeable drop in the buckling load (compare the φ = 180° curve in Fig. 4a) whereas in other cases the local load did not seem to influence the carrying capacity until the local load reached much higher values. (Curve for φ = 90°). Actually, in one case (curve φ = 270° in Fig. 4e), application of the local load actually increased the carrying capacity of the cylinder. In this case buckling was noted to initiate 180° away from the local load until very large loads were applied indicating the presence of large imperfections or irregularities at the boundary. Since the local load introduces tensile stresses in the meridional direction and in this specific case caused a beneficial redistribution of the stress field such that a slight increase in the buckling load resulted. This observation indicates that there cannot be a complete correspondence between a local imperfection and a deflection caused by a local load. In the cylinder case, however, assumption of such a correspondence appears to be on the safe side.

Figs. 4b and 4f show the buckling load of the two specimens as a function of a point load applied at x = l/4 along the meridian. The load reduction is smaller than when the load acts at the shell midpoint, which is of course a consequence of the fact that the deflections caused by the local force are smaller.

A local bending moment acting along the circumferential coordinate causes buckling load reductions according to Fig. 4c. The effect is very similar to that found in the point load case.

Finally, the effect of a bending moment acting along the meridian is shown in Fig 4d. In this case a comparatively large moment had to be applied before a reduction in the carrying capacity was noted. This may be explained by the fact that the deflection caused in this case by the moment does not resemble the buckling mode of the cylinder and is thus considerably less critical.

45

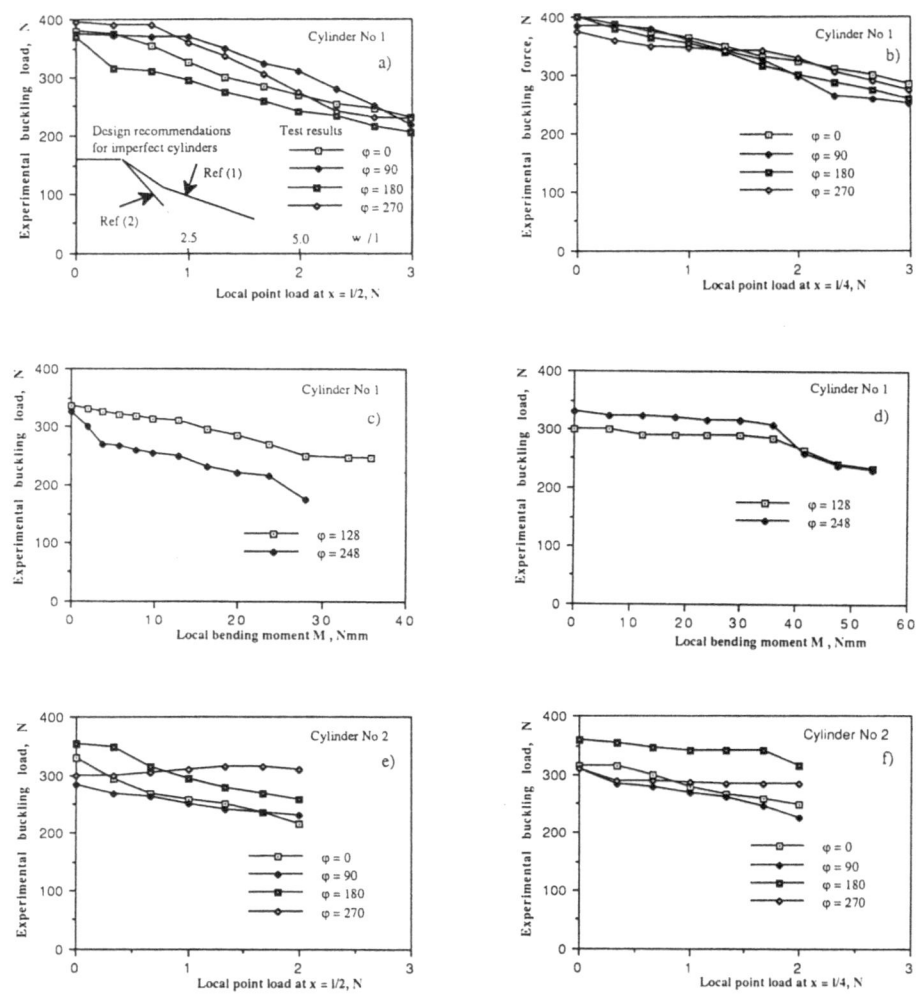

Fig. 4. Experimental results for test specimens 1 and 2.

Deflections caused by point loads on shell surfaces

Shell structures have limited capacity for loads acting perpendicular to the middle surface. Also, the analysis of deflections and stresses in the vicinity of local disturbances is fairly complex even in the case when a linear analysis is adequate. A fairly large number of investigations have been published. Here [9] through [11] were chosen to represent research investigation related to this particular class of shell problems.

Cylindrical shells under radial point loads

Bijlaard, [9] developed a theory for calculation of deflections and stresses in a cylindrical shell subjected to a local load in the lateral direction. A computer program was written for the specific purpose of determination of a load P^* yielding a deflection w_r corresponding to the relative displacement $w_r/l_r = 0.01$. The results are shown in Fig. 5

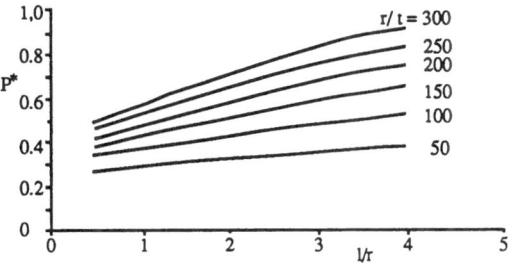

Fig. 5 Relative displacements in a cylinder subjected to local loads. $P^* = P\sqrt{rl}/(Et^3)$

Similar diagrams may be developed for deflections caused by local bending moments.

Spherical shells

The representation of a characteristic deflection is somewhat simpler in this case since the only geometry parameter of importance is the r/t ratio. The result, obtained by use of the BOSOR4 computer code, [12] is shown in Fig. 6.

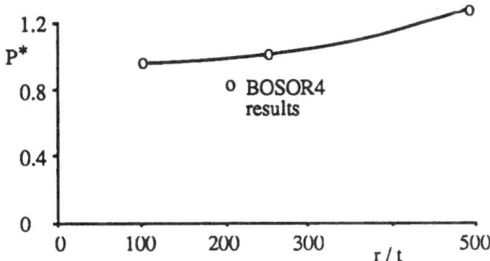

Fig. 6 Nondimensional point load on a spherical shell giving a deflection $w_r = 0.5t$. $P^* = Pr/Et^3$.

Numerical analyses

Cylindrical shells

Finite element analyses of the buckling behavior of cylindrical shells under a constant point load and simultaneously subjected to axial compression were carried out . This case is very difficult and expensive to analyze numerically and therefore only a single geometry and a few load combinations could be treated. The results are shown in Fig. 7.

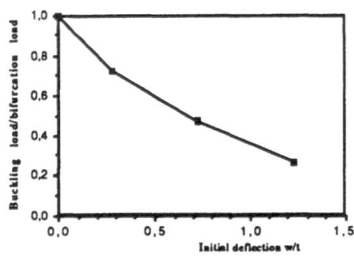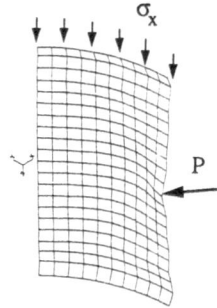

Fig. 7 The reduction of the buckling load of the test cylinder under axial
compression due to a local radial load at the shell midpoint.

A number of models were tested in order to find an appropriate mesh for the analysis. The
bifurcation load was calculated and was compared to the classical value given above. The
results showed that a mesh of 20x40 elements was required in order to simulate the bifurcation
load to within 5 per cent. In the nonlinear collapse analyses summarized in Fig. 7, a slightly
less accurate model was used as included in the figure. The bifurcation load of this specific
model was 830 N or 25 per cent higher than the classical value. The collapse loads given in the
figure are referred to the higher bifurcation load of the model.

Spherical shells

Spherical shells were studied in [10] and [11].In the present investigation, a number of collapse
analyses were carried out by use of BOSOR4, [12] in order to evaluate the influence of a point
load in reference to the tolerance limits currently used in the design of spherical shells ([1] and
[5]). The shell geometry is given in Fig. 8 where the deflection caused by a local load P is
included. A point load of 600 N gives a relative deflection of 0.2t and reduces the carrying
capacity to 0.583 of the classical value. The corresponding figures for a point load of 1200 N
are: Initial deflection = .37t and load reduction to 0.358 of the classical critical pressure.
Finally, a point load of 2500 N produces a relative displacement of 1.05t and a collapse
pressure of only 0.12 times the classical value. The load reduction caused by deflections from
point loads thus appears to be more severe than that caused by initial deflections of a
corresponding magnitude, see Fig. 8.

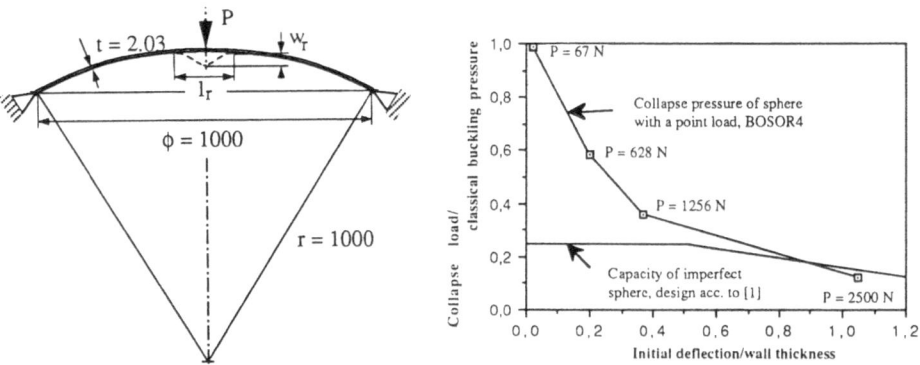

Fig. 8 Geometry of spherical cap under external pressure and a local load
acting at the apex. Included are numerical results and design values
for an imperfect shell according to [1].

48

Typical load-deflection curves are shown in Fig. 9 where the lateral inward deflection at the apex of the cap is plotted against the external pressure ratio p / p_{cr}.

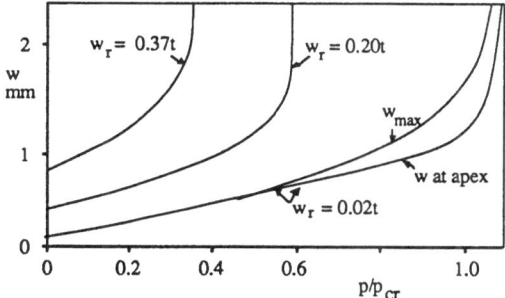

Fig. 9 Lateral deflection w at the apex of a spherical cap as a function of the external pressure. $r = 1000$ mm, $t = 2.03$ mm, $\psi_0 = 30°$, $E = 210000$ MPa.

Discussion

The assumption of [1] regarding the equivalence between initial imperfections and deflections caused by local radial loads appears to be somewhat conservative as is shown in Fig. 4 for cylinders under axial compression. However, in the case of spherical shells, Fig. 8 indicates that the point load causes a more severe disturbance than an initial imperfection of he same magnitude. The difference in behavior between the two cases may be attributed to the difference in the stress fields caused by the local load. The design recommendation given in [1] for cylindrical shells is, therefore on the safe side and the allowable magnitudes of the local forces may even be increased provided additional testing on steel and aluminum test specimens verify the results presented in this report. A tentative procedure may be proposed as demonstrated in Fig. 9:

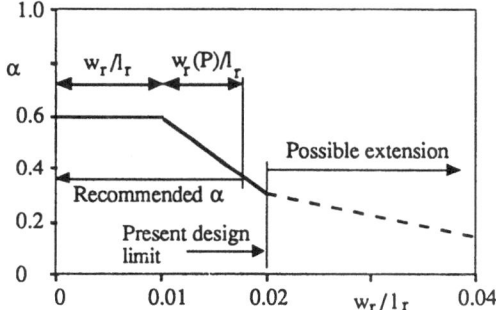

Fig. 10 Suggested design procedure for imperfect shells subjected to additional disturbances such as point loads.

Assume that the tolerance level of 0.01 is valid for a cylindrical shell under axial compression. The reduction factor is then given by α_0 according to ECCS, [5].

A local load P_0 causes an increase of the initial deflection. Using Fig. 5, a value of P^* can be chosen corresponding to a relative displacement $w_r/l_r = 0.01$. Application of the full load P^* implies that the equivalent imperfection amplitude would be equal to the sum of the displacements or 0.02. The allowable stress would then be defined by the reduction factor $0.5\alpha_0$.

Similarly, a smaller value of P leads to a higher value of the reduction factor whereas higher values are not acceptable according to the present rules.

Since interaction between the point load and the most severe imperfection is unlikely to occur, it was suggested in [1] that the effective tolerance level w_{ref}/l_r be determined from:

$$w_{ref}/l_r = \frac{1}{l_r}\sqrt{w_r^2 + w_r(P)^2}$$

Concluding remarks

Up till now, little information has been available on the effect of local loads, acting on a shell surface, on the load carrying capacity of the shell when subjected to buckling. In particular, current codes give no advice for design.

The assumption that the deflection caused by a local disturbance may be regarded in the same way as an initial imperfection leads to a fairly simple rule for design of shells subjected to local loads. The results of this method were compared with test results and finite element analyses for cylindrical and spherical shells.

For cylindrical shells under axial compression, assumption of equivalence between an imperfection and the deflection caused by a local point load appears to be slightly conservative. In the case of spherical shells under external pressure, indications are that the effect of the point load is slightly more severe than that of an imperfection of the same magnitude.

On the other hand, if the correspondence between initial imperfections and deflections caused by local loads etc could be established, testing similar to that presented in the present paper would give results which could be used for establishing more accurate tolerances on initial imperfection levels than those currently in use.

The results of the investigation show that simple design rules may be applied in order to estimate the effect of local disturbances on the buckling capacity of different types of shells. The scope of the present investigation was, however, rather limited and additional tests and finite element analyses are required for a final verification.

Acknowledgements

The author wishes to recognize the careful work performed by Mr Henrik Christiansson, student at the Royal Institute of Technology, Stockholm, who carried out the experiments reported in the present paper, and Mr Peter Dillström, the Swedish Plant Inspectorate, who carried out the finite element analyses.

References

[1]   Samuelson, L. Å., Ed.: The Shell Stability Handbook. Stockholm 1990. (In Swedish)

[2]   Eggwertz, S.F, Samuelson, L. Å.: Buckling Strength of Spherical Caps. The Swedish Plant Inspectorate, R & D Report No 88/04, 1988.

[3]   Tennyson, R.C.: The Effects of Unreinforced Cutouts on the Buckling of Circular Cylindrical Shells under Axial Compression. J. Eng. Ind., ASME, Vol 90, Nov 1968, pp 541-546.

[4]   Brush,D.O., Almroth, B. O.: Buckling of Bars, Plates and Shells. McGraw-Hill, New York 1975.

[5]   ECCS: European Recommendations for Steel Construction. Buckling of Shells, 1988.

[6]   Starnes, J.H.: The Effect of a Circular Hole on the Buckling of Cylindrical Shells. Ph D Thesis, Caltec, Pasadena, Calif. 1970.

[7]   Miller, C.D.: Experimental Study of the Buckling of Cylindrical Shells with Reinforced Openings. Chicago Bridge & Iron Co., CBI-5388, July 1982.

[8]   Gunnarsson, J., Sandberg, D.: Buckling of Cylindrical Shells under Local Loads. Dept. of Steel Constr. Luleå Technical University, Rep. No 037E 1987. (In Swedish)

[9]   Bijlaard, P.P.: Stresses from Radial Load in Cylindrical Pressure Vessels. Welding Journal, Vol 33, Dec 1954, pp 615s-623s.

[10]   Forsberg, K., Flügge, W.: Point Load on Shallow Elliptic Paraboloid.J. Appl. Mech., Sept 1966, pp 575-585.

[11]   Loo, T.C., Evan-Iwanowski, R.M.: Deformations and Collapse of Spherical Domes Subjected to Uniform Pressure and Normal Concentrated Load at the Apex. Proc. World Conf. of Shell Structures, Oct 1962, San Francisco. Techn Editor: R.W. Spangler, pp 297-304.

[12]   Bushnell, D.: Stress Stability and Vibration of Complex Branched Shells of Revolution: Analysis and User's Manual for BOSOR4. NASA CR-2116, 1972.

A STUDY OF BUCKLING IN COLUMN-SUPPORTED CYLINDERS

J.G. TENG and J.M. ROTTER

Department of Civil Engineering
University of Edinburgh, Scotland, U.K.

Abstract
This paper presents a finite element study of the non–linear buckling behaviour of column–supported steel cylinders. Both perfect and imperfect shells are examined. High compressive stresses develop in the vicinity of the column support and these can lead to buckling of the shell at a load much lower than that for a uniformly supported shell. Imperfections are shown to play an important role in reducing the buckling load even when the structure is supported on a small number of columns.

1. Introduction
Large elevated silos are generally supported on a number of columns (Fig. 1). The discrete supports of the columns give rise to high stresses adjacent to the column terminations. In particular, very high meridional compressive stresses arise above the column termination and these can lead to buckling of the shell at a load much lower than that for a uniformly supported shell.

A few studies have investigated the linear prebuckling stress distributions in column–supported silos [1–7], but none appears to have investigated the buckling behaviour of these structures.

Despite the extensive research efforts on shell buckling in the last few decades, only a few studies [8–13] have examined the buckling behaviour of shells under circumferentially varying axial loads. A simple general conclusion from this work on perfect shells might be that

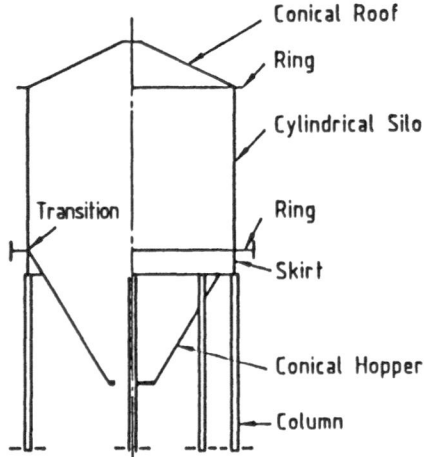

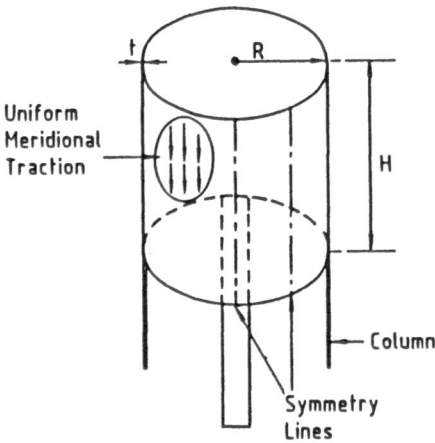

Fig. 1. Typical Column–Supported Silo   Fig. 2. Column–Supported Cylinder

buckling occurs under a circumferentially non-uniform distribution of axial stress when the maximum stress is similar to the classical elastic critical value for uniform axial compression. Libai and Durban [13] gave simple expressions which describe the increase in buckling stress above this simple rule, but the strength gains are generally small. These classical solutions are based on linear bifurcation analyses of perfect shells, though shells under local loads generally behave in a non-linear manner. All the above authors only dealt with perfect shells, even though shell buckling under axial compression is normally acutely imperfection-sensitive. Ory and Reimerdes [2] suggested that column-supported cylinders might be 20-30% less imperfection-sensitive than uniformly supported cylinders. Rotter [14] proposed that the imperfection-sensitivity should steadily decrease as the stress regime becomes more localised, leading to vanishing sensitivity for very local stresses. However, the above were all speculative suggestions, no calculations or experiments on imperfect cylinders are known.

In the study presented here, finite element analyses were carried out to explore the stresses in and stability of column-supported steel cylinders (Fig. 2). The study is a first step towards a better understanding of the more complicated buckling problem in practical column-supported silos, where interaction between the cylindrical shell, the ring and the other shell segments must be considered. Even for this simplified structure, many parameters need to be considered in a detailed exploration. These include the radius-to-thickness ratio, the height-to-radius ratio, the column width, the boundary conditions, the number of columns and the shape, position and amplitude of possible imperfections. The study presented here is thus restricted only to a single shell geometry with a different number of column supports (Fig. 2). Both perfect and imperfect shells are considered. An axisymmetric geometric imperfection, which was previously used by the authors [15] to represent a local weld depression, is introduced to simulate a practical fabrication imperfection. This form and amplitude of imperfection has been repeatedly observed in prototype welded steel silos [16, 17].

2. Finite Element Modelling

The finite element results presented in this study were obtained using an elastic large deflection analysis with the semi-loof curved thin shell element available in the LUSAS system [18]. Details of the semi-loof shell element may be found in [18] and [19].

The columns of this study were assumed to terminate at the lower edge of the shell. The columns were modelled as rigid supports, in which all displacements of the shell lower edge in contact with the column were rigidly restrained. To model the frictional force imposed on the silo wall by the stored bulk solid, a uniformly distributed axisymmetric downward meridional traction was applied to the cylinder (Fig. 2). For an axisymmetrically loaded cylinder on n columns, there exist 2n axes of symmetry for the circular cross-section, so only $1/(2n)$ of the cylinder (Fig. 3a) needs to be modelled. This use of symmetry conditions excludes buckling (failure) modes which do not satisfy the same symmetry conditions. For many practical column-supported silos, the number of column supports is small and this assumption is likely to be valid. A very fine mesh is required, especially in the area near the column support, to model the local nature of stresses and deformations accurately. Away from the shell edge which is supported by the columns, the top and bottom edges were assumed to be restrained against radial and circumferential displacements and against meridional rotations, but free to move vertically. The boundary conditions for the two vertical (meridional) edges of the modelled section were determined by symmetry considerations.

The geometry of the example cylinder (Fig. 2) was defined by the radius-to-thickness ratio $R/t=350$, the height-to-radius ratio $H/R=2$ and the column-width-to-radius ratio $d/R=0.2$. The steel was represented by a Young's modulus $E=2\times10^5$ MPa and a Poisson's ratio $\nu=0.3$. The imperfect surface of the shell was represented by superimposing a local axisymmetric inward depression on a perfect cylinder at an appropriate height above the column supports.

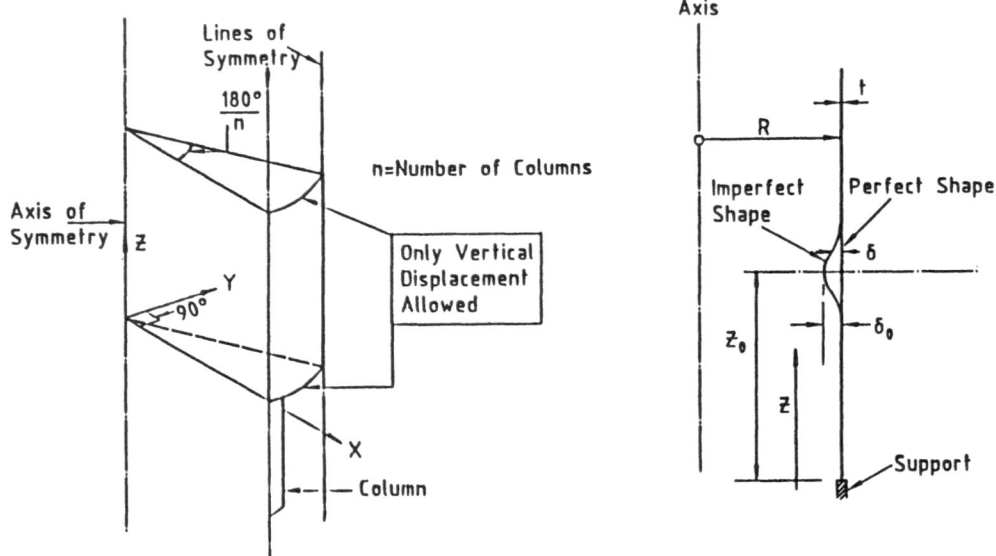

(a) Finite Element Model   (b) Meridional Profiles

Fig. 3 Finite Element Model

In this study, the centre of the imperfection was located at $Z_0/t = 60$. The imperfection shape (Fig. 3b) was taken to be that of the Type A local imperfection used by Rotter and Teng [15] to represent a weld depression. This imperfection is given by:

$$\delta = \delta_0\, e^{-\pi z/\lambda}[\ \sin(\pi z/\lambda) + \cos(\pi z/\lambda)\ ] \qquad (1)$$

$$z = Z - Z_0 \qquad (2)$$

where δ is the imperfection amplitude at height z from the centre of the imperfection, δ_0 is the characteristic amplitude at the centre of the imperfection and λ is the linear elastic bending half wavelength of the cylinder ($\approx 2.444\sqrt{Rt}$). In this study, a local depression with a characteristic amplitude of one wall thickness was used ($\delta_0=t$).

3. Stresses in Column–Supported Cylinders

The stresses in a column–supported cylinder are complicated and not easy to generalise [4,20]. No attempt is made here to explore the stresses in column–supported cylinders in general. Instead, only the membrane stress distributions in cylinders on 4 columns are described to illustrate the patterns which may be typically found in these structures. The bending and shear stresses have been studied elsewhere [21]. Here, comparisons are first made between the stresses in a perfect cylinder and those in an imperfect one with a local depression centred at $Z_0/t=60$.

Figure 4a shows the variation of meridional membrane stresses with the horizontal tangential coordinate Y (Fig. 3a) at different heights in a perfect cylinder subject to a uniform mèridional traction $p_z = 0.02$ MPa. Because the column is modelled as a rigid support, the reaction from the column at the bottom edge of the cylinder (Z/t=0) is far from uniform.

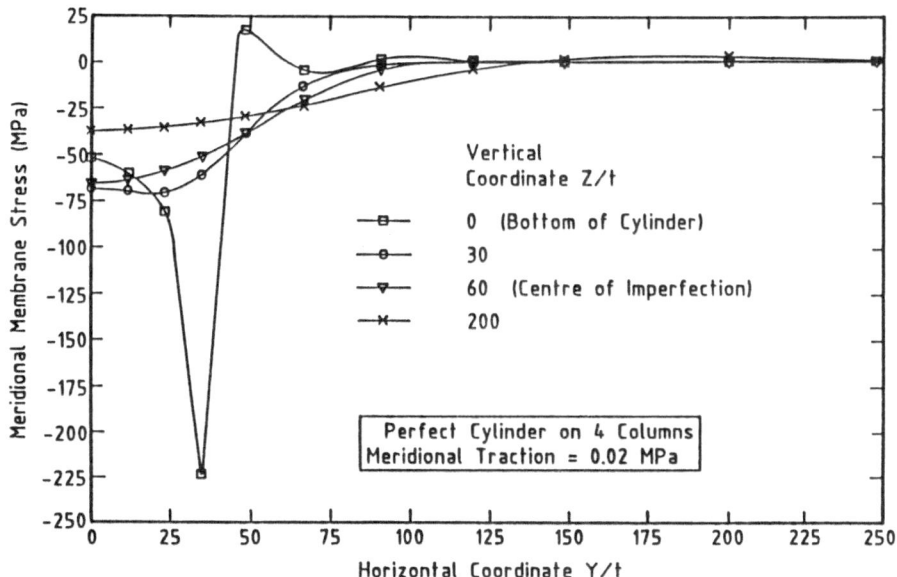

(a) Perfect Cylinder

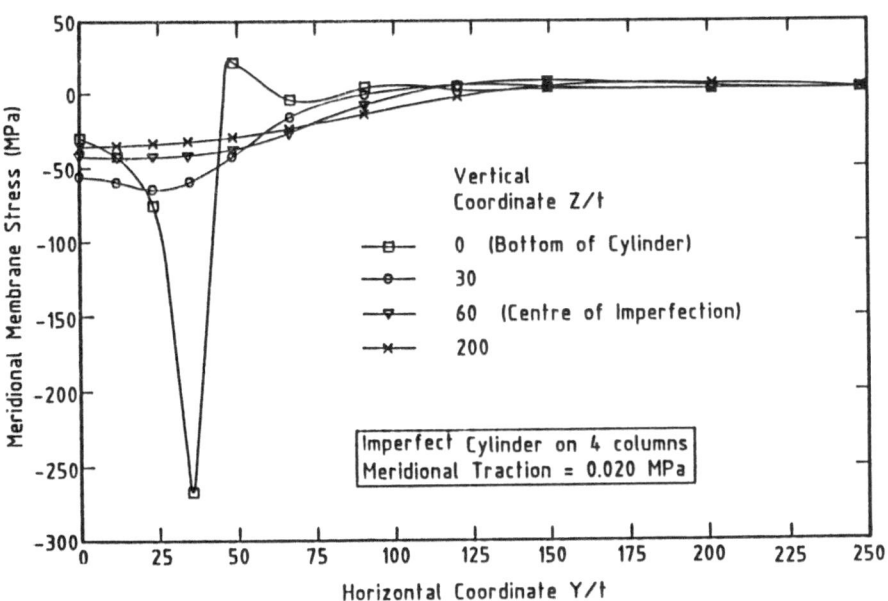

(b) Imperfect Cylinder

Fig. 4 Meridional Membrane Stresses

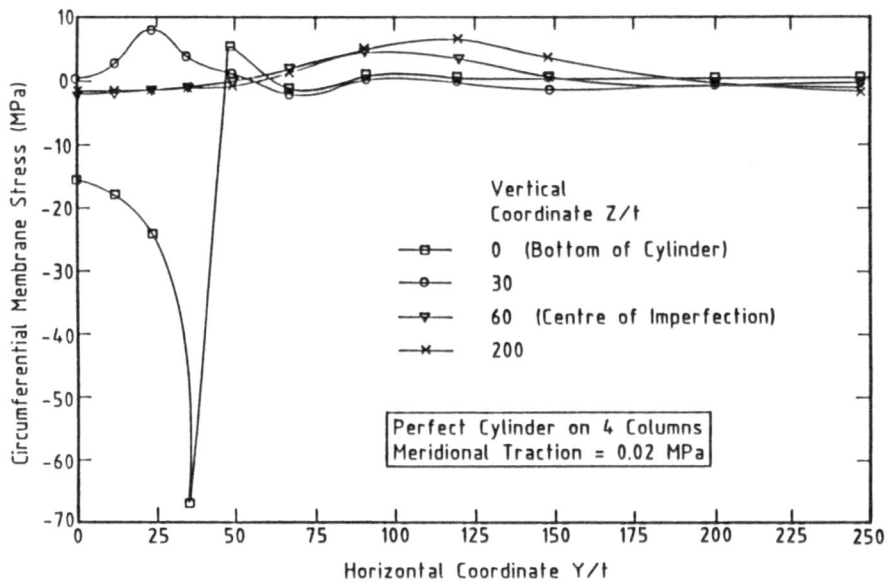

(a) Perfect Cylinder

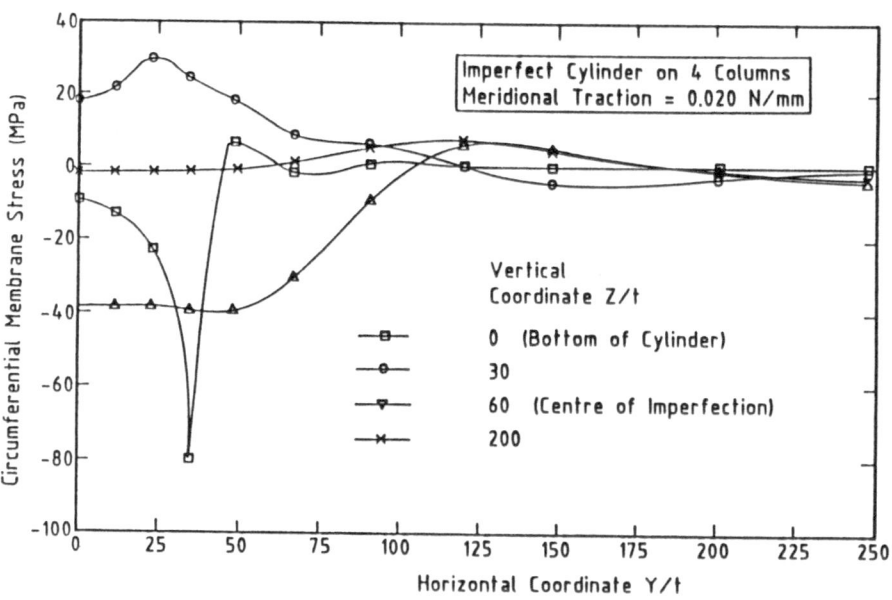

(b) Imperfect Cylinder

Fig. 5 Circumferential Membrane Stresses

Very high compressive stresses develop towards the edge of the column (Y/t=35), whilst much lower but relatively uniform stresses occur at the column centre. At the base of the cylinder, no meridional membrane stresses should exist away from the column, and the small non—zero stresses shown in Fig. 4a are a result of finite element approximations. At a short distance above the base (Z/t=30), the stress distribution is fairly uniform in the zone above the column. Higher up the cylinder, the highest compression quickly moves to the centre line above the column support (eg at Z/t=60). At Z/t = 200, the meridional compression is spread over a large zone with the maximum meridional membrane stress occurring above the centre of the column. For this structure, the highest meridional membrane stress, which occurs at the column edge, is about 16 times that in a uniformly supported shell. The variation of the meridional membrane stresses up the cylinder can be found in [21]. The meridional membrane stresses decrease rapidly above the bottom of the cylinder in the region above the column support. The largest meridional membrane stress occurs on the meridian above the column centre except in the immediate vicinity of the column. The very high stresses adjacent to the column are a local phenomenon and disappear quickly at a short distance above the support.

Figure 4b shows the circumferential variation of meridional membrane stresses in the corresponding imperfect cylinder under the same load. The stress distributions are very similar to those in the perfect shell (Fig. 4a), but the meridional stresses at the cylinder base are larger at the column edge and smaller at the column centre than those in the perfect shell. This change in the reaction distribution is due to the larger deformations in the imperfect shell which has a reduced stiffness because of the imperfection. At the centre of the imperfection (Z/t=60), the stresses in the imperfect shell are smaller and more uniform than those in the perfect one (Fig. 4b).

Returning to the perfect shell, the variation of circumferential membrane stresses with the horizontal tangential coordinate Y is shown in Fig. 5a. The circumferential membrane stresses at the base of the cylinder follow the same pattern as that of the meridional membrane stresses due to Poisson effects. Large compressive circumferential membrane stresses only occur in a very small region immediately above the column. Slightly higher above the column support, the circumferential stresses change to a significant tension at Z/t=30 and then to a very small compression higher up the cylinder (Z/t>60). Near the quarter—span (Y/t=134), tensions develop higher up the cylinder. Overall, the circumferential membrane stresses are much smaller than the corresponding meridional membrane stresses in this perfect shell. Thus, the buckling strength of the perfect shell is likely to depend only on the meridional membrane stresses.

For the imperfect shell, the variation of the circumferential membrane stresses with Y coordinate is shown in Fig. 5b. The circumferential membrane stresses in the imperfect shell are very different from those in a perfect cylinder. At the base of the cylinder (Z/t=0), the circumferential stresses again follow the pattern of the meridional membrane stresses due to Poisson effects. At Z/t = 30, large tensions develop over a significant zone above the column. This is followed by large circumferential compressive stresses at the centre of the imperfection at Z/t=60, which are almost uniform in the region over the column. It should be noted that these stresses are as large as the meridional membrane stresses at this height. Thus, the buckling strength of the imperfect cylinder can be expected to be sensitive to the amplitude of a critically located imperfection. Away from the centre of the imperfection, the circumferential stresses quickly decrease and are similar to those in a perfect shell.

4. Behaviour of Perfect Cylinders
Perfect cylinders on 4, 8, 12 and 20 columns were first analysed using the geometrically non—linear elastic analysis. The load deflection curves are shown in Fig. 6. The plotted displacement was chosen as the normal displacement at a height of Z/t=60, where the centre of imperfection is located in the imperfect shell. Although the LUSAS program includes the arc—length method [22] to follow an unstable postbuckling load deflection path, the program was unfortunately unable to follow this path in the problems considered here, so the analyses were carried out using the Newton—Raphson method (either real or modified).

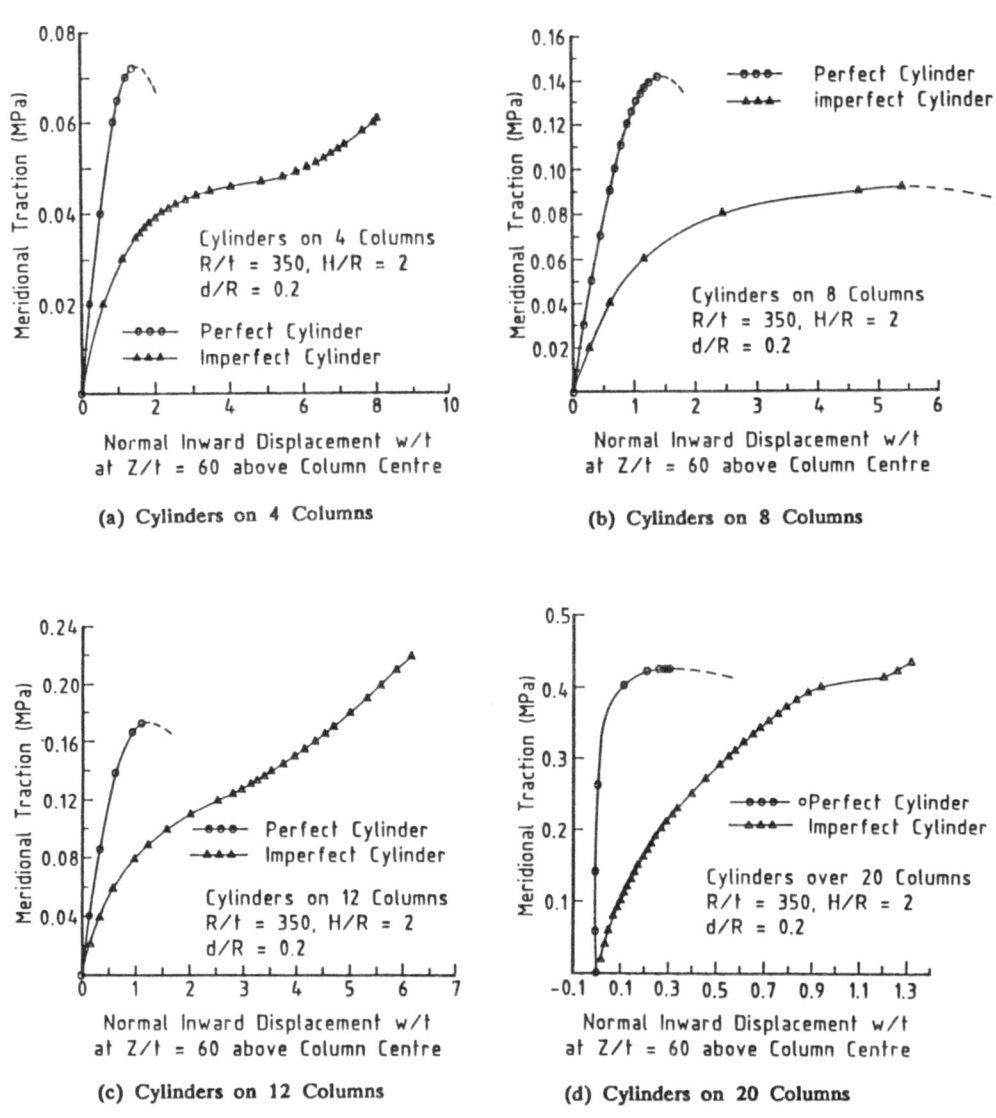

(a) Cylinders on 4 Columns

(b) Cylinders on 8 Columns

(c) Cylinders on 12 Columns

(d) Cylinders on 20 Columns

Fig. 6 Load Deflection Curves

All four analyses failed to converge after a sign change in the determinant of the stiffness matrix was detected, indicating an unstable postbuckling response. The meridional traction at the maximum load for a perfect shell on 4 columns is only 0.072 MPa. This induces a maximum compressive meridional membrane stress of 869 MPa at the column edge, though the mean stress above the column is only 396 MPa. In most practical structures, the stress distribution, and possibly the strength, would thus be affected by yielding of the steel. If the shell had been uniformly supported the meridional stress at the bottom edge would have been

50.4 MPa. The classical elastic critical stress for a uniformly compressed cylinder is 346 MPa. The mean stress just above the column support at the maximum load in this case is thus slightly higher than the classical elastic critical stress. As the number of columns of the same width is increased, the meridional traction at the maximum load rises rapidly (Fig. 6). The deformation mode of the perfect cylinder on 4 columns is shown in Fig. 7a. The deformations at failure are dominated by long wave bending, with larger amplitude short wave deformations in the lower region of the cylinder over the column support, where the meridional compressive stresses are large. The deformations at mid-span are very small.

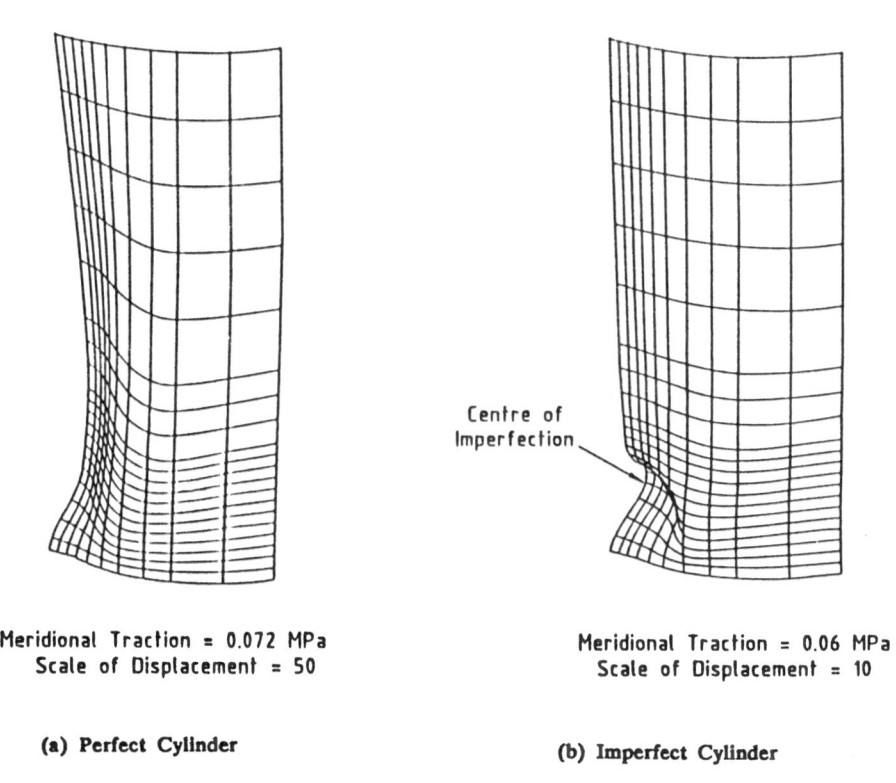

Meridional Traction = 0.072 MPa
Scale of Displacement = 50

(a) Perfect Cylinder

Meridional Traction = 0.06 MPa
Scale of Displacement = 10

(b) Imperfect Cylinder

Fig. 7 Deformed Shapes of Cylinders on 4 Columns

5. Behaviour of Imperfect Cylinders

It was demonstrated earlier that large compressive circumferential stresses can arise due to an appropriately located imperfection. It might be expected that this circumferential compression would lead to a significant reduction in strength, as it does in a uniformly compressed cylinder [15].

All the imperfect shells were found to be much more flexible than their perfect counterparts (Fig. 6). The postbuckling behaviour was unstable for the imperfect cylinder on 8 columns, but stable for the other imperfect shells. It is difficult to decide on an appropriate criterion of failure for some of these imperfect shells. In some analyses, a change of sign in the determinant of the stiffness matrix was detected after significant stable postbuckling deformations (e.g. Figs 6a and 6c). This may indicate an unstable path following the initial stable postbuckling behaviour.

The deformation mode of the imperfect cylinder on 4 columns is shown in Fig 7b. The deformations are localised around the the initial local depression and spread over a circumferential distance slightly wider than the column width. This localisation is due to the local high circumferential compressive stresses caused by the imperfection.

Although the 'failure' loads of the imperfect shells are difficult to define precisely, large reductions in these loads were found due to the imperfection for cylinders on 4, 8 and 12 columns. However for cylinders on 20 columns, the strengths were very similar for both perfect and imperfect shells. A close examination of the deformed shapes shows that the deformations of the perfect shell are localised towards the centre of the column, but those of the imperfect shell are almost axisymmetric [21]. This may be caused by two factors: (a) the imposed symmetry condition may have suppressed a possible bifurcation mode at a lower load, or (b) the location of the imperfection may not be the most detrimental. As a test of the former, a linear bifurcation analysis was carried out for the imperfect structure, and the bifurcation load was predicted to be at $p_z = 0.318$ MPa, significantly lower than the result from the nonlinear analysis. This suggests that the results of these large deflection analyses constitute an upper bound to the strength, when the number of columns is large.

6. Conclusions
This paper has explored the buckling behaviour of steel cylinders on discrete column supports. Both perfect and imperfect shells have been examined. It has been shown that very high compressive membrane stresses develop in the vicinity of the column support and these can lead to buckling of the shell at a load much lower than that for a uniformly supported shell. A criterion of failure for imperfect shells is difficult to define, but imperfections can evidently reduce the useful strength significantly.

7. References
1. H.Ory, H.G.Reimerdes and W.Tritsch. Beitrag zur bemessung der schalen von metallsilos. Stahlbau. 8, pp 243–248 (1958).

2. H.Ory and H.G.Reimerdes. Stresses in and stability of thin walled shells under non-ideal load distribution. Proc. Int. Colloquium on the Stability of Plate and Shell Structures, Gent, Belgium, 6–8 April, ECCS, pp 555–561 (1987).

3. Z.Bodarski, E.Hotala and H.Pasternak. Untersuchung des spannungszustande im mantel von metallsilos". Proc., 10th Internationaler Kongress uber Anwendungen der Mathematik in den Ingenieurwissenschaften, Weimar. (1984).

4. P.L.Gould, S.K.Sen, R.S.C.Wang and D.Lowrey. Column–supported cylindrical conical tanks. J. Struct Divn, ASCE. 102(ST2), pp 429–447 (1976).

5. J.M.Rotter. Analysis of ringbeams in column–supported bins. Proc., Eighth Australasian Conference on the Mechanics of Structures and Materials, University of Newcastle, Aug. (1982).

6. J.M.Rotter. Analysis and design of ringbeams. in Design of Steel Bins for the Storage of Bulk Solids, edited by J.M. Rotter, Univ. Sydney, March, pp 164–183 (1985).

7. L.A.Samuelson. Design of cylindrical shells subjected to local loads in combination with axial or radial pressure. Proc. Int. Colloquium on the Stability of Plate and Shell Structures, Gent, Belgium, 6–8 April, ECCS, pp 589–596 (1987).

8. D.Abir and S.V.Nardo. Thermal buckling of circular cylindrical shells under circumferential temperature gradients. Jnl of the Aerospace Sciences. 26(12), pp 803–808 (1958).

9. D.L.Bijlaard and R.H.Gallagher. Elastic instability of a cylindrical shell under arbitrary circumferential variation of axial stresses. Jnl of the Aerospace Sciences. 27(11), pp 854-858 and 866 (1959).

10. N.J.Hoff, C.C.Chao and W.A.Madsen. Buckling of a thin-Walled circular cylindrical shell heated along an axial strip. Jnl of Appl. Mech., ASME. 31, pp 253-258 (1964).

11. D.J.Johns. On the linear buckling of circular cylindrical shells under asymmetric axial compressive stress distribution. Jnl of the Royal Aeronautical Society, Dec. pp 1095-1097 (1966).

12. A.Libai and D.Durban. A method for approximate stability analysis and its application to circular cylindrical shells under circumferentially varying loads. Jnl Appl. Mech., ASME. 40, pp 971-976 (1973).

13. A.Libai and D.Durban. Buckling of cylindrical shells subjected to nonuniform axial loads. Jnl Appl. Mech., ASME. 44, pp 714-720 (1977).

14. J.M.Rotter. The analysis of steel bins subject to eccentric discharge. Proc., Second International Conference on Bulk Materials Storage Handling and Transportation, Institution of Engineers, Australia, Wollongong, July, pp 264-271 (1976).

15. J.M.Rotter and J.G.Teng. Elastic stability of cylindrical shells with weld depressions. Journal of Structural Engineering, ASCE. 115(5), pp 1244-1263 (1989).

16. M.J.Clarke and J.M.Rotter (1988) A technique for the measurement of imperfections in prototype silos and tanks. Research Report R565, School of Civil and Mining Engineering, University of Sydney, March, (1988).

17. J.M.Rotter. Calculated buckling strengths for the cylindrical wall of 10,000 tonne silos at port kembla", Investigation Report S663, School of Civil and Mining Engineering, University of Sydney, June, (1988).

18. FEA, LUSAS User's Manual and LUSAS Theory Manual, Version 9, Finite Element Analysis Ltd, Surrey, U.K., (1989).

19. B.M.Irons. The semi-loof shell element. in Finite Elements for Thin Shells and Curved Members, eds. Ashwell and Gallagher, Wiley, (1976).

20. A.Kildegaard. Bending of a cylindrical shell subject to axial loading. Proc., Second Symposium on Theory of Thin Shells, IUTAM, Copenhagen, Sept 1967, Springer, pp 301-315 (1969).

21. J.G.Teng and J.M.Rotter. The nonlinear and buckling behaviour of column-supported cylinders. Research Report 90.9, Department of Civil Engineering, University of Edinburgh, U.K., 1990.

22. M.A.Crisfield. A fast incremental/iterative solution procedure that handles snap-through. Computers and Structures. 13, pp 55-62 (1981)

THE SUPPORT OF CYLINDRICAL VESSELS ON RIGID AND FLEXIBLE SADDLES
- AN IMPROVED ANALYSIS

FAISAL A MOTASHAR and ALWYN S TOOTH

Department of Mechanical Engineering
University of Strathclyde, Glasgow G1 1XJ
United Kingdom

Abstract

This paper examines the distribution of the interface forces which occur between the support and a cylindrical vessel under fluid loading. In the early work, by the authors, it was assumed that these pressures were of constant magnitude across the width of the saddle. The validity of this assumption is examined by extending the existing analysis to derive interface pressures across the width of the support for both rigid and 'flexible' saddles. From these values the stresses, which occur in the critical regions of the vessel, can be derived with more accuracy than previously. An illustrative example of a large storage vessel is presented.

1. Notation

B	Distance of saddle centre profile from vessel end
b_1	Distance of a discrete area centre from vessel end ($x = 0$)
C	Half saddle width
D	Extension rigidity $= Et/(1 - v^2)$
E	Modulus of elasticity
i, j	Discrete areas in the ϕ and x directions
k, l	General discrete areas in the ϕ and x directions
K	Bending rigidity $= Et^3/12(1 - v^2)$
L	Length of vessel (tan/tan length)
$N_x N_\phi N_{x\phi} M_x M_\phi M_{x\phi}$	Stress resultants
NA, NC	Total number of discrete areas in the x direction and ϕ direction
P_x, P_ϕ, P_r	Externally applied loading in the x, ϕ and radial directions
$P_{xmn}, P_{\phi mn}, P_{rmn}$	Loading and displacement coefficients, in the x, ϕ and radial
u_{mn}, v_{mn}, w_{mn}	directions employed in the Fourier series
R	Mean radius of the cylindrical vessel
t	Wall thickness of vessel
u, v, w	Mid surface displacements in the x, ϕ and radial directions
x, ϕ, z	Coordinates in the axial, circumferential and radial directions
β, γ	Half discrete area size in the ϕ and x directions
Δ	Vertical upward rigid body displacement of saddle with respect to end of the vessel
λ	$m\pi R/L$
v	Poisson's ratio
ϕ_i	Angular distance of a discrete area centre from ($\phi = 0°$)

2. Introduction

The design of horizontal vessels supported on twin saddles has been dealt with by several authors over the years. However, the approach given in the Pressure Vessel Standards BS5500 and ASME are essentially the work of L.P. Zick [1], who used a modified beam and ring analysis so that the mathematical model for the vessel predicted values which agreed with the experimental results he had available. More recent work by Tooth, et al [2,3] had indicated that Zick's treatment for the vessel full of fluid predicts stresses which are in reasonable agreement with the experimental values when a flexible saddle is employed. However, when the saddle is rigid, compared with the vessel, Zick's treatment underestimates the peak stresses which actually occur at the horn.

The key to understanding the behaviour of such problems lies in deriving the interface forces which occur between the support and the vessel. The magnitude and distribution of these forces depends upon the vessel flexibility and the rigidity of the support. In the earlier analytical work by Tooth et al [4], the configuration of the support was found to have a crucial effect on the stress in the vessel - primarily in the 'horn' region of the saddle. For example, when a flexible saddle is employed, the vessel stresses can be reduced by up to 50%.

In order to determine the interface pressures between the saddle and the vessel, the saddle contact area is divided into a number of discrete areas, each of which is subject to unknown uniformly distributed pressures in both the radial and tangential direction. For ease of calculation, the early work [2-5] assumed that these pressures were of constant magnitude across the saddle width. That is, the saddle had an element of radial flexibility across the width to avoid pressure high spots.

When the saddles are of rigid construction (i.e. no radial or tangential flexibility) compared to the vessel, such as with a GRP vessel on a steel or concrete support, the above assumption will not be valid. However, the majority of steel vessels are placed on fabricated steel saddles with a central web. These do contain radial flexibility across the width. This work examines the validity of this assumption by deriving the interface forces across the width of the saddle for both rigid and flexible supports used in a large cylindrical storage vessel.

3. Fourier Expansion Solutions

The behaviour of the thin-walled circular cylindrical shell can be described using the set of governing differential equations proposed by Sanders [6]. The surface loading with symmetry about the nadir, $\phi = 0°$ can be represented as a double Fourier series expansion as follows:-

$$\begin{bmatrix} P_r \\ P_x \\ P_\phi \end{bmatrix} = \sum_{m=0}^{\infty} \sum_{n=0}^{\infty} \begin{bmatrix} P_{rmn} & \cos(n\phi) & \sin(\lambda x/R) \\ P_{xmn} & \cos(n\phi) & \cos(\lambda x/R) \\ P_{\phi mn} & \sin(n\phi) & \sin(\lambda x/R) \end{bmatrix} \tag{1}$$

where $\lambda = m\pi R/L$

An appropriate solution for the displacements can be written:-

$$\begin{bmatrix} w \\ u \\ v \end{bmatrix} = \sum_{m=0}^{\infty} \sum_{n=0}^{\infty} \begin{bmatrix} w_{mn} & \cos(n\phi) & \sin(\lambda x/R) \\ u_{mn} & \cos(n\phi) & \cos(\lambda x/R) \\ v_{mn} & \sin(n\phi) & \sin(\lambda x/R) \end{bmatrix} \tag{2}$$

By substituting equations (1) and (2) into the governing shell equations a matrix equation, involving coefficients of displacements and loadings, is obtained. This can be evaluated using Cramer's rule or the Co-factor method of matrix inversion. From this u_{mn}, v_{mn} and w_{mn} are expressed as functions of P_{rmn}, P_{xmn} and $P_{\phi mn}$. Using the stress resultant - displacement relations the values of N_x, N_ϕ, M_x and M_ϕ can be obtained for each of the loading cases (radial P_{rmn}, tangential $P_{\phi mn}$ and longitudinal P_{xmn}). The only unknowns in the double Fourier series expressions, which arise from the above approach, for w, v, u and the stress resultants are the loading coefficients P_{rmn}, $P_{\phi mn}$ and P_{xmn}. These can readily be related to the applied loading P_r, P_ϕ and P_x.

4. The Saddle Supported Vessel

The cylindrical vessel of length L, mean radius R and of constant wall thickness t is supported on twin rigid saddles which are located on equal distance B, from the vessel ends. It is assumed that the loading on the cylinder is symmetrical about the vertical plane through the cylinder centre line. In this treatment the saddle is assumed to be welded to the vessel. The loose saddle can be tackled in a similar manner by forcing the outward radial forces on the vessel to be zero and introducing a tangential frictional force.

The Interface Force System

The saddle/vessel contact area is assumed to be divided into a number of equal size discrete areas $2\beta \times 2\gamma$ in size, as shown in Figure 1. These can be graded to provide smaller areas at the top and/or the edges of the saddle. However, in this present case equal sizes are used in both the axial and circumferential directions. The discrete areas in the axial direction are identified as 'j' with a total number equal to NA, and in the circumferential direction as 'i' with a total number of NC on both sides of the vessel nadir ($\phi = 0°$).

Each discrete area is loaded with a uniform radial pressure and tangential shear. For example, on area 'ij' a radial pressure of p_{ij} and tangential shear of t_{ij} is assumed to act. The radial and tangential displacements of the vessel at a general point 'kl' due to p_{ij} and t_{ij} are given by:-

$$w_{kl} = t_{ij} (w_t)_{ij,kl} + p_{ij} (w_r)_{ij,kl}$$

$$v_{kl} = t_{ij} (v_t)_{ij,kl} + p_{ij} (v_r)_{ij,kl} \tag{3}$$

where $(w_t)_{ij,kl}$ and $(v_t)_{ij,kl}$ are the radial and tangential displacements of point 'kl' due to unit tangential shears applied over area 'ij' and $(w_r)_{ij,kl}$ and $(v_r)_{ij,kl}$ are the radial and tangential displacements of point 'kl' due to unit radial pressures applied over area 'ij'.

The total radial and tangential displacements of point 'kl' on the surface of the vessel due to all the interface forces are then given by:-

$$W_{kl} = \sum_{j=1}^{NA} \sum_{i=1}^{NC} t_{ij}(w_t)_{ij,kl} + \sum_{j=1}^{NA} \sum_{i=1}^{NC} p_{ij}(w_r)_{ij,kl}$$

$$V_{kl} = \sum_{j=1}^{NA} \sum_{i=1}^{NL} t_{ij}(v_t)_{ij,kl} + \sum_{j=1}^{NA} \sum_{i=1}^{NC} p_{ij}(v_r)_{ij,kl} \tag{4}$$

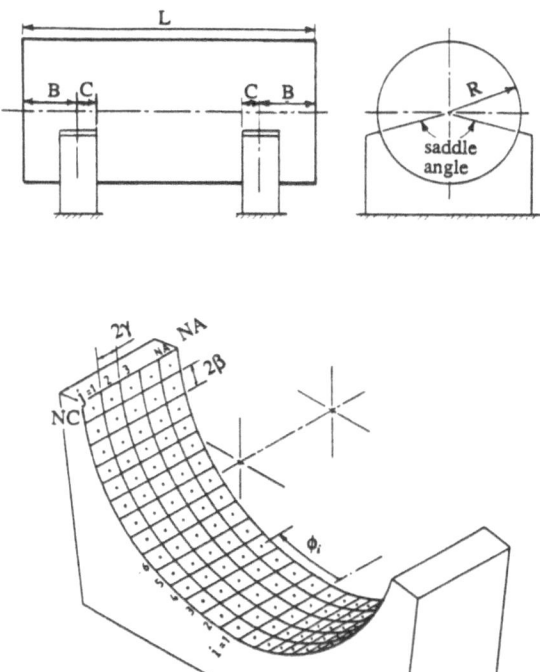

Fig 1   The saddle/vessel contact surface of a twin saddle supported cylinder.

These expressions are valid for the NA discrete areas along the saddle width and the NC areas around each half attachment about $\phi = 0°$. They can be rewritten in matrix form:-

$$[W] \ = \ [WT][T] \ + \ [WR][P]$$

$$[V] \ = \ [VT][T] \ + \ [VR][P] \tag{5}$$

The elements of the flexibility matrices [WR], [VR], [WT] and [VT] are given by the series forms of the displacements w and v in terms of the loading functions.

The loading coefficients P_{rmn} and $P_{\phi mn}$ can be found for the four areas detailed above (i.e. an area each side of the two saddles), where P_r and $P_\phi = 1$ in the loaded region and zero elsewhere. In addition to the reactive interface forces the vessel is subjected to the applied loading which is a combination of hydraulic pressure, internal surcharge, and the self weight of the vessel. The loading coefficients for these can also be derived from which the radial and tangential displacements at the centres of the discrete areas of the support can be obtained. These are written in matrix form as [WHSW] and [VHSW], respectively.

65

Compatibility and Equilibrium

Compatibility at the centre of the discrete areas is enforced by the following:-

$$[WR][P]+[WT][T]+\Delta[CS]+[WHSW]= -[WTS][T]-[WRS][P]$$

$$[VR][P]+[V][T]-\Delta[SN]+[VHSW]= -[VTS][T]-[VRS][P] \tag{6}$$

where Δ is a rigid body movement in the vertical direction, and [WTS], [WRS], [VTS] and [VRS] are the saddle flexibility matrices, obtained from say a finite element method (for the rigid case these are set to zero).

The final set of equations are those to establish equilibrium. In this the components of the interface pressures must balance the applied load.

Determination of Vessel Stresses and Displacements

Using the compatibility and equilibrium equations it is possible after some manipulation to determine the unknown interface forces [T] and [P] in terms of the known applied loading. Combining these with the gravity and pressure loadings in the vessel the total loading coefficients $P_{rmn}, P_{\phi mn}$ can be obtained. These are used to determine the displacements and stress resultants at all points in the vessel. Certain techniques have been devised to reduce the computing time required to generate the vessel flexibilities and to derive the subsequent distribution of stresses.

5. Illustrative Example

To illustrate distribution of the interface pressure across and around the saddle a selection of typical theoretical results are presented for a large gas receiver. The vessel is 3658 mm mean diameter, 54860 mm barrel length, 26.6 mm constant wall thickness. The saddles are 762 mm wide of an angle 162° and located 6858 mm from the ends of the barrel length. Both rigid and flexible saddle designs are examined. The saddle flexibilities were obtained using a finite element package. The case when the vessel is full of water is considered, the weight of the vessel is ignored in these calculations.

A range of cases were examined where the saddle width was divided into 3,5,7,9 and 11 discrete areas and up to 30 in the half circumferential arc length.

It was found that the resulting stresses in the vessel, when a flexible saddle was used, were insensitive to the number of discrete areas employed. Thus for the sake of illustration the case of 20 areas in the half saddle arc length, NC = 20, and 3 areas across the saddle width, NA = 3, i.e. (20 x 3) is presented. In the case of the rigid saddle the stresses were higher when more discrete areas were considered. The case of 20 x 3 presents typical values. For comparison the 20 x 1 case (uniform pressure across the width) is also presented. The results for the interface pressures are shown in Figures 2(a), (b) and (c).

In the rigid case, (a), high values of radial and tangential interface pressures exist at the horn. Both of these are so disposed as to cause local moment loading. These occur in (i) a plane at right angles to the vessel axis and (ii) a plane containing the vessel outer surface. Both of these are required to satisfy compatibility of the welded vessel rigid support interface which occurs in this case. In the flexible case, (c), the pressures are much reduced and the local moment loading is almost totally absent. Furthermore it is also noted that the forces in the immediate horn region are more nearly uniform across the width in the flexible saddle case than when the saddle is assumed rigid. The uniform nature of these forces is particularly noticeable in the 10 x 3 and 10 x 5 cases (not presented).

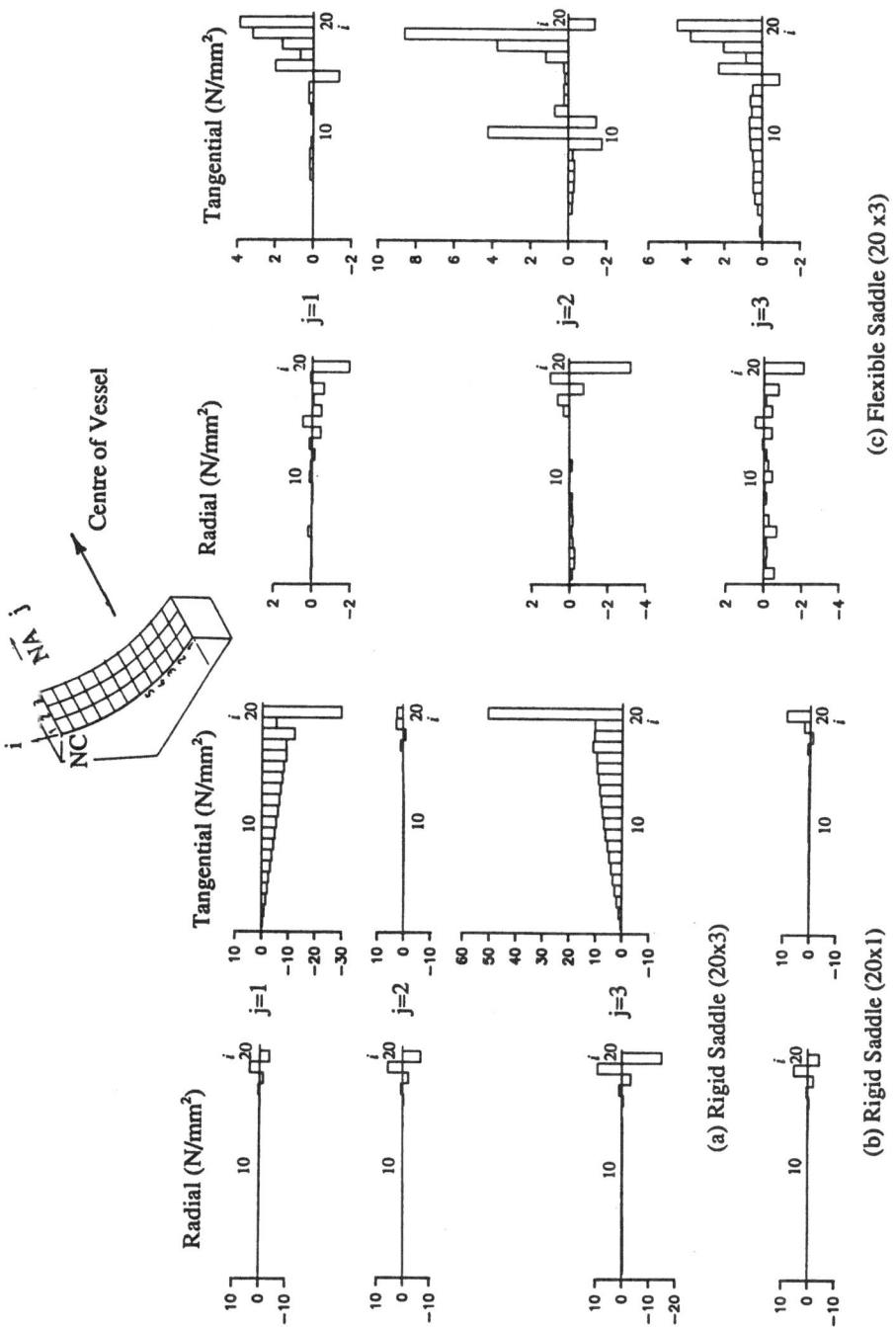

Figure 2   Radial and Tangential Interface Pressures for Rigid and Flexible Saddles

67

The resulting maximum stresses, which occur on the outer surface in the circumferential direction, are shown in Figures 3 and 4, plotted around the vessel and along the generator in the horn region. The advantage of the flexible saddle is clearly noted since the peak stress in dramatically reduced. It is almost one third that which occurs when a rigid saddle is used. When the (20x1) rigid case is assumed for a flexible configuration a conservative value is obtained.

6. Concluding Comments

The analysis presented enables an assessment to be made of the peak stresses which occur in a vessel in the region of a saddle support. The superiority of the flexible support is clearly demonstrated. The assumption of a uniform load across the width of a rigid saddle can be safely assumed for those cases where the saddle is of a fabricated construction with a central web, i.e. with inbuilt flexibility across the width.

Acknowledgements

The authors wish to acknowledge the financial assistance they have received from The Institution of Mechanical Engineers, who made available a Senior Engineers Conference Grant to enable A S Tooth to present the paper and attend the IUTAM meeting in Prague September 1990.

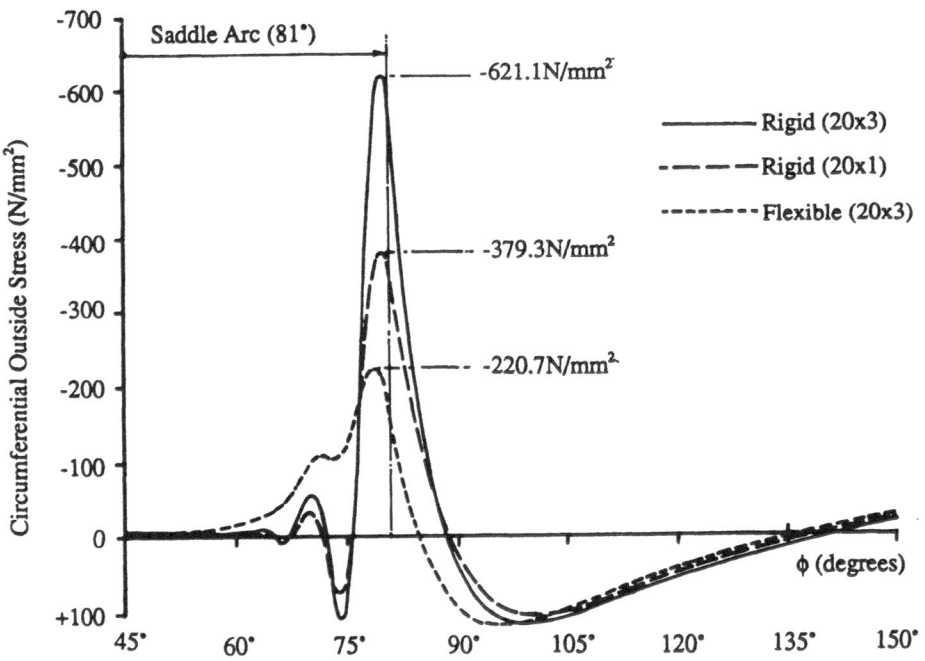

Figure 3   Distribution of the outside surface circumferential stress around the vessel in the profile containing maximum stress for rigid and flexible saddles

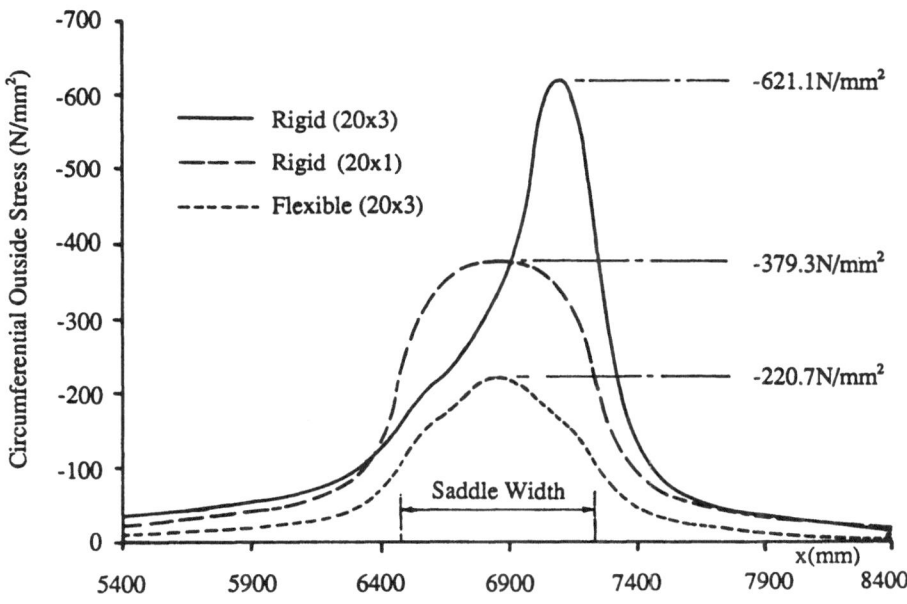

Figure 4   Distribution of the outside surface circumferential stress across
the saddle in the horn region for rigid and flexible saddles

References

1.   Zick, L.P., Stresses in large horizontal cylindrical pressure vessels on two saddle supports.
Welding Res Supplement, 1951, Vol30, No. 9, 435-445s.

2.   Duthie, G. and Tooth, A.S. The analysis of horizontal cylindrical vessels supported by
saddles welded to the vessel - a comparison of theory and experiment, 3rd Int. Conf. on
Pressure Vessel Tech, Tokyo, 1977, pp 25-38.

3.   Tooth, A.S., Duthie, G., White, G. C. and Carmichael, J., Stresses in horizontal storage
vessels - a comparison of theory and experiment, J. of Strain Analysis, 1982, 17, No. 3,
169-176.

4.   Wilson, J. D. and Tooth, A. S., The support of unstiffened cylindrical vessels, 2nd Int.
Conf. on Pressure Vessel Tech, San Antonio, 1973, pp 67-83.

5.   Duthie, G., White, G. C., and Tooth, A. S., An analysis for cylindrical vessels under local
loading - application to saddle supported vessel problems, J of Strain Analysis, 1982, 17,
No. 3, 157-167.

6.   Sanders, J. L., Jr, An improved first approximation theory for thin shells. NASA, TR-R24,
1959.

EXPERT LIKE SYSTEM FOR CYLINDRICAL SHELL SUPPORT

P. VOKROJ

VÍTKOVICE - Institute of Applied Mechanics,
611 00 Brno 11; box 32,
Czechoslovakia

Abstract

The submitted contribution describes expert-like SAUDES program
system based on data base of 288 cases of contact problem solutions. The data base consists of extreme values of longitudinal
and circumferential stresses at the saddle horn. These values
are functions of 3 parameters that are shell geometry, dimensions
of loaded area and non-dimensional distance of saddle from stiffening element (plate, rib or drumhead). The solution is based
on semibending shell theory for semi-infinite length of cylindrical shell, for line load contact forces. Both ends influence is
done as superposition of two single side influence. The basis
of task is 3-dimensional linear interpolation of 3-parameters
function. The stresses obtained are evaluated by help of ASME
stress categorisation. The procedure suggested is compared with
six examples of theoretical and experimental solutions, known
from technical literature.

1. Introduction

Expert like system SAUDES is based on database of results of
288 cases of contact problem solutions. The database consists
of extreme values of longitudinal and circumferential stresses
at a horn of a rigid saddle of cylindrical shell as a function
of 3 parameters:
a) shell geometry together with the depth of saddle support in
   direction of axis of symetry of shell,
b) circumferentia l angle of saddle in plane perpendicular to
   symetry axis,
c) distance between saddle center profile and shell drumhead or
   rib or plate.

The stress results are obtained by help of space linear interpolation that consists of two two-dimensional and one one-dimensional linear interpolations.

2. Method Used

The contact problems were analysed by help of conditions of continuation of deformation at individual points of contact region
together with condition of statical equilibrium of contact forces with respect to saddle reaction applied. The contact forces
were introduced as line ones using division of contact region
in circumferential direction by one degree, Lit. |5|. An influen-
ce lines for contact problems were introduced for semibending

state of cylindrical shell of semi-infinite length. The accele-
ration of used series was introduced for reduction of computa-
tion time required, Lit. |6|. The scheme used was leading to
system of linear simultaneous equations, solved by Gauss-Jordan
elimination. The extreme values were obtained by help of superposition of contact forces influence. Both ends effect is done
as double use of one end influence.

$$k = \frac{b}{r} \sqrt{\frac{t}{r}} \qquad \bar{a} = \frac{a}{r} \sqrt{\frac{t}{r}}$$

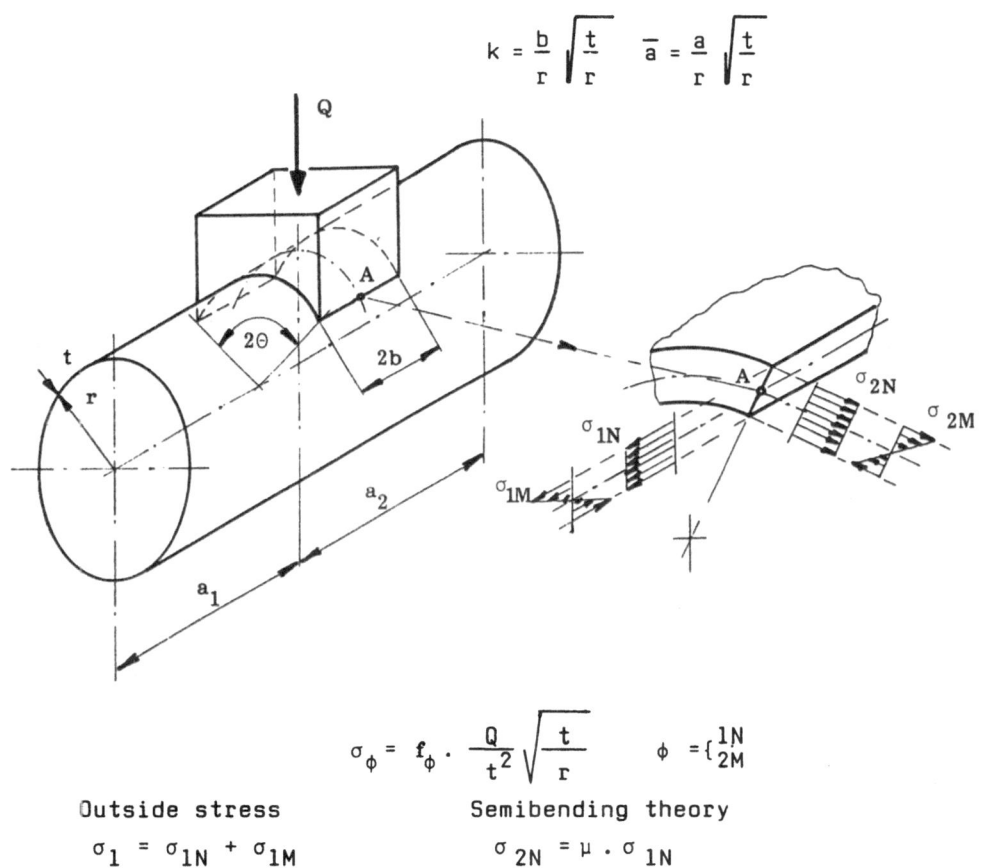

$$\sigma_\phi = f_\phi \cdot \frac{Q}{t^2} \sqrt{\frac{t}{r}} \qquad \phi = \{^{1N}_{2M}$$

Outside stress

$$\sigma_1 = \sigma_{1N} + \sigma_{1M}$$

$$\sigma_2 = \sigma_{2N} + \sigma_{2M}$$

Semibending theory

$$\sigma_{2N} = \mu \cdot \sigma_{1N}$$

$$\sigma_{1M} = \mu \cdot \sigma_{2M}$$

Notation:

r - mean radius of shell
t - thickness of cylindrical wall
$\bar{a}_1$ - non-dimensional distance of saddle central profile from 1.end
$\bar{a}_2$ - non-dimensional distance of saddle central profile from 2.end
b - saddle width
k - non-dimensional shell parameter
θ - half saddle angle
Q - saddle support reaction

71

k	Θ	$\bar{a}$
0,0005	5°	0,005
0,001	15°	0,02
0,0025	30°	0,05
0,005	45°	0,1
0,01	60°	0,2
0,025	75°	0,5
0,05		1,0
0,1		

3. Results

The four stresses are obtained in both longitudinal and circumferential directions and on outside and inside surfaces of cylindrical shell. These stresses obtained are evaluated by help
of ASME stress categorisation. From material yield and strength
stress limits is established allowable one. Such allowable stress
is compared with the amplitude of principal stresses, composed of
a) local membrane stresses,
b) summation of local membrane and local bending stresses,
to fulfill the ASME criteria of safe service.

4. Verification

For verification of suggested procedure, six examples were compared with SAUDES system, that are known from technical literature.
Most of them were published together with results of measurements,
that complete the verification in best way; see Lit. |1| - |4|.

Vessels:

1 - CRIF report MT113
2 - Forbes, Tooth 1968
3 - Duthie, Tooth 1977 - case 3
4 - Duthie, Tooth 1977 - case 4
5 - Lakis, Dore 1978
6 - Duthie, Tooth 1977 - case 1, data see 5

Geometrical data:

Vessel	1	2	3	4	5
R [mm]	952,75	142,88	453,35	452,7	1817,3
t [mm]	2,5	0,71	3,3	4,6	25,4
a1 [mm]	815,0	246,8	1410,0	1410,0	9140,0
a2 [mm]	2645,0	1099,4	5910,0	5910,0	45720,0
2b [mm]	400,0	44,7	102,0	102,0	760,0
Θ [deg]	51,5	72,0	75,0	75,0	75,0
Q [N]	55000,0	431,7	23630,0	23570,0	2846000

Comparision table:

OUTSIDE STRESS [MPa]

Ves.	longitudinal			circumferential		
	theory	experim.	SAUDES	theory	experim.	SAUDES
1	-48,3	-26,0	-40,2	-129,6	-116,4	-120,0
2	-5,3	-4,1	-5,4	-20,3	-17,9	-14,9
3	-71,7	-56,5	-96,5	-194,6	-172,9	-251,8
4	-56,5	-21,7	-59,8	-128,6	-85,7	-155,1
5	-261,7	-	-233,9	-636,4	-	-620,8
6	-285,8	-	-233,9	-608,2	-	-620,8

The program system SAUDES was composed for IBM PC compatible computers and series of tests were done for comparison with published analytical and experimental results of contact problems solved. It is possible to suppose, that system SAUDES will be convenient tool for designers of pipelines and storage tanks.

5. References

1. R.Beeckman-D.Van Leuwen. Programmes de calcul per ordinateur, CRIF report MT 113, Sept. 1976, Bruxelles.

2. P.V.Forbes-A.S.Tooth. An Analysis for Twin Saddle Supported Unstiffened Cylindrical Vessels, In.Recent Advances in Stress Analysis, Royal Aeronautical Society, March 1968, London.

3. G.Duthie-A.S.Tooth. The Analysis of Horizontal Cylindrical Vessel Supported by Saddles Welded to the Vessel, In. 3-rd Int. Conf. on Pressure Vessel Technology, Tokyo April 1977.

4. A.A.Lakis-R.Doré. General Method for Analysing Contact Stress on Cylindrical Vessels, Int.J. Solid Structures Vol. 14, pp. 499-516 (1978).

5. V.Křupka. The Contact Between a Rigid or Flexible Saddle Support and Thin Elastic Shell, Arch. Budowy Maszyn XXIV (1977).

6. P.Vokroj. Local Load on Cylindrical Shell Analysis, Thesis, VÍTKOVICE-Institute of Applied Mechanics, Brno 1987 (in Czech).

DISCUSSIONS

LOAD CARRYING CAPACITY OF SHELLS
WITH REGARD TO CUMULATION OF DAMAGES

M. T. ALIMZHANOV

Laboratory of Mechanics deformable rigid body,
Institute of Mathematics and Mechanics of AS of KazSSR,
Pushkin str., 125, 480021 Alma-Ata, USSR

It is known, the degree damaged of the material is characterized by the size of measure damage. Most simple and widely extended from the phenomenological hypothesis is the hypothesis of the linear summation of damages /1 - 3/. The essence of this hypothesis is contained in that the damage, evoking given cycle of the stresses, is assumed not depending from the condition of the construction at present moment and from the previous history of the loading and simple is summed with the damages, caused by the previous cycles. The measure of the damage in input condition of the construction is assumed equal zero, and in moment of the destruction equal unit

In the message rather different approach to the calculation of stress-deformed condition (SDC) of elements of thick-walled constructions with regard of accumulation of damages establishment of their optimal amounts /4/ is considered.

As far as the elements of the engineering structures and machines not infrequently have the form of thick-walled cylindrical shells that in the capacity of example consider the tick-walled cylinder, lying in action uniform inner and outer pressures, where $P_b > P_a$. The process of the accumulation of damages in the body of thick-walled cylinder in result of strong effect (increase of outer pressure) reflects in lowering stability characteristic his material in zone of inelastic deformations (ZID). If introduce designations: K - meaning of stability characteristic entire material, K_0 - minimal meaning stability characteristic of defective material, that correspond raised level of accumulation damages on inner surface cylinder under maximal development of ZID (i. e. when this zone goes out on outer surface), ρ_0 - dimensionless outer radius of (ZID), defined in the process of solution the problems, $\alpha = a/b$, a and b - respectively inner and outer radius of the cylinder, $\rho = r/b$ - dimensionless current coordinate, then starting from distribution of stresses by thickness cylinder, the regularity of lowering of stability characteristic of the material of ZID can be approximate by the function

$$K(\rho) = K + (K_0 - K) \frac{\alpha^n}{\alpha^n - 1} \left[1 - (\frac{\rho_0}{\rho})^n \right] , \qquad (1)$$

where n - the parameter of approximation.

On Figure 1 according to expression (1) is shown the regularity of change stability characteristic of the material $K(\rho)$ of ZID depending on parameters n and ρ_0. As is obvious from this figure on measure increase of ZID (that connected with the increase of outer pressure) the mean of stability characteristic on inner surface cylinder is dropped, that allow to consider the formation and development micro and microsplits in the process of deformation. In limiting case, when $\rho_o \rightarrow 1$, the function $K(\rho)$ and inner surface cylinder tends to its limit minimal meaning K_o. Thus, the presence in expression (1) parameter ρ_o, characterizing stress condition of the cylinder allow to consider the process of accumulation of damages on measure of development of ZID.

In the capacity of development of transfer material in condition of inelastic deformation we take the condition of plasticity Tresca-Saint Venant or Coulomb-Mohr.

Solving elastic -plastic problem about ZID of thick-walled cylinder under the condition plasticity of Tresca-Saint Venant with regard to expression (1), is obtained, that the boundary ρ_o between the zones elastic and inelastic deformations is derived from the following relation:

$$2K\ln\frac{\rho_o}{\alpha} + 2(K_o-K)\frac{\alpha^n}{\alpha^n-1}\left[\ln\frac{\rho_o}{\alpha} + \frac{1}{n} - \frac{1}{n}(\frac{\rho_o}{\alpha})^n + \right.$$

$$+ K(1-\rho_o^2) - P_b + P_a = 0 \qquad (2)$$

with characterize the regularity of change radius ρ_o of ZID as degree of increase outer pressure on cylinder.

On figures 2, 3 represent the graphs this function for different meanings α, n and β ($\beta=K_o/K$ - parameter characterizing the level of damaged).

Note, that under $K_o=K$ the relation (2) turns into well-known solution /5/, corresponding on graphs of curved $\beta = 1$. The boundary ρ_o between the zones of elastic and inelastic deformations in case of condition plasticity of Coulomb-Mohr is defined from the equation:

$$\rho_o^{\alpha_2}\left\{P_a\alpha^{-\alpha_2} + ctg\,\varphi\left[K(\alpha^{-\alpha_2} - \rho_o^{-\alpha_2}) + \right.\right.$$

$$+ (K_o-K)\frac{\alpha^n}{\alpha^n-1}(\alpha^{-\alpha_2} - \rho_o^{-\alpha_2} - \rho_o^n\frac{\alpha_2}{\alpha_2+n}$$

$$\left.\left.(\alpha^{-\alpha_2-n} - \rho_o^{-\alpha_2-n}))\right]\right\} + \frac{(P_b+Kctg\,\varphi)\sin\varphi}{1-\rho_o^2\sin\varphi}(1-\rho_o^2) - $$

$$- P_b = 0, \qquad (3)$$

$$\alpha_1 = \frac{1+\sin\varphi}{1-\sin\varphi}, \qquad \alpha_2 = \frac{2\sin\varphi}{1-\sin\varphi}, \qquad \varphi - \text{angle of inner friction}$$

The graphs, constructed in accord of expression (3), qualitative coincide with the graphs on figures 2 and 3. Note, that under $K_o = K$ the relation (3) turns into solution, previously obtained by the author /6/.

78

The analysis of the relations (2) and (3) show, that the dependence of radius ϱ_0 of ZID from outer pressure qualitative seem the following form (Fig.4) On graph, the point C, corresponding the beginning of not falling part on this curve corresponds to the moment of loss carrying capacity of cylinder In addition the steepness of not falling part characterize the level of damaged material of ZID. As far as not difficult to prove the existence and uniqueness of maximum point, then we have the second equation for determination of optimal thickness of cylinder. Thus, if the value of uniform (hydrostatic) pressure is known, the meaning of outer radius cylinder b, his stability characteristics, parameters defining the degree of damaged material of ZID, then optimal (least) thickness of cylinder will find on formula:

$$d = b(1 - \alpha),$$

where the parameter α is defined from the system of two transcendental equations

$$f(\varrho_0, \alpha) = 0, \quad \frac{\partial}{\partial \varrho_0} f(\varrho_0, \alpha) = 0.$$

Thus in case of condition Tresca-Saint Venant this system takes the view:

$$\ln \frac{\varrho_0}{\alpha} + (\beta-1) \frac{\alpha^n}{\alpha^n - 1} \left[\ln \frac{\varrho_0}{\alpha} + \frac{1}{n} - \frac{1}{n} (\frac{\varrho_0}{\alpha})^n \right] +$$

$$\frac{1 - \varrho_0^2}{2} - \frac{P_b - P_a}{2K} = 0,$$

$$\frac{1}{\varrho_0} - \varrho_0 + (\beta-1) \frac{\alpha^n}{\alpha^n - 1} \frac{1}{\varrho_0} - \frac{\varrho_0^{n-1}}{\alpha^n} = 0, \qquad (4)$$

and in case of Coulomb-Mohr the view:

$$\frac{1 - \varrho_0^2 \sin\varphi}{1 - \sin\varphi} \left\{ (\frac{\varrho_0}{\alpha})^{\alpha_2} - 1 + (\beta - 1) \frac{\alpha^n}{\alpha^n - 1} \left[(\frac{\varrho_0}{\alpha})^{\alpha_2} - 1 + \right. \right.$$

$$\left. \left. + \frac{\alpha_2}{\alpha_{2+n}} (1 - (\frac{\varrho_0}{\alpha})^{\alpha_{2+n}}) \right] \right\} + \frac{\alpha_2}{2} (1 - \varrho_0^2) + \frac{P_b}{K} \operatorname{tg}\varphi = 0,$$

$$\frac{1 - \varrho_0^2 \sin\varphi}{1 - \sin\varphi} \left\{ \frac{\varrho_0^{\alpha_2 - 1}}{\alpha^{\alpha_2}} + (\beta-1) \frac{\alpha^n}{\alpha^n - 1} \frac{\varrho_0^{\alpha_2 - 1}}{\alpha^{\alpha_2}} \quad 1 - (\frac{\varrho_0}{\alpha})^h \right] \right\} -$$

$$- \varrho_0 \left\{ (\frac{\varrho_0}{\alpha})^{\alpha_2} + (\beta-1) \frac{\alpha^n}{\alpha^n - 1} \left[(\frac{\varrho_0}{\alpha})^{\alpha_2} - 1 + \frac{\alpha_2}{\alpha_{2+n}} \right. \right.$$

$$(1 - (\frac{\varrho_0}{\alpha})^{\alpha_{2+n}}) \right] \right\} = 0, \qquad (5)$$

Solving the system (4) we will find the optimal (least) thickness of cylinder. Numerical results are reduced in Table 1. In tables for comparison is reduced also the limit meaning of thickness cylinder d_1 (not considering process irreversible microsplit formation), corresponding the known solution elasto--plasticity problem /5/.

Table 1: Optimal (minimal thickness of cylinder depending on coefficient ß, characterizing the level of damaged material of ZID, under n = 5, $(P_b-P_a)/K = 0,4$.

	ß = 1	ß = 0,5	ß = 0,4	ß = 0,3	ß = 0,2	ß = 0,1
q_o	1,000	0,893	0,891	0,883	0,877	0,871
d_*	0,181	0,198	0,199	0,201	0,202	0,203
d_1				0,181		

As is obvious from the Table 1, with decrease of parameter ß and n the level of irreversible damages are increased and by the same token, the meaning of optimal (least) thickness of cylinder is increased. The analogous results are obtained and under solving the system (5). Note in addition, that with decrease of angle of inner friction φ , as parameters ß and n, the meaning of optimal thickness of cylinder also is increased.

Besides, the influence of parameters ß, n and φ is investigated on optimal (least) thickness of cylinder, obtained starting from the theory of stability elasto-plasticity balance of bodies /6/.

The comparison of results calculation of thickness of thick--walled cylinder from the point of view the loss as carrying possibility, thus the stability of his balance showed, that for the determination more safe thickness it should be calculated her on the stability /6/.

LITERATURE

1. Bolotin V. V., Ermolenko A. Ph.:The investigation models of accumulation fatique damages. In the book: The calculations on stability. M.: Machine-building, 1979, output. 20, p. 3-29.

2. Bolotin V. V.: The prediction resource of machines and constructions. M.: Machine-building, 1984, p. 312

3. Gusev A. S., Svetlitsky V. A.: The calculation of constructions under random effects. M.:Machine-building, 1984, 240 p.

4. Alimzhanov M. T., Adaijbekov R. M.: On calculation of thick--walled shells, lying in situated under uniform pressure. Herald AS of KazSSR, 1986, N4, p. 54-62.

5. Malinin N. N.:Applied theory of plasticity and creep. M.: Machine-building, 1968, 400 p.

6. Alimzhanov M. T.: Stability of balance bodies and the problems of mechanics mining rocks. Alma-Ata: Science, 1982, p. 272.

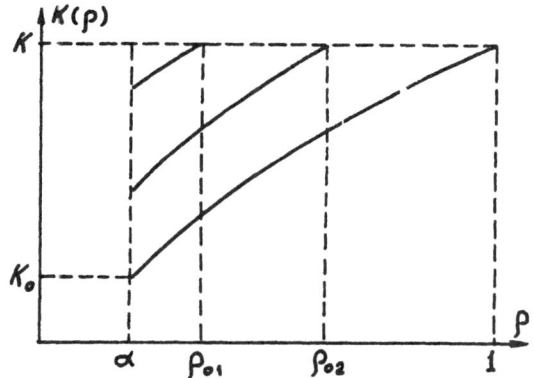

Fig. 1:

The regularity of decrease stability characteristics of the material on thickness cylinder on measure development of the zone inelastic deformations ς_0 (increase the level of irreversible damages).

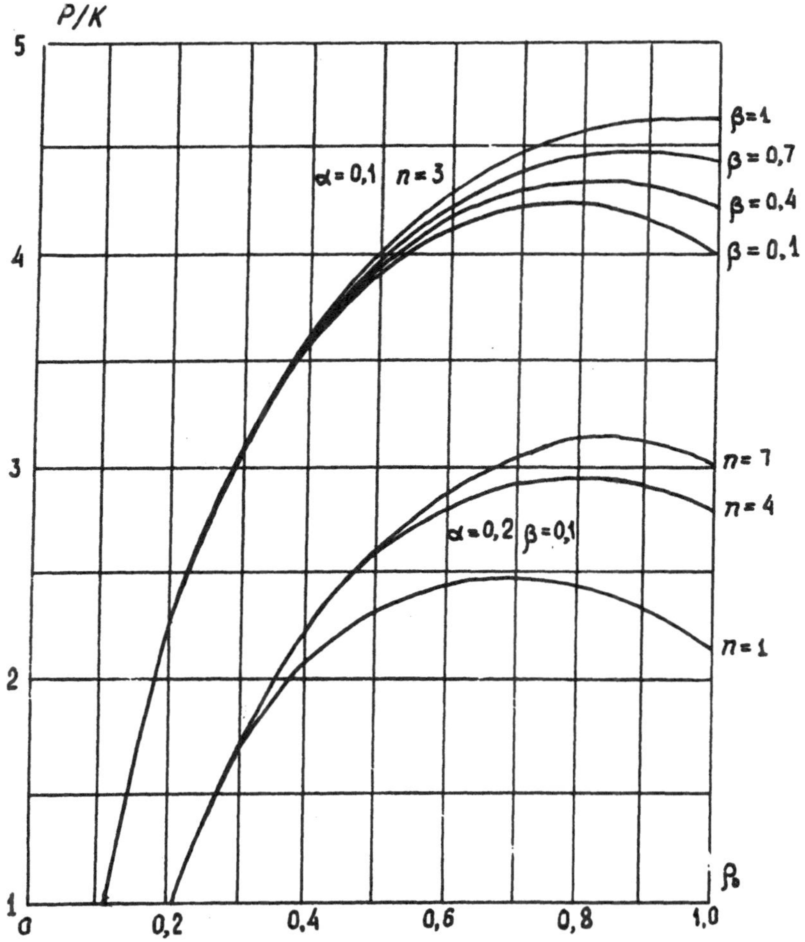

$\beta = 1$
$\beta = 0.7$
$\beta = 0.4$
$\beta = 0.1$

$\alpha = 0.1$ $n = 3$

$n = 7$
$n = 4$

$\alpha = 0.2$ $\beta = 0.1$

$n = 1$

Fig. 2: The regularity between pressure on thick-walled cylinder and by radius of ZID under a/b = 0,1; n = 3; ß = 0,1; 0.4; 0,7; 1,0 and a/b = 0,2; ß = 0,1; n = 1; 4; 7.

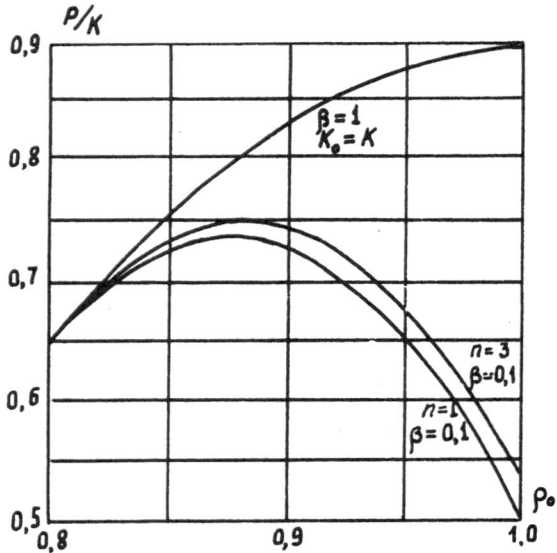

Fig. 3: The regularity between pressure on thick-walled
cylinder and by radius of ZID under a/b = 0,8;
ß = 0,1 (n = 1; 3) and ß = 1,0.

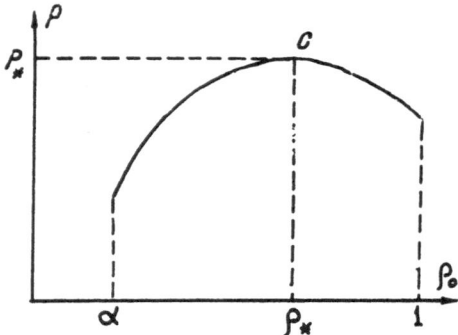

Fig. 4: The regularity between pressure on thick-walled
cylinder and by radius of ZID in the presence of
cumulative damages (ß ≠ 1).

EFFECTS OF FIBRE ORIENTATIONS ON FAILURE MODES OF COMPOSITE
SHELLS SUBJECTED TO UNILATERAL CONSTRAINTS

A. MUC

Technical University of Cracow,
Institute of Mechanics and Machine Design,
ul. Warszawska 24,
31-155 Kraków,
Poland

Abstract

The main purpose of the present work is to estimate maximal
loads which can be carried out by laminated FRP axi-symmetrical
cylindrical shells restrained by an outer rigid walls and to
find optimal fibre orientations. Various failure criteria in
the stress space will be applied herein in order to determine
the lowest failure load for the identical fibre orientations in
individual layers.

1.   Introduction

Due to the influence of fibre orientations and stacking
sequences on contact domains and pressures the analysis of
unilaterally constrained laminated shells leads to much more
problems than in the considerations of isotropic structures.
However, on the other hand the anisotropy of fibre reinforced
plastics (FRP) allows to design such fibre orientations in
shell structuresthat their weakest places in shell structures
(from the viewpoint of the assumed failure criteria) may be
reinforced during the production process.
In the previous works concerning the analysis of laminated
shell structures under unilateral boundary conditions,
attention was mainly focused on the stress analysis only.
Pielekh and Sukhorolski [1] investigated anisotropic shells but
under the assumption of transversly isotropic or orthotropic
materials. The works [2,3] dealt with laminated shell theory
(i.e. included the orthotropy of materials) but ,in fact, were
restricted to the analysis of specified fibre orientations. In
the area of laminated unilaterally constrained shell structures
the optimisation of the strength (or strains) with respect to
fibre orientations, obviously important to designers, still
remains in its infancy. Of course, some attempts have been made

in the area of strength optimisation of laminated shell
structures subjected to bilateral boundary conditions. However,
the applicability of obtained results is strongly limited due
to some auxiliary assumptions; for instance Chao et al. [4]
prescribed in advance a linear or constant displacement field
throghout the mid-surface.

2. Basic relations

In the present considerations we restrict our analysis to
the frictionless axisymmetric contact problem, so that
unilateral boundary conditions can be described by the
following relations:

$$\text{if} \quad v_3 < \varepsilon \quad \text{then} \quad S_N(x) = 0 \text{ for } x \notin S$$

$$\text{if} \quad v_3 = \varepsilon \quad \text{then} \quad S_N(x) + f(x) = 0 \text{ for } x \in S \quad (1)$$

where S_N denotes the normal reaction of foundation, x is the
shell longitudinal coordinate, v_3 is the shell normal
displacement and ε means the nonzero gap between the shell and
the rigid founadtion. When the contact occurs on the contact
area S (uknown in advance), the normal reaction S_N is
determined by an uknown function $f(x)$.

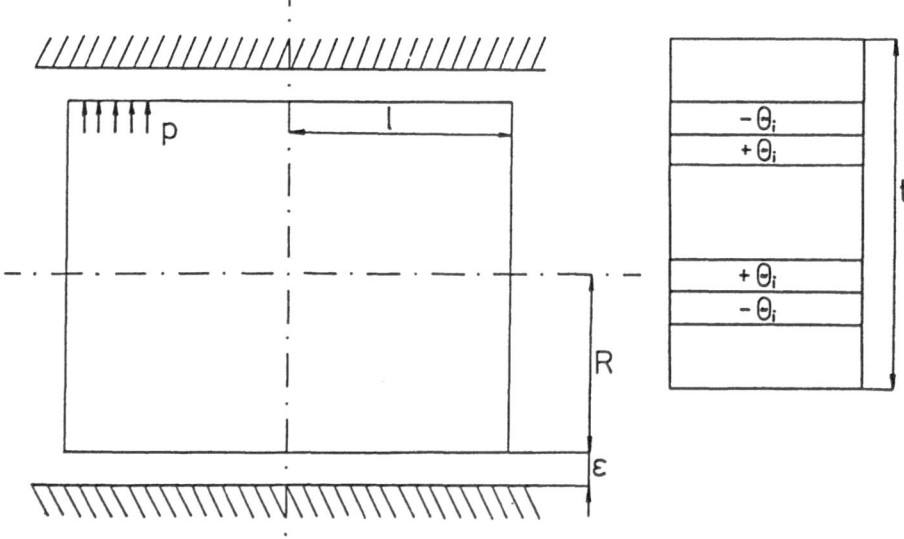

Fig.1 Shell geometry and lamination.
For laminated shells the geometric relations are analogical to
those for isotropic shells - see Reddy [5]. However, the form
of the assumed kinematical hypothesis is strongly connected
with the mechanical properties of the considered structures.
Since the characteristic feature of composite materials is a
very low shear-to-Young's modulus ratio E_1/G_{13} (varying in the
interval [20,50]) the use at least five-parametrical shell is
theories are required. Let us notice that the identical to the

84

above condition should be also satisfied with respect to the accuracy of description of contact forces - see Galimov [6].
In our considerations we use five-parametrical geometrically linear shell theory. The strain-displacement relations applied here are analogical to those presented in [7] and we shall not dwell on them. We analyse a laminated cylindrical shell consisting of a finite number of layers of FR materials having the same mechanical properties and thicknesses except fibre orientations. The lamination is assumed to be symmetric with respect to the mid-surface of the shell and the layers are stacked alternately with $+\vartheta_i$ and $-\vartheta_i$ angles (angle-ply orientation). The shell geometry and the lamination are presented in Fig.1. The assumed lamination simplifies significantly constitutive equations for laminated structures which in the analysed case take the following form:

$$N_{ij} = A_{ijkl}\,\varepsilon_{kl}, \quad M_{ij} = D_{ijkl}\,\varkappa_{kl}, \quad Q_{13} = A_{44}\,\varepsilon_{13} \qquad (2)$$

In the above equations i, j = 1, 2 ; N_{ij}, Q_{13} and M_{ij} are the extensional, transverse shear and moment stress resultant; ε_{kl}, ε_{13} and $\varkappa_{kl}$ are the corresponding extensional, shearing and bending strains, respectively. A_{ijkl}, A_{44} and D_{ijkl} are the extensional, shearing and bending coefficients. Their explicit form can be found eg. in [8] — p.222. It is necessary to emphasize here that in the distinction to isotropic materials

the stiffness coefficients are functions not only of material properties but are also dependent on fibre orientations in individual layers and their positions (a stacking sequence) with respect to the shell mid-surface. Let us notice also that in Eq.(2) the coupling (extensional - bending) B_{ijkl} terms of the stiffness matrix (see [8]) are equal to zero due to the prescribed fibre orientations (angle - ply). In addition, the element A_{44} of the stiffness matrix is independent on the fibre orientations because we assume the equality of the out - of - plane shearing moduli G_{13} and G_{23}.
With the help of the relations (1),(2), one can formulate now the functional J of total potential energy of unilaterally constrained axisymmetric angle-ply cylindrical shells:

$$J = \int_0^1 [\tfrac{1}{2}(N_{ij}\varepsilon_{kl} + M_{ij}\varkappa_{kl} + Q_{13}\varepsilon_{13}) - (p - S_N)v_3]R\,dx \qquad (3)$$

As usual all terms in (3) can be expressed as the functions of displacements v_1, v_3 and angle of rotation γ. Similarly as for isotropic bodies, the solution of the considered contact problem for laminated shells can be determined as the infinum (not the minimum) of the functional J on the field of kinematically admissible displacements defined by the Eq.(1). From the mathematical point of view, it is equivalent to the variational inequality. The proof of the above and of the existence of the solution is identical to that presented in [7]. This identity is possible thanks to the prescribed angle - ply fibre orientations in the laminate what eliminates entirely the B_{ijkl} coupling terms in the stiffness matrix. The nonzero B_{ijkl} terms may be both positive and negative, and the existing mathematical theorems of the existence of solutions of contact problems for elastic plate or shell structures base mainly on the nonnegativity of all terms in the stiffness matrix.

3. Failure criteria

The majority of the existing and applied in calculations failure criteria is summarised and discussed for instance in Tsai's book [9]. In general, the used criteria for laminated structures are a simple transformation of the wellknown criteria of elastic carrying capacity for anisotropic bodies. It is worth to mention that identically as for plasticity the quadratic criteria in the stress space give the better approximation of failure loads (lower) than linear ones. However, there is no a general rule in this subject and additionally the variety of the quadratic failure criteria make the problem more complicate. This is the main reason that our intention is to consider some commonly applied failure criteria and then to decide which one of them gives the minimal failure load. The obtained in this way failure load will be further optimised with respect to fibre orientations ϑ_i.

In our analysis we use three criteria formulated in the stress space, i.e. two quadratic :
- Tsai - Wu's [10]:

$$\frac{(\sigma'_{11})^2}{X_t X_c} - \frac{\sigma'_{11}\sigma'_{22}}{\sqrt{X_t X_c Y_t Y_c}} + \frac{(\sigma'_{22})^2}{Y_t Y_c} + \sigma'_{11}(\frac{1}{X_t} - \frac{1}{X_c}) + \sigma'_{22}(\frac{1}{Y_t} - \frac{1}{Y_c}) = 1$$

(4)

- Hoffman's [11]:

$$\frac{(\sigma'_{11})^2 - \sigma'_{11}\sigma'_{22}}{X_t X_c} + \frac{(\sigma'_{22})^2}{Y_t Y_c} + \sigma'_{11}(\frac{1}{X_t} - \frac{1}{X_c}) + \sigma'_{22}(\frac{1}{Y_t} - \frac{1}{Y_c}) = 1 \quad (5)$$

and one linear in the form of maximal stresses

$$\max(\frac{\sigma'_{11}}{X_{t(c)}}, \frac{\sigma'_{22}}{Y_{t(c)}}, \frac{\sigma'_{12}}{S}, \frac{\sigma'_{13}}{Z}, \frac{\sigma'_{23}}{T}) = 1 \qquad (6)$$

where σ'_{11}, σ'_{22}, σ'_{12}, σ'_{13}, σ'_{23} and $X_{t(c)}$, $Y_{t(c)}$, S, Z, T are the longitudinal, transverse, in-plane shear and transverse shear stresses and strength, respectively determined in the local coordinates system (denoted by prime over the symbols) associated with fibres directions. Since for composite materials allowable maximal strengths are different for tension and compression we distiguish them by the subscripts t and c, respectively. In the global coordinates system the stresses $\sigma_{\alpha\beta}$ (α, β = 1, 2, 3) are calculated from the distributions of the stress resultants in the following way:

$$\sigma_{ij} = \frac{N_{ij}}{t} + \frac{12M_{ij}z}{t^3} \ , \qquad \sigma_{13} = \frac{3Q_{13}}{2t}\left(1 - \frac{4z^2}{t^2}\right) \qquad (7)$$

where t is the shell thickness and z the distance from the shell midsurface.

The initial failure of the FRP shell structure is predictable with the aid of the criteria (4)-(6) on the inner – and on the outer – most layers (the bending effects are taken into account) and at the shell midsurface (the maximum of out-of-plane shearing forces). Generally, after the initial failure of one layer (so-called first ply failure), the structure may be able to carry additional loads. However, with the emphasis being placed on the application of the results of the optimisation, only the initial not ultimate failure is considered in this study.

In each layer the angle between the fibres and the longitudinal shell axis (the global coordinate system) is denoted by ϑ_i. The basic mechanical properties of the single layer, such as Young's E_1, E_2, Kirchhoff's G_{12}, G_{13} moduli, Poisson's ratio ν_{12} and the ultimate admissible strengths $X_{t(c)}, Y_{t(c)}, S, Z, T$ are determined experimentally in the local coordinate system where 1-axis is parallel to the fibres at the point of interest. The transformation of the mechanical properties of FRP to the global coordinate system results in the appearance of the terms A_{ijkl}, D_{ijkl} being functions of ϑ_i. After the solution of the problem described by the functional (3), one can obtain the stresses σ_{ij}, σ_{13} – Eq.(7) in the global coordinate system. Since the ultimate strength $X_{t(c)}, Y_{t(c)}, S, Z$, T are known in the local system it is necessary to retransform the stress components $\sigma_{\alpha\beta}$ back to the local coordinate system by using the usual coordinate transformation law:

$$
\begin{bmatrix} \sigma'_{11} \\ \sigma'_{22} \\ \sigma'_{12} \\ \sigma'_{13} \\ \sigma'_{23} \end{bmatrix} =
\begin{bmatrix}
m^2 & n^2 & 2mn & 0 & 0 \\
n^2 & m^2 & -2mn & 0 & 0 \\
-mn & mn & m^2-n^2 & 0 & 0 \\
0 & 0 & 0 & m^2 & n^2 \\
0 & 0 & 0 & n^2 & m^2
\end{bmatrix}
\begin{bmatrix} \sigma_{11} \\ \sigma_{22} \\ 0 \\ \sigma_{13} \\ 0 \end{bmatrix}
\qquad (8)
$$

$$m = \cos\vartheta_i \ , \quad n = \sin\vartheta_i$$

Let us notice that the axisymmetry of the contact problem considered leads automatically to the existence of three nonzero components of the stress tensor (i.e. σ_{11}, σ_{22}, σ_{13}) which appear in Eq.(8). This results also in the independence on fibres orientations and the equality of the out-of-plane shear stresses σ'_{13} and σ'_{23} in the local coordinate system. However, the appearance of the nonzero stresses σ'_{13} and σ'_{23} in the strength criteria (4),(5) reduces the failure strength of shell structures in the comparison with the cases where the stres component σ_{13} is neglected in the analysis (ie. as three-parametrical shell theory is applied in considerations).

4. Numerical results and discussions

First of all let us briefly summarise the basic steps of numerical procedures used in calculations:
- finite difference discretisation of the functional (3) and searching for the infinum of it-this procedure is described in details in [7].
- determination of the stresses in the global coordinate system with the use of Eqs. (2),(7).
- retransformation of the stresses - Eq.(8) and searching for such a values of external loadings that one of the failure criteria (4)-(6) is fulfilled on the inner or outer shell surface.
- optimisation (maximisation) of external loadings with respect to fibres orientations θ_i by using the golden section search.

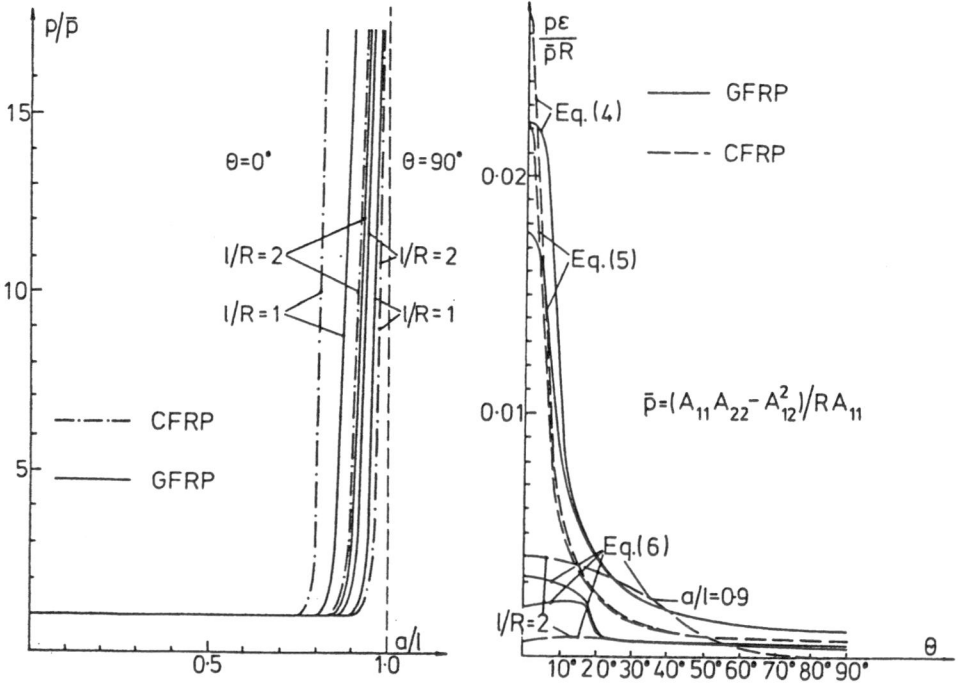

Fig.2 Variations of internal pres-   Fig.3 Dimensionless failure
sure vs the length of first contact   pressures $(R/l=1, a/l= 0.1)$.
zone.

As the numerical example let us analyse the case of laminated pressure vessel loaded by an uniform internal pressure p and surrounded by a rigid wall — see Fig.1. In addition, we assume that a cylindrical part of the vessel is closed by very stiff domed ends which enable to prescribe bilateral boundary conditions at the ends of the cylindrical part in the form of clamped edges, ie. :

$$v_3 = 0, \ \gamma = 0 \text{ and } N_{11} = 0 \text{ for } x=1 \tag{9}$$

For the simplicity of the numerical analysis we assume also the following form of the uknown contact pressures $f(x)$ - Eq.(1) :

$$f(x) = a_1 \exp(a_2 x) + a_3 \exp(a_4 x) + a_5 \tag{10}$$

where $a_1, \ldots, a_5$ are uknown constants determined numerically with the aid of Newton-Raphson's procedure. The above relation is identical to that given in [7].

Numerical examples are carried out for 3-ply and 7-ply (it means that the total number of layers is equal to 6 and 14, rsp.) symmetric angle-ply glass/ (GFRP) and carbon/ (CFRP) epoxy laminated cylindrical shells. The material constants are given in [9] :

	E_1	E_2	G_{12}	ν_{12}	X_t	X_c	Y_t	Y_c	S
	in	GPa			in	MPa			
CFRP	203	11.2	8.4	0.32	3500	1540	56	150	98
GFRP	38.6	8.27	4.14	0.26	1062	610	31	118	72

Table 1

For a 3- and 7- ply laminated shell the numerical results dealing with the optimal fibre orientations are identical. The maximal value of the internal pressure corresponding to the failure in the sense of the criteria (4) or (5) or (6) is obtained for the identical value of ϑ angles in each of layers (ie. as $\vartheta_i = \vartheta$, i = 1,...,N). Therefore, the results (Figures 2, 3) present the variations of the internal pressure p versus fibre orientation ϑ, not taking into account the total number of layers N or stacking sequences. As clearly indicated in Fig.2 the value of the internal pressure p equivalent to the appearance of the first contact domain having the logitudinal length a/l is a function of the shell geometry l/R and fibr orientations ϑ but is independent on the value of ε/R. It should be pointed out that the growth of a/l to 1 reduces the number of contact areas to one only and is associated simultaneously with the increase of the referential internal pressure to infinity for a/l =1.

Figure 3 presents the distributions of dimensionless failure pressures versus fibre orientations ϑ for various failure criteria (4)-(6). First of all let us notice that for all cases considered dimensionless failure pressures are much lower than those corresponding to the appearance of the first contact zone ($p/\bar{p} \geq 1$ - see Fig.2). However, in Fig.3 the $p/\bar{p}$ ratio is multiplied by the coefficient ε/R, so that it shows clearly that the contact as well as the failure are possible not for all values of the ε/R ratio. The limit values of ε/R one can simply determine by the comparison of the results plotted in Figs. 2,3. As it may be seen in Fig.3 the failure pressures obtained with the use of the qudratic criteria (4),(5) are much higher than those calculated with the aid of Eq.(6). This is caused by the different mechanisms of failure which are analysed by both (linear and quadratic) failure criteria. The concentration of stresses may occur at two points : x=a (as a/l > 0.8 - one contact zone only) and x=l (as a/l < 0.8 - more than one contact zone). In the first case the maximal stresses occur at the middle layer due to shearing forcesQ_{13}-see Eq.(7), whereas in the second case due to shearing forces at the middle layer ($\sigma_{11}=\sigma_{22}=0$-Eq.(7)) or at the outer- or inner- most layers due to bending effects ($Q_{13}=0$). Thus, the quadratic failure criteria can be used in calculations as bending effects prevail shearing. The results drawn in Fig.3 show that Hoffman's criterion (5) gives lower values of failure pressures than Tsai-wu's (4) but it is obvious that the failure occurs at the middle layers of cylinders due to shearing forces caused by the assumed boundary conditions of the type (9). Let us notice that the latter has also its reflection in the optimal fibre orientation. The quadratic failure criteria shows that the optimum occurs for $\vartheta = 0°$, ie. when fibres are oriented along the shell axis (in the global coordinate system). The use of the criterion (6) changes the positions of the optimal fibre orientation to the angle $22°$ but the differences between the failure pressures for $\vartheta = 0°$ and $\vartheta = 22°$ are very low.

References

1. B.L. Pielekh and M.A. Sukhorolski, Contact problems in elastic anisotropic shells theory, Naukova Dumka,Kiev (1980 in Russian).
2. A.S.Tooth,W.M.Banks and D.H.A.Rahman, The specially orthotropic GRP multilayered cylindrical pressure vessels — A theoretical approach,J.Composite Str.9, 53-68 (1988).
3. A.S.Tooth,W.M.Banks and D.H.A.Rahman, The specially orthotropic GRP multilayered cylindrical pressure vessels - The radial patch load, ibidem ,69-83.
4. C.C.Chao,C.T.Sun and S.L.Koh,Strength optimization for cylindrical shells of laminated composites,J.Comp.M.9, 71-85 (1975).
5. J.N.Reddy,Exact solution of moderately thick laminated shells,J.Eng. Mech.110,794-809(1984).
6. K.Z. Galimov. Shell theory with transverse shear effects, Izd. Kaz. Univ., Kazan (1977 in Russian).
7. A.Muc,Theoretical and numerical aspects in contact problems for shells, ZAMP 35,890-905 (1984).
8. J.R.Vinson,T.W.Chou,Composite materials and their use in structures, Appl.Sc.Publ.Ltd., London(1975).
9. S.W.Tsai, Composite design, Dayton, Think composites(1987).
10. S.W.Tsai,E.M.Wu, A general theory of strength for anisotropic materials, J.Comp.M.5, 58-80(1971).
11. O.J. Hoffmann, J.Comp.M.1,200-214(1967).

CONTACT PROBLEM IN THE THEORY OF ANISOTROPIC CYLINDRICAL SHELLS

B.L.Pelekh, N.N.Shcherbina, D.D.Matieshyn

Institute of   Applied Problems
of Mechanics and Mathematics,
AS Uk SSR
290601 Lvov
USSR

Abstract

The contact problem of deformable bodies (one of which may
be an absolute solid) is a vital problem in estimation of
their stiffness and strength as well as of occurring local
effects with probable occurence of plastic strains   in the
neighbourhood of transition zones. Since it is an attribute
of this class of problems that application of Kirchoff-Lyava
classical shell theory results in the contradictions in
their physical nature, the question of the choice of correct
mathematical models for describing the state of shell-type
thin-walled elements with reference of the contact problem
is of paramount importance.

1. Solvable Equations of Thin Anisotropic Cylindrical Shells

Contact   problems   for thin-walled elements of designs
(pivots,plates and shells) have specific character   of
organization and solution. It is connected with the fact
that Herts hypothesis is incompetent in this case and
well-developed ways of solution of three-dimentional contact
problems can't be   applied   to   plates   and   shells.
Investigations in this direction found their reflection in
monographs [1-3]. As stated earlier the result of solving
two-dimensional contact problems depends essentially on
accepted mathematical models of thin-walled element states
(Kirchoff-Lyava, Timoshenko and others). Choice of this or
that   model   includes   its   solvability requirements in
engineering calculations and the correspondence of the
results obtained to experimental data. Proceeding from this
position, a generalized theory [3] is assumed as a basis of
investigations which allows to take account of finite shift
stiffness and anisotropy of   physical   and   mechanical
properties   of   the   shell   and,   finally, to obtain
not-contradictory solutions. Reduction of three-dimentional
problem of the shell theory to the two-dimentional one is
implemented by the way of decomposition of   unknown
quantities relatively to normal co-ordinate.
      Approximations of problems are constructed on the basis
of stress approximation to finite numbers of polynoms when
performing exactly bound conditions on external   surfaces.

The following representations are accepted for stresses in the bounds of $\{m,n\}$ approximation

$$\sigma_\alpha = \sum_{\kappa=0}^{m} \frac{2\kappa+1}{2} N_{\alpha\kappa} P_\kappa(z) \ (\alpha \to \beta), \quad \tau_{\alpha\beta} = \sum_{\kappa=0}^{m} \frac{2\kappa+1}{2} N_{\alpha\beta\kappa} P_\kappa(z),$$

$$\tau_{\alpha z} = \sum_{\kappa=0}^{n+2} \frac{2\kappa+1}{2} N_{\alpha z\kappa} P_\kappa(z) \ (\alpha \to \beta), \tag{1}$$

$$\sigma_z = \sum_{\kappa=0}^{n+1} \frac{2\kappa+1}{2} N_{z\kappa} P_\kappa(z)$$

and for displacements

$$U_\alpha = \sum_{\kappa=0}^{m} U_{\alpha\kappa} P_\kappa(z) \ (\alpha \to \beta, \ m \leqslant n+1),$$

$$U_z = \sum_{\kappa=0}^{n} U_{z\kappa} P_\kappa(z), \tag{2}$$

where $P_\kappa(z)$ – Lezhandr polynoms.

Equations of shell theory based on Timoshenko and Kirchoff-Lyava hypotheses follow as private cases from equations [3] shown for some variants of contact loading.Note, that within the scope of discrete-continual approach non-traditional contact problems can be solved for flaky cylindrical shells under the influence of patterns or local loadings on the basis of a generalized shell theory [4-6].

Some ways of solving contact problems are known. In [3] method of reducing contact problem to Volterra integral equations in effectively used for definition of contact pressure under the pattern. Solutions of integral equations are constructed with the help of Laplas conversion. It is necessary to note that solution of contact problems for flaky shells within the scope of discrete approach (kinematic hypotheses are used for each layer separately) is connected with definite difficulties.In [6] matrix approach is developed for solution of such class of problems. Proposed in [4] models allow to consider different kinds of contact interactions of design layers (friction,slipping and others), taking account of exfoliation formation which is important for definition of interphase destruction.

2. Inversely-Symmetrical Contact Problems

A N- flaky shell is considered under the influence of stiff patterns or distributed loading (particularly concentrated) and even internal pressure. Shell layers are connected without slipping. Let's denote R_i and $2h_i$ – radius and thickness of the i-th layer ($i = 1$ corresponds to an upper layer), $\sigma_i^{\pm}$ and $\tau_i^{\pm}$ – normal and tangent stresses on upper and lower surfaces of the l-th layer correspondently. Solvable equation system for a given case is the following:

$$\frac{d^2 F_i}{d\varphi^2} + F_i = \frac{R_i^2}{B_i}\left(\sigma_i^+ - \sigma_i^-\right),$$

$$\frac{d^2 \gamma_i}{d\varphi^2} + \frac{(z_i^2 - 1)}{R_i}\frac{dF_i}{d\varphi} = -\frac{R_i}{B_i}\left[(z_i^2 - 1)(\tau_i^+ - \tau_i^-) + \frac{2h_i}{R_i}\left(\tau_i^+ + \tau_i^-\right)\right],$$

(3)

$$\frac{d^2 W_i}{d\varphi^2} + W_i - \left(1 + S_i^2\right)F_i + R_i\frac{d\gamma_i}{d\varphi} = -\frac{R_i^2}{B_i}\left[S_i^2\left(\sigma_i^+ - \sigma_i^-\right) + \right.$$

$$\left. + \frac{h_i}{3R_i}\frac{d}{d\varphi}\left(\tau_i^+ + \tau_i^-\right)\right],$$

where $F_i = \frac{dv_i}{d\varphi} + W_i$ is a solvable function; $S_i^2 = \frac{6}{5}\frac{B_i}{\Lambda_i}$, $z_i^2 = 1 + 3\left(\frac{R_i}{h_i}\right)^2$ – are undimentioned parameters.

Normal and tangential point shifts of external surfaces are defined as follows:

$$U_z^{(i)}(\pm h) = W_i,$$

$$U_z^{(i)}(\pm h) = v_i \pm \frac{5h_i}{6}\left(\gamma_i - \frac{1}{5R_i}\frac{dW_i}{d\varphi} + \frac{v_i}{5R_i}\right) +$$

(4)

$$+ \frac{1}{G_i'}\left[\left(\tau_i^+ \pm \frac{\tau_i^+}{2}\right) - \left(\tau_i^- \mp \frac{\tau_i^-}{2}\right)\right].$$

In the absence of slipping between layers and coming unstuck under the pattern the following conditios are performed:

$$U_z^{(i)}(-h_i) = U_z^{(i+1)}(+h_{i+1}), \quad U_z^{(i)}(-h_i) = U_z^{(i+1)}(+h_{i+1}),$$

$$\sigma_i^- = \sigma_{i+1}^+, \quad \tau_i^- = \tau_{i+1}^+, \quad i = \overline{1, N-1}$$

(5)

$$w_1(+h_1) = f(\varphi) \ , \quad (0 \leqslant \varphi \leqslant \varphi_o) \tag{6}$$

where $f(\varphi)$ – is a given function (describes a pattern profile), the angle φ_o characterizes contact scope.

In a general case the inversely-symmetrical contact problem consists in solution of equation system (3) taking account of correlations (5), (6) and symmetry conditions

$$\frac{d w_i}{d\varphi} = 0 \ , \quad \mathcal{V}_i = 0 \ , \quad \mathcal{T}_i = 0 \ .$$

$$\left(\varphi = 0 \ , \quad \varphi = \frac{\pi}{2} \ ; \quad i = \overline{1,N} \right) \tag{7}$$

For the shell of an open profile instead of (7) corresponding bound conditions are given.

For one- and two-layer shells analytical solution is constructed with the help of Laplas integral conversion. Its representation in the form of integrals of curtail type allows to solve effectively a number of problems for an arbitrary symmetrical shell loading including the effect of patterns. Realization of contact condition, which represents equality of curvatures of shell and pattern contact surfaces in the field of contact results in Volterra integral equation of the second kind of relatively unknown contact pressure $q(\varphi)$. From condition of pattern balance as a stiff body

$$P = -2B \int_0^\theta q(\varphi) \cos\varphi \, d\varphi \tag{8}$$

we get a transcedent equation for the definition of an unknown contact field θ . In [5] dependence of sagging on the parameter has been investigated, characterizing material pliability for shift. With its increase sagging is growing. Sagging forms of a flaky shall are also constructed and laws of changing inter-layer stresses have been obtained at the boundary of layer division with different geometrical and elastic characteristics. Calculations performed [3,4] concerning contact pressure distribution in contact region testify that when decreasing material pliability for shift contact pressure concentrates more at the edges of contact region. Dependence of loading on contact region quantity has nonlinear character.

3. Axissymmetrical Contact Problems

Let's consider a contact problem for an infinitely long N – layer shell in conditions of axissymmetrical loading (Fig. 1). The problem consists in definition of contact

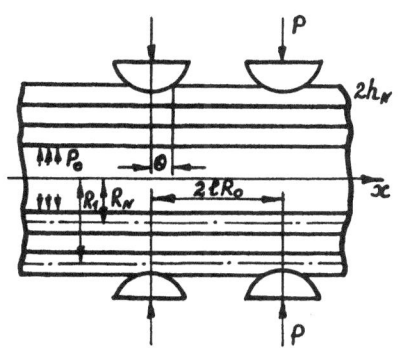

stresses as well as in normal and tangential inter-layer stresses. In the case given equilibrium equations of the i-th layer are as follows:

$$\frac{d}{dx} N_1^{(i)} = - R_i \left(\tau_i^+ - \tau_i^- \right),$$

$$\frac{d}{dx} M_i - R_i Q_i = -h_i R_i \left(\tau_i^+ + \tau_i^- \right), \quad (9)$$

$$\frac{d}{dx} Q_i - N_2^{(i)} = R_i \left(\sigma_i^+ - \sigma_i^- \right).$$

Fig. 1. THE SCHEME OF
AXISSYMMETRICAL LOADING

Correlations of elasticity, normal and tangential shifts of external surfaces of the layer are defined according to formulas [3]. Conditions of conjugation between layers are such as:

$$U_z^{(i)}(-h_i) = U_z^{(i+1)}(+h_{i+1}), \quad U_1^{(i)}(-h_i) = U_1^{(i+1)}(+h_{i+1}),$$

$$\sigma_i^- = \sigma_{i+1}^+, \quad \tau_i^- = \tau_{i+1}^+ .$$

Besides, the conditions of periodicity have to be performed (interactions with the same equally moved away bandages)

$$U_i = 0, \quad \gamma_i = 0, \quad \frac{dw_i}{dx} = 0 \quad (x = 0, \ x = l)$$

or corresponding bound conditions (for finite, infinite, semi-infinite shells).

4. Analysis of Some Results for Anisotropic Cylindrical shell under circular pattern effect of arbitrary contour

Solution of the given contact problem is being constructed by analogy with the described scheme with the help of Laplas integral conversion [3] and allows to obtain analytical solutions for a finite, infinite and semi-infinite shell.
The condition of contact is as follows:

$$W = \beta - \frac{R^2}{2R_o} \left(\alpha - \theta_o \right)^2, \quad \alpha = \frac{x}{R},$$

where R_o - is radius of pattern curvature, quantity θ_o fixes its position, β - is the pattern settling. For a semi-finite shell external loading is defined as follows:

$$\sigma(\alpha) = \begin{cases} 0 & , & 0 \leqslant \alpha < \theta_1 \\ q(\alpha), & \theta_1 \leqslant \alpha \leqslant \theta_2 & , \\ 0 & , & \theta_2 < \alpha < \infty \end{cases}$$

where θ_1, θ_2 are unknown boundaries of contact region. Contact problem in this case is reduced, in the long run, to the system of eight non-linear equations. Analysis of calculations performed for a semi-infinite shell shows that influence of the free edge of the shell on distribution of contact pressure is revealed considerably if free of the edge shell region is less than the quantity of contact region.

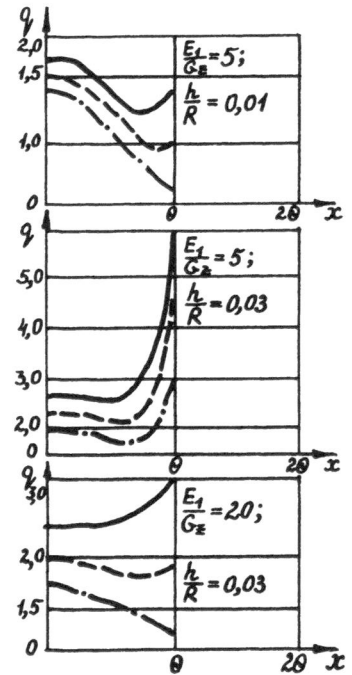

In [3] regularities of local stresses at local loading as well as influence of a distance between bandages on distribution of contact pressure and on stress-strained state have been investigated.

On the basis of strained solutions local stresses in an elastic infinite cylindrical shell in contact zones with a stiff body depending on parameters which characterize orthotropy and pliability for a shift. Some results are presented in Fig.2.

Their analysis allows to come to the conclusion that dangerous gradients of contact stresses are reached either in the neighbourhood of the edges or in the centre of contact region depending on the combination of parameters of being thin-walled and anisotropy of shell materials.

Regulation of these properties is especially important when designing a thin-walled construction made of composition materials.

Fig. 2 .DISTRIBUTION OF CONTACT PRESSURE DE-PENDING ON ORTHOTROPY

96

1. V.M.Alexandrov, S.M.Mkhitaryan, Contact Problems for Bodies with Thin Coverings and Layers. Nauka Press, Moscow (1983).
2. E.I.Grigolyuk, V.M.Tolkachev, Contact Problems of Theory of Plates and Shells. Mashinostroyenie Press, Moscow (1980).
3. B.L.Pelekh, M.A.Sukhorolsky, Contact Problems of Theory of Elastic Anisotropic Shells. Naukova Dumka Press, Kiev (1980).
4. B.L.Pelekh, A.V.Maximuk, I.M.Korovaichuk, Contact Problems for Flaky Elements of Designs and Bodies with Coverings. Naukova Dumka Press, Kiev (1988).
5. B.L.Pelekh, N.N.Shcherbina. Contact Stiffness of Flaky Cylindrical Shells. 1.Analitical Solution. J. Mechanics of Composite Materials. 4, 663-668 (1983).
6. B.L.Pelekh, A.V.Maximuk, N.N.Shcherbina.Contact Stiffness of Flaky Cylindrical Shells. 2.Matrix Method of Solving Contact Problems for Multilayer Cylindrical Shells. J. Mechanics of Composite Materials. 2, 276-280 (1986).

PART II.

PLATED STRUCTURES

MODELLING AND EXPERIMENTAL INVESTIGATION OF PLASTIC RESISTANCE AND LOCAL BUCKLING OF H OR I STEEL SECTIONS SUBMITTED TO CONCENTRATED OR PARTIALLY DISTRIBUTED LOADING

J.M. ARIBERT, A. LACHAL and M. MOHEISSEN

Structures Laboratory, Institut National des Sciences Appliquées (I.N.S.A.)
20, avenue des Buttes de Coësmes - 35043 RENNES CEDEX
FRANCE

Abstract

The ultimate resistance of H and I rolled beams submitted to symmetrical local compression is analyzed by means of both numerical and analytical approaches. Theoretical results deduced from these approaches and also from recent formulae of the literature are compared statistically with experimental data of thirty three tests on European beams. A new formula, which appears simple and more accurate, is proposed for calculation of the design resistance.

1. Introduction

This paper deals with a synthesis of works carried out on the last fifteen years by the Structures Laboratory of INSA in Rennes. The topic is the resistance in symmetrical local compression of normal rolled I or H members loaded on both flanges, for steel grades going from FeE 235 to FeE 460 ; for instance,such situation is met for internal symmetrically loaded beam-to-column connections.

Although this topic may appear elementary, it is in fact relatively complicated because failure can result from two different phenomena : first, the reaching of plastic limit resistance of the profile web, and secondly the elastic or elasto-plastic web buckling on the whole web depth.

Another aspect may be considered adding to this complexity, which is the mechanical concept used to investigate the problem concerned : either, the limit state is only based on the ultimate web resistance which is obviously the prevailing component of the total resistance and may be sufficient for design, or the participation of the profile flanges is also taken into account to expect a more accurate determination of the total resistance.

In this paper, these two concepts are developped by means of different types of theoretical approach : a numerical one to analyse the plastic web resistance, which is followed by a mechanical analogy to transform it into ultimate resistance ; and an analytical approach by plastic hinged mechanism to evaluate the ultimate resistance when there is interaction between flanges and web. Then, the formulae deduced from the previous theoretical approaches and other formulae mentioned in recent literature and codes are compared with a data base including thirty three experimental tests on European sections ; in this comparison, the authors try to bring out the most accurate formula. Finally a new design formula is recommended, which covers both the crushing and buckling ultimate resistances and can be applied to all the steel grades considered with the same reliability as that adopted in Eurocode 3.

101

2. Modelling of the ultimate resistance of the web

2.1 Plastic resistance of the web

2.1.1 Plastic resistance criterion

Our first experimental works on this subject have allowed to establish a well defined criterion for the determination of the plastic collapse of the web alone [1, 2]. This criterion corresponds to the limit state "when there is equality between the rigid-plastic resistance of the web and the external load F applied on each flange" ; the rigid-plastic web resistance is defined as the product of the yield strength f_{yw} and the length l_p really yielded along the web-to-flange fastening line (called "k line"), also multiplied by the web thickness t_w. Thus, the real elasto-plastic web behaviour can be expressed in an equivalent rigid-plastic behaviour.

A such criterion requires an accurate elasto-plastic analysis to determine the real stress distribution along the k line ; this analysis may be an experimental or numerical one.

2.1.2 Experimental analysis

Two experimental methods were used in ref. [1] to determine the plane strain tensor along the k line : first, the photoelastic coating method which gives a global visualisation of the strain field ; secondly, the electric strain gauge technique associated with stress computer calculation, which provides local values of stress at successive steps of loading ; for that, the incremental Prandtl-Reuss law is used, possibly with an isotropic hardening effect.

2.1.3 Numerical model simulation

Now, it is presented briefly a numerical model which remains relatively simple and nevertheless supplies significant distributions of the normal stress along the web-to-flange fastening line. In this model, each flange is considered as an elastic beam lying on a continuous elasto-plastic support with variable rigidity $C(x)$ which corresponds to the web (figure 1). The flange deflection $y(x)$ is governed by the classical equation :

$$\frac{d^2 y}{dx^2} = - \frac{M(x)}{EI(x)} \qquad (1)$$

where E is the elastic modulus, $I(x)$ the moment of inertia of the flange cross-section and $M(x)$ the bending moment which is depending on the distributed reaction $R(x)$ and the external discretized loads $F(x_j)$ applied on the flange. Two boundary conditions have to be associated with equation (1) :

$$T(0) = M(0) = 0 , \qquad (2)$$

where $T(x)$ is the vertical shear force.

Using a geometrical discretization of the flange with small intervals Δx, the non-linear problem must be solved by an iterative procedure ; for information, the R reaction of each interval Δx is assumed to be of elasto-perfectly plastic type, with stiffness C and plastic resistance R_y ; so, the R reaction is equal to :

$$R(x) = \begin{cases} - C \cdot y(x) & \text{in elastic behaviour} \\ R_y = f_{yw} \cdot t_w \cdot \Delta x & \text{in plastic behaviour} \end{cases} \qquad (3)$$

In the particular case of the figure 1, where the web yielding is assumed only for interval $x_k \leqslant x \leqslant x_l$, $M(x)$ is given by the relationship :

$$M(x) = - \int_x^{x_k} R(t) (t-x) \, dt - \int_{x_l}^{l} R(t) (t-x) \, dt - \sum_{x_j > x} F_j(x_j - x) - \sum_{x_k \leqslant x_j \leqslant x_l} R_y(x_j - x) \qquad (4)$$
$$\text{if } x \leqslant x_k \; ;$$

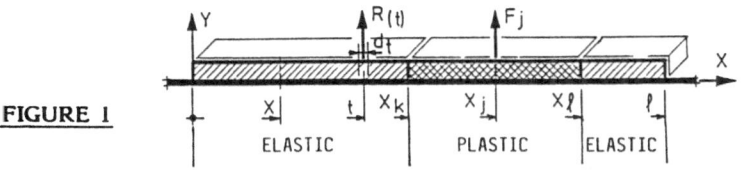

FIGURE 1

ELASTIC | PLASTIC | ELASTIC

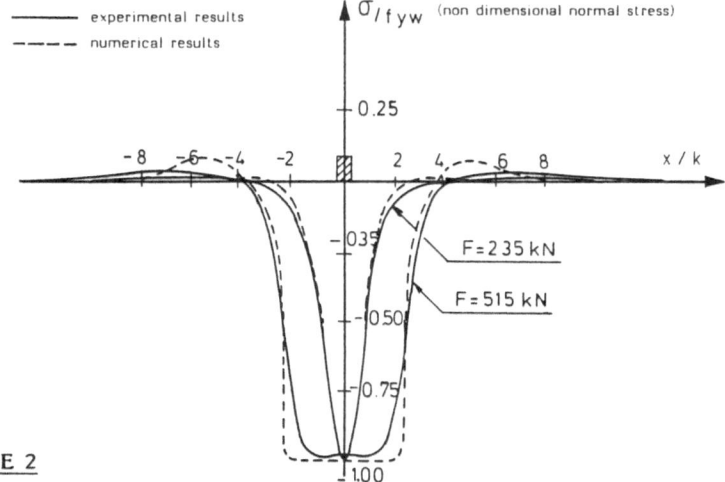

experimental results
numerical results

σ / f_{yw} (non dimensional normal stress)

F = 235 kN
F = 515 kN

x / k

FIGURE 2

Yielded length	Applied load F
x/k	235 kN
	515 kN

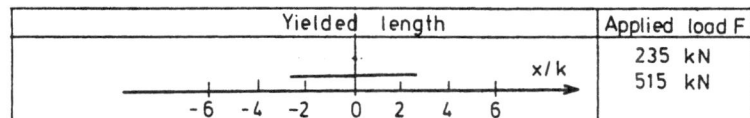

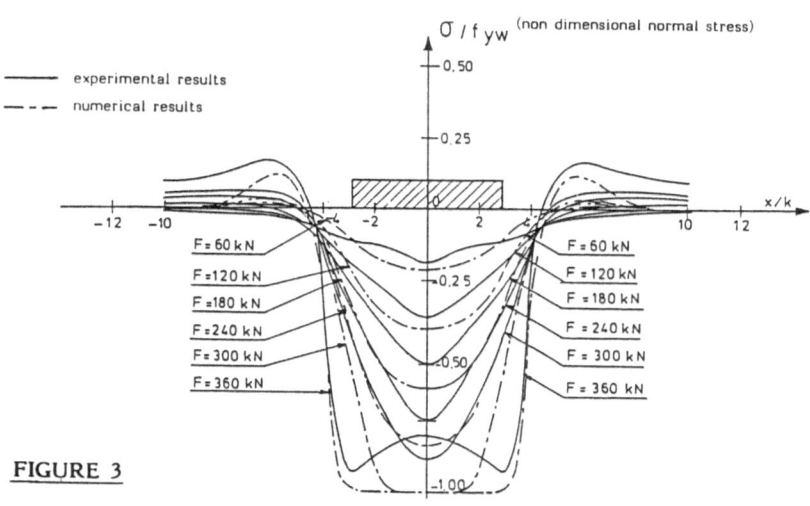

experimental results
numerical results

σ / f_{yw} (non dimensional normal stress)

F = 60 kN
F = 120 kN
F = 180 kN
F = 240 kN
F = 300 kN
F = 360 kN

F = 60 kN
F = 120 kN
F = 180 kN
F = 240 kN
F = 300 kN
F = 360 kN

x/k

FIGURE 3

Yielded length	Applied load F
x/k	320 kN
	360 kN
	380 kN
	460 kN

there are similar relationships for $x_k \leqslant x \leqslant x_\varrho$ and $x \geqslant x_\varrho$

In reference [2], a more detailed presentation is given about the matrix formulation of the problem and the numerical algorithms used to solve it. To be noted also that the stiffeness C of each support has been adjusted numerically (collocation method) in such way that the median support reaches its value R_v when a single concentrated load above this support is equal to the well-known experimental value of local compression at the elastic limit state [1].

2.1.4. Examples of comparison between experimental and numerical results

In the case of an HEB 200 profile subjected to a single point load on each flange, **figure 2** shows the experimental and numerical normal stress distributions along the k line at the experimental elastic and plastic limit states of the web. Also, in the case of an IPE 360 profile subjected to a distributed load by means' of a rigid piece of limited length, **figure 3** illustrates the evolution of the two types of stress distribution up to the plastic web resistance (experimentally equal to 370 kN and numerically to 380 kN).

2.1.5 Synthesis of the numerical and experimental results [2,3]

Table I collects all the formulae of plastic resistance established from the numerical simulator presented here above ; most of them have been proved by experimental tests.

PLASTIC RESISTANCE IN SYMMETRICAL LOCAL COMPRESSION		
TYPE OF LOADING		PRACTICAL FORMULATION
	Point loading	$F_{pw} = 5\,k\,t_w\,f_{yw}$ with $k = t_f + r$ (5)
	Point loading on intermediate plate	$F_{pw} = [5k + 2t_e]\,t_w \cdot f_{yw}$ (6)
	Point loading with an eccentric stiffener	$F_{pw} = 5kt_w f_{yw} + \alpha S_r f_{ys}$ S_r : total cross-section area of the two stiffeners $\alpha = \dfrac{h/4-e}{h/4}$ if $0 \leqslant e \leqslant \dfrac{h}{4}$ $\alpha = 0$ if $e > \dfrac{h}{4}$ (7)
	Uniformly distributed loading	$F_{pw} = [d + 2pk]\,t_w\,f_{yw}$ with $p = \dfrac{2.5}{1+2(\frac{d}{5k})}$ (8)
	Two point loads	$F_{pw} = [d + 2pk]\,t_w\,f_{yw}$ for $d \leqslant 5k$ and $p = 2,5\left[1 - 1,2\,\dfrac{d}{5k}(1 - \dfrac{d}{5k})\right]$ (9) $F_{pw} = 10\,k\,t_w\,f_{yw}$ for $d > 5k$

TABLE I

As commentary of **Table I**, we point out that formulae **(5)** and **(6)** differ from EC3 **[4]** and AISC 86 **[5]** which both introduce a possible effect of the thickness "t" of the element (plate or beam flange) ; in addition, EC3 takes into account the throat thickness "a" of the fillet weld connecting the beam flange and the column flange (or possibly the end plate bolted on this column flange).

But our investigation has demonstrated that these two parameters are not fully significant from statistical point of view with regard to experimental results.

As for the case of an uniformly distributed load and the case of two point loads acting simultaneously on each flange, formulae **(8)** and **(9)** express that the spreading plastic slope is function of length d (see **figures of Table I**) ; to be noted that formula **(8)** has been adopted recently in the last draft of EC3 (paragraph 5.7.3). Another field for application of formulae **(8)** and **(9)** could concern welded joints between hollow section braces and I or H section chord ; so, two of the present authors have proposed [3] a more reliable design formula than that of EC3 (annex K) concerning a bracing member in compression.

2.2 Ultimate resistance of the web

a/ The ultimate resistance of the web can be considered as the previous plastic one only if buckling collapse does not occur before. To take into account the favourable effect of post-critical resistance in buckling, it is possible to base our approach on a Von Karman type formulation and to propose for the web ultimate resistance :

$$F_{uw} = \sqrt{F_{cr}^{(w)} \cdot F_{pw}} \leqslant F_{pw} \qquad (10)$$

where the critical resistance $F_{cr}^{(w)}$ is calculated here in the case of double punching :

$$F_{cr}^{(w)} = \frac{\pi E t_w^3}{3(1 - \nu^2)h_w} \qquad (11)$$

For plastic resistance F_{pw} determination, formulae given in **table I** have to be used.

b/ In fact, it is well-known that initial geometrical defects and residual stresses may affect more or less the ultimate strength ; so, adopting the point of view of Winter formula, the present authors prefer to propose the following modified formulation :

$$F_{uw} = F_{pw} \left[\frac{1}{\overline{\lambda}}(1 - \frac{0,22}{\overline{\lambda}}) \right] \leqslant F_{pw} \qquad (12)$$

where $\overline{\lambda}$ is the non-dimensional web slenderness given by :

$$\overline{\lambda} = \sqrt{\frac{F_{pw}}{F_{cr}^{(w)}}} \qquad (13)$$

3. Modelling with the flange participation

The approach adopted here to introduce the web-to-flange coupling consists in developing a mechanism model with plastic hinge lines ; obviously, with a such model, the analysis of web buckling requires to consider second order geometrical effects, which may lead to go beyond the validity conditions of Limit Analysis in Plasticity. Two types of collapse are distinguished, like in paragraph 2 : crushing and buckling of the web.

3.1 Crushing resistance

The crushing model is presented in **figure 4** and looks like other models such those proposed by BERGFELT [6] and WARDENIER [7] ; however, our model takes into account the effect of the root radius r of the profile, so the crushing resistance is given by :

$$F_{cw} = (t + 2 \ell_1) f_{yw} \cdot t_w \qquad \text{with} \qquad \ell_1 = 2 (\frac{M_{pf}}{f_{yw} \cdot t_w})^{1/2} \qquad (14)$$

where M_{pf} is the plastic bending moment of the flange including the part due to the profile roots :

$$M_{pf} = (\frac{b_f \, t_f^2}{4} + Z_r) \, f_{yw}$$

with :

$$Z_r = \frac{rt_f}{2}(t_w + 0.43 \, r) + (\frac{t_w}{2} + 0.1 \, r) \, r^2 - \frac{r^2}{4 \, b_f} (t_w + 0.43 \, r)^2 \; .$$

In the presence of intermediate plate, expression of ℓ_1 have to be multiplied by $(1 + \frac{M_{pte}}{2 M_{pf}})$, where M_{pte} is the plastic bending moment of the intermediate plate.

3.2 Buckling resistance

In addition to the previous flange mechanism, it is considered a second plastic hinged mechanism in the web , as shown in **figure 5** , which occurs only when the web transversal deflection has some value, characterized by angle θ_o on the figure. Assuming that the load which is deduced from the combined mechanism corresponds to the ultimate load, its value can be calculated from the simple relationship :

$$F_{bwf} = q \, f_{yw} \, t_w \qquad (15)$$

where q represents the yielded length due to normal stresses along the k line on each side under the applied load :

$$q = \frac{8 \ell_2 + 2 \, t}{1 + \frac{f_{yw} \cdot t_w}{2 \, m_{pw}} \sin \theta_o} \qquad (16)$$

m_{pw} is the plastic moment of the web per unit length.

Geometrically, we have :

$$\sin \theta_o = 2 \sqrt{\frac{\delta}{h_w}} \qquad (17)$$

where δ is the maximum flange deflection which can be evaluated in the same way as ROBERTS and ROCKEY [8]

$$\delta = \frac{M_{pf} \cdot \ell_2^2}{6 \, EI} \qquad (18)$$

By minimizing the internal work of the combined mechanism, we find also :

$$\ell_2 = \frac{h_w}{2} (\sqrt{\alpha} + \sqrt{1 + \alpha}) \, , \text{ with } : \; \alpha = (\frac{M_{pf}}{m_{pw}})^2 \, \frac{M_{pf}}{6 \, EI \, h_w} \quad ; \qquad (19)$$

in the presence of intermediate plate, only the expression of α is modified by changing :

$$(\frac{M_{pf}}{m_{pw}})^2 \quad \text{by} \quad (\frac{M_{pf} + 0.5 \, M_{pte}}{m_{pw}})^2$$

3.3 Ultimate resistance

It is logical to propose :

$$F_{uwf} = \text{the smaller of } (F_{cw}, F_{bwf}) . \qquad (20)$$

4. Other recent formulae

4.1 Eurocode 3 (paragraph 5.7 and annex J.1.1.)

In the case of a point load applied to one flange and transferred through the web directly to the other flange, the last draft of Eurocode 3 specifies that the web resistance should be taken as the smaller of :

- the crushing resistance $(s + 5 k) t_w f_{yw}$ where :
$$s = t + 2 a \sqrt{2} + 2 t_e \qquad (21)$$

- the buckling web resistance obtained by considering the web as a virtual compression member with an effective breadth :

$$b_{eff} = [h^2 + s^2]^{1/2} , \qquad (22)$$

and by using the buckling curve C.

4.2 German formula

Recently, an empirical formula probably obtained by statistical method from some data base [9, 10] has been proposed in Germany to determine the ultimate resistance under local compression :

$$F_{uwf}^* = [24 \, (\frac{235}{f_{yw}})^{0,28572} \, t_w + 2 \, t_e] . \, t_w . \, f_{yw} \qquad (23)$$

To be noted that such formula does not allow to distinguish the failure mode.

5 - Comparison between all the mentioned formulae and experimental data

In **table II**, characteristics and failure loads F_u^{exp} of thirty three tests are plotted. These tests concern European rolled H and I sections, for steel grade going from FeE 235 to FeE 460, loaded in symmetrical local compression. The web slenderness ratio $\frac{h_w}{t_w}\sqrt{\frac{f_{yw}}{235}}$ is also mentioned.

In this table, we can do a comparison between experimental ultimate loads and theoretical ones F_{ui}^{th} given by formulae (12), (20), (21 - 22) and (23). To make this comparison easier, we have used the following statistical quantities:

- the mean value :

$$\bar{b} = \frac{1}{n} \sum_{i=1}^{n} \frac{F_{ui}^{exp}}{F_{ui}^{th}} \qquad (24)$$

where n is the number of tests ;
- the standard deviation of the error terms $\delta_i = \dfrac{F_{ui}^{exp}}{\bar{b} \, F_{ui}^{th}}$, i.e. :

$$s_\delta = \sqrt{\frac{1}{n-1} \, (\sum_{1}^{n} \delta_i^2 - n \, \bar{\delta}^2)} \qquad (25)$$

with obviously : $\bar{\delta} = \dfrac{1}{n} \sum_{i=1}^{n} \delta_i = 1$;

- the correlation coefficient between experimental and theoretical values :

REFERENCES		SECTION	NOMINAL CHARACTERISTICS				EXPERI-MENTAL FAILURE LOAD	CALCULATED LOADS			
Laboratory	TEST N°		f_{yw} (MPa)	$\frac{h_w}{t_w}\sqrt{\frac{f_{yw}}{235}}$	t (mm)	t_e (mm)	F_u^{exp} (kN)	AUTHORS F_{uw} (12) (kN)	AUTHORS F_{bw} (20) (kN)	EC3 (21, 22) (kN)	GERMAN FORMULA (23) (kN)
INSA	L1	HEB 140	320	13,1	10	-	365	291	279	253	345
INSA	L2	HEB 200	320	14,9	10	-	770	504	475	439	570
INSA	L3	HEB 260	320	17,7	10	-	870	674	668	571	703
INSA	L4	HEB 140	320	13,1	20	-	375	314	301	255	345
INSA	L5	HEB 200	320	14,9	15	-	780	518	489	439	570
INSA	L6	HEB 200	320	14,9	20	-	825	533	504	440	570
INSA	L7	HEB 260	320	17,7	20	-	880	696	700	572	703
INSA	N1	HEB 160	275	13,0	15	-	550	341	334	294	404
INSA	T1	HEB 200	265	14,9	15	10	760	477	462	385	569
INSA	T2	HEB 200	265	14,9	15	15	800	500	497	389	605
INSA	T3	HEB 200	265	14,9	15	20	840	525	543	393	641
INSA	T4	HEB 200	265	14,9	15	30	940	572	579	405	712
INSA	M1	IPE 140	303	23,8	10	-	175	125	105	127	176
INSA	M2	HEA 260	335	23,6	15	-	608	423	472	346	442
INSA	M3	IPE 220	284	30,2	10	-	300	189	189	177	248
INSA	M4	IPE 360	326	37,4	15	-	530	347	382	284	491
S.T. Delft	1,1	IPE 240	367	33,2	40	-	380	265	277	183	298
S.T. Delft	2,1	IPE 240	425	37,7	40	-	320	290	300	190	331
S.T. Delft	3,1	HEA 240	317	21,9	40	-	483	407	422	313	393
S.T. Delft	4,1	HEA 300	357	26,5	40	-	630	561	580	411	549
S.T. Delft	5,1	HEA 500	286	32,5	40	-	980	783	837	618	934
INSA	MH1	HEA 140	484	23,4	10	-	365	280	297	230	307
INSA	MH2	HEA 160	481	24,3	10	-	530	377	403	315	422
INSA	MH3	HEA 160	475	24,3	10	-	522	367	386	305	404
INSA	MH4	HEA 200	542	28,8	10	-	760	549	602	443	607
INSA	MH5	HEA 200	542	28,8	10	-	740	568	620	458	623
INSA	MH6	HEAA 200	610	34,1	10	-	402	346	370	235	462
INSA	MH7	HEAA 300	544	38,8	10	-	588	442	498	267	475
INSA	MH8	IPE 240	566	42,9	10	-	454	303	345	198	393
INSA	MH9	IPEA 360	524	63,4	15	-	490	295	328	160	436
INSA	MH10	HEA 160	481	24,3	10	10	580	421	418	322	519
INSA	MH11	HEA 160	481	24,3	10	15	620	439	429	326	567
INSA	MH12	HEA 160	481	24,3	10	20	664	458	439	332	616

buckling cases are underlined

TABLE II

$$\rho = \frac{\sum_1^n F_{ui}^{exp} . F_i^{th} - n \overline{F_u^{exp}} . \overline{F^{th}}}{n \, S_F^{exp} \, S_F^{th}} \qquad (26)$$

where $\overline{F_u^{exp}}$ and $\overline{F^{th}}$ are the mean values respectively of experimental and theoretical ultimate loads, and S_F^{exp} and S_F^{th} the corresponding standard deviations.

If ρ is greater than 0.9, then the correlation may be considered sufficient for a statistical interpretation with the theoretical formula concerned. To be noted that the introduction of partial safety factors like γ_{M0} and γ_{M1} in the above-mentioned theoretical formulae does not affect the two main quantities ρ and s_δ for the statistical interpretation. The results are the following ones :

Formula	$\overline{b}$	s_δ	ρ
Authors (12)	1,406	0,113	0,911
Authors (20)	1,381	0,138	0,883
EC3 (21, 22)	1,823	0,181	0,883
German (23)	1,201	0,112	0,916

As commentary, EC3 design formulation gives the highest standard deviation and the lowest correlation coefficient. A detailed examination of results shows that, for all the tests, the buckling load (22) is systematically lower than the crushing resistance (21) ; therefore, the buckling formula (22) is probably too conservative. On the other hand, it appears that theoretical values calculated from empirical formula (23) and formula (12) proposed by the authors lead to the best adjustment to experimental data, each formula giving practically the same accuracy.

It may be surprising that the mechanism model with web-to-flange coupling (20) is less efficient for adjustment than simplified approach (12) and empirical formula (23). A possible explanation of this remark lies in the fact that the real phenomenon may involve several modes of mechanism which are not imperatively discontinuous ones with yielded lines, even if the mechanism adopted here seems relatively sophisticated ; however, the introduction of initial defects in the web profile might improve the adopted model, as the good adjustment shown by formula (12) (but not the Von Karman formula) tends to demonstrate such assumption.

6. Conclusion

Among all the formulae presented in this paper, formulae (12) and (23) appear the simplest and the most accurate to calculate the ultimate load under local compression.

Nevertheless, formula (12) shows several advantages in comparison with formula (23) :

- it is based on a mechanical concept whereas (23) is fully empirical ;
- it allows to distinguish the failure mode (crushing resistance or buckling).

Consequently, formula (12) seems more suitable for some generalization. For instance, by using another data base concerning 26 American rolled sections [11], we have observed a better adjustment with formula (12) (s = 0,093 and ρ = 0,957) than formula (23) (s = 0,120 and ρ = 0,956). Also, there is good reasons for believing that formula (12) could be generalized easily to welded beams. (in this case, we suggest to replace the root radius r by the quantity $a_c \sqrt{2}$, where a_c is the throat thickness of the weld).

In order to bring formula (12) to take a design form, a specific statistical treatment [12] has been carried out [11] according to the reliability index of Eurocode 3 (β = 3.8) ; the corresponding partial safety factor so obtained has been found exactly equal to γ_{M1} = 1.1. Finally, the best formula proposed by the present authors has to be written :

$$F_{ud} = 1,1\ F_{pw} \left[\frac{1}{\overline{\lambda}} (1 - \frac{0,22}{\overline{\lambda}}) \right] / \gamma_{M1} \leqslant 1,1\ F_{pw} / \gamma_{M1} \qquad (27)$$

where F_{pw} and $\overline{\lambda}$ are calculated from nominal characteristics.

7. References

[1]   J.M. Aribert and A. Lachal. Elasto-plastic analysis of the local compressive stresses on a web. Construction Métallique n° 4, 1977, p. 51-66.

[2]   J.M. Aribert, A. Lachal and El Nawawy. Elasto-plastic modelling of the strength of a rolled section under local compression. Construction Métallique, n° 2, 1981, p. 3-26.

[3]   J.M. Aribert, F. Ammari and A. Lachal. Influence of the application mode of local compression loading on the plastic strength of a section web. Case of tubular connections. Construction Métallique, n° 2, 1988, p. 3-30.

[4]   Eurocode n° 3 (projet). Design of steel structures. Vol. 1 and 2, final draft, novembre 1989.

[5]   A.I.S.C. . Load and resistance factor design specification for structural steel building 400 Northe Michigan Avenue, Chicago, Illinois 60611-4185 - september 1986.

[6]   A. Bergfelt. Patch loading on a slender web, influence of horizontal and vertical web stiffeners on the load carrying capacity. Sweden 1979.

[7]   J. Wardenier, P. Stolle, P.W. Dool. The strength and behaviour of plate to I chord connections March 1982.

[8]   T.M. Roberts and K.C. Rockey. Method for predicting the collapse load of a plate girder when subjected to patch loading in the plane of the web. Construction Métallique, n° 3, 1978, p. 3.13.

[9]   G. Sedlacek and W. Hensen. Statistical analysis of strength functions for welded H section joint with respect to available experimental data (Aachen). ARBED. Research Project FeE 460 november 1989.

[10]   Background documentation. For design rules specific for high strength steels according to EN 10113. Annexe D. Document D-01, november 1989.

[11]   J.M. Aribert, A. Lachal and M. Moheissen. Interaction between buckling and web plastic resistance of a rolled beam submitted to symmetrical local compression (for steel grades going to FeE 460).Construction Métallique n° 2, 1990, p. 3-23.

[12]   F.S.K. Bijlaard, G. Sedlacek and J.W.B. Stark. Procedure for the determination of design resistance from tests. Background report to Eurocode 3, IBBC, TNO, report n° B1-87-112, june 1988.

FIGURE 4

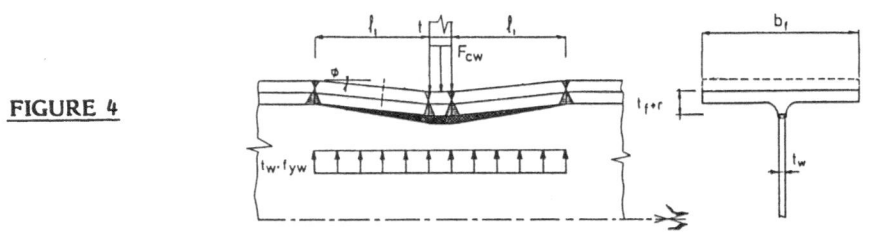

FIGURE 5

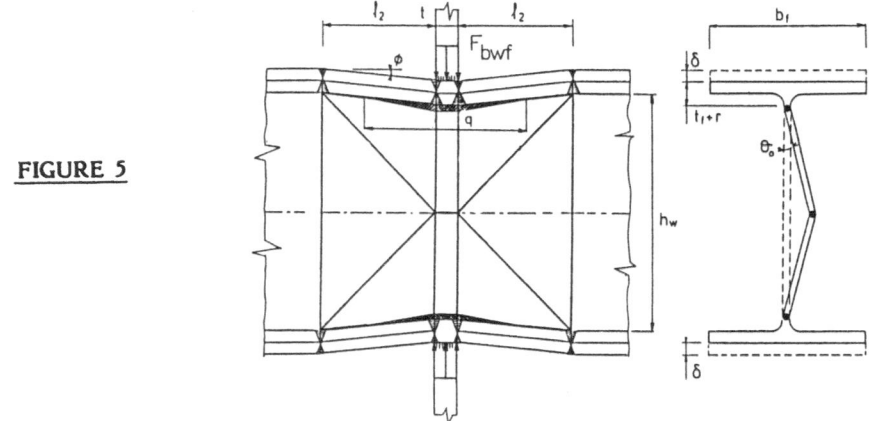

NON - STIFFENED STEEL WEBS WITH FLANGES UNDER PATCH LOADING

M. DRDÁCKÝ

Central Laboratory for Experimental Mechanics,
Institute of Theoretical and Applied Mechanics,
Vyšehradská 49
128 49 Prague 2
Czechoslovakia

Abstract

The paper presents selected results of experimental research into the behaviour of steel webs subjected to partial edge
loadings. The deformations of webs as well as their stress state including plastification are studied. A general method for
derivation of empirical forms for predicting the load-carrying
capacity is presented.

I. Introduction

The behaviour of slender plate-girder webs subjected to
partial edge loadings has been investigated experimentally in
the Institute of Theoretical and Applied Mechanics of the Czechoslovak Academy of Sciences for 25 years. The results have
enriched knowledge obtained through other experiments which
have been performed for more than half a century. From the
engineering point of view, experimental data can be utilized in
two ways. They are vital for improvements of theoretical models
describing the structural behaviour of plate girders under
loading or they serve as a basis for construction of empirical
or semi-empirical formulas estimating the load-carrying capacity of plate-girder webs. The paper deals with the authors
experimental research into the behaviour of non-stiffened steel
webs with flanges taking into account the both above mentioned
purposes.

Different steel web panels were tested. Their typical geometrical, loading and material charakteristics are denoted in
Fig. 1.

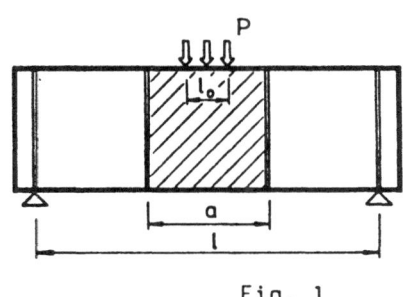

J_{fb}...bending rigi-
dity of the
loaded flange

J_{ft}...torsional rigidity of the
loaded flange

f_{yw}...yield stress
of web

f_{yf}...yield stress
of flange

w_o ...initial web
deflection

Fig. 1

2.   Mechanical behaviour of webs with flanges /1/

2.1. Development of failure mechanism

The behaviour of webs constrained with flanges of a plate girder is rather complicated. Nevertheless, in practical cases, the loss of the load carrying capacity generally occurs after a crippling of the web under the loaded flange regardless to an extent of a previous plastification in the web. The flange thus loses its support and penetrates into the deformed web. In this process a kinematic mechanism originates in the web and in the loaded flange determined by a system of plastic hinges (Fig. 2). Plastic hinges in the flange occur below the

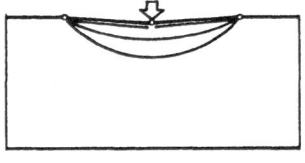 

load and at the beginning of the segment hinges in the web, as if the web had fictious vertical stiffeners there.

The length of the hinges in the web is considerably influenced by the dimensions of the flange, the flexural rigidity of which is most

Fig. 2

marked. The form of the segment hinges in the web retains an identical character regardless of the shape and the dimensions of the flange.

Plastic hinges of the described failure mechanism are rather plastic or elastic - plastic areas depending on the geometrical characteristics of the web, flanges and spacing of

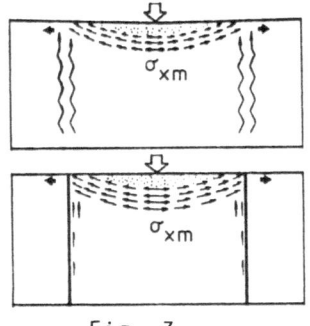

Fig. 3

vertical stiffeners and usually only the line hinge just under the loaded flange is fully plastified. Even in cases of very thin webs because of the stabilization effect of membrane stresses forming a curvilinear tension band in the web acting as a set of ropes hanging between two supports (Fig. 3).

Supposing a rigid plastic mechanism, (Fig. 2), the unit length plastic hinges of the web can sustain the reduced plastic moment

$$m_{pl,r} = \frac{1}{2} t_w^2 f_{yw} \left[1 - (\frac{\sigma_{ym}}{f_{yw}})^2 \right] \qquad (1)$$

From the equilibrium of this mechanism the concentrated load at collapse can be expressed in the general form

$$P_{failure} = const . t_w^2 . f_{yw} . f(...) \qquad (2)$$

where the function f(...) expresses the influence of plastic hinges length, plastic moment reduction due to membrane stresses, web deflections atc. The contribution of plastic moments in the loaded flange is neglected for webs without vertical stiffeners.

112

2.2. Web deformation

The deflections of webs during the course of loading increase from the very beginning, at first stage more or less linearly,(in Fig. 4 at loads lower than the criti- cal one),

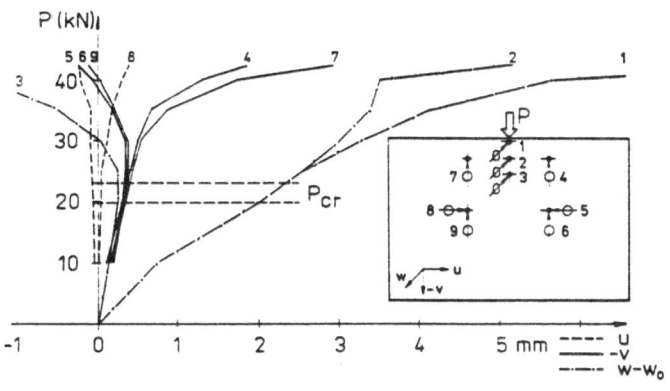

Fig. 4

further in the major part of the web they grow monoto- nously until the moment of the web failure. The in-plane web displace- ments exhibit a similar character and they show the remarkable location of the deformation

into the proximity of the loaded flange, (Fig. 4). The typical
buckling shape of the web in the vicinity of the applied load
is conserved also for webs of very low slenderness ratio, (Fig.5). Other overall web deflections are very strongly influ- enced by initial

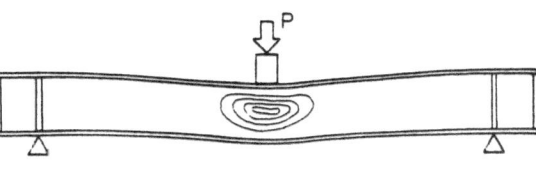

Fig. 5

deflections which often represent the dominant portion of web
deflections at failure, for this type of loading. For illustration, Fig. 6 shows the ratios of the initial and maximum web

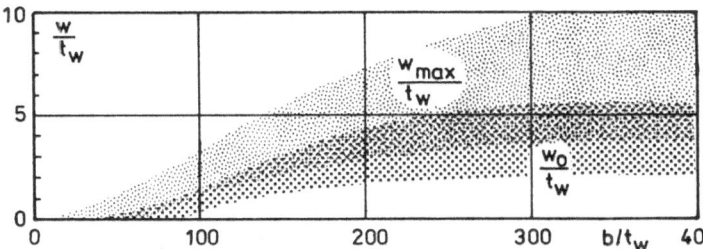

Fig. 6

deflections plotted against the web thickness. Further, the
magnitude of the initial deflections depends on the web
slenderness ratio to about the value of $b/t_w=200$, then it
becomes stationary and oscillates about 1% of the web depth.
Maximum overall deflections vary about 2,5% of web depth./2/

The web deflections in the vicinity of the loaded flange

113

are strongly influenced by an increase of eccentricity of the applied load at only little dependence on the initial web deflections. The dependency is linear.

2.3. Stress in the web

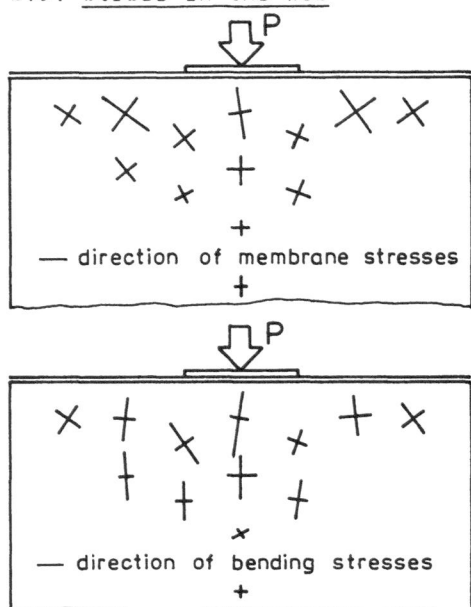

— direction of membrane stresses

— direction of bending stresses

Fig. 7

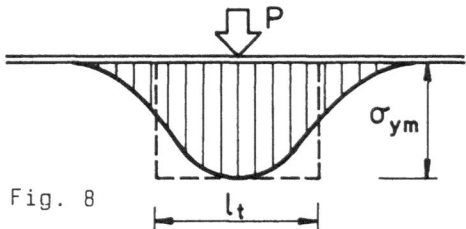

Fig. 8

The stress state in the web is a result of superposition of membrane stresses due to the compressive external loading and flexural stresses due to buckling of a thin web. Because of boundary restraints the web cannot bend freely and in some areas the directions of principal membrane and bending stresses do not coincide and hence the resulting stress state is nonhomogeneous - the resulting principal stresses rotate along the thickness of the web, (Fig. 7).

From the characteristic membrane stress distribution in a web the idea of Girkmann s transmission length l_t follows, see Fig.8. He has calculated this length taking into account the flexural rigidity of the loaded flange in the form

$$l_t = 3.27 \ t_w \left(\frac{J_{fb}}{t_w^4}\right)^{0.333} \tag{3}$$

from which yield for maximum membrane stress under the load

$$\sigma_{ym,max} = \frac{P}{l_t \cdot t_w} \tag{4}$$

Then, in the case of "nonbuckling" webs, the ultimate compresive load P_{uc} can be calculated as

$$P_{uc} = 3.27 \ t_w^2 \cdot f_{yw} \left(\frac{J_{fb}}{t_w^4}\right)^{0.333} \tag{5}$$

The change of stresses in the web is influenced by an eccentricity of the applied load and the increase of web surface stresses under the loaded flange can be estimated as

$$\Delta\sigma = \pm \ e \ \frac{P \cdot t_w \cdot a \ \cdot}{0,75 J_{ft} \cdot b} \tag{6}$$

114

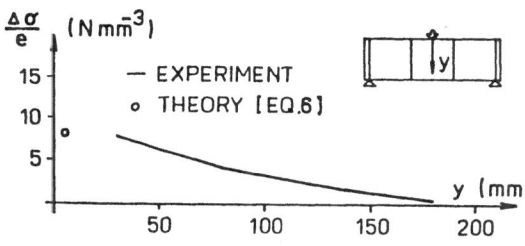

The relative increase of surface stresses plotted against the distance of a measured point from the loaded flange is shown in Fig. 9. /4/

Fig. 9

2.4. Plastification of the web

A web in the vicinity of the applied load is highly stressed beyond the elastic material limit even at early stage of loading. Fig. 10 presents experimentally ascertained beginnings of plastification for webs of various slenderness ratios, Fig. 11 for different loading lengths l_o. They involve measurements by accoustic emission method and by strain gauges, where the attainment of plasticity surface was checked according to Huber - Mises - - Hencky hypothesis. /3/

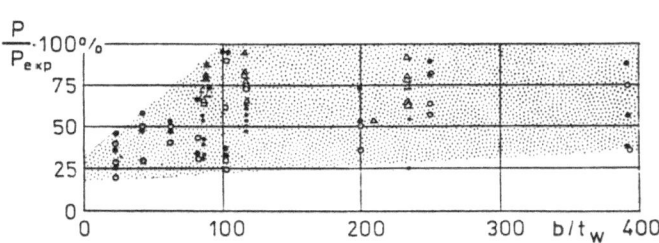

○ Onset of yielding on the surface of unstiffened web.
△ Onset of yielding on the surface of stiffened web.
● Onset of membrane yielding in the unstiffened web.
▲ Onset of membrane yielding in the stiffened web.
• Undirectly signalized plastification of the structure.

Fig. 10

It is seen that a part of the web can be in a plastic state already under loads corresponding with 25 % of the maximum load, particularly in the case of webs of lower slenderness and narrow loading lengths. Newertheless, the plastification is always a very local phenomenon; therefor it seems adequate

$l_o = 0\,mm$

| P = 35.6% | 56.8% | 75.8% | 94.7% | 98.5% P_{max} |

$l_o = 100\,mm$

| P = 51.3% | 64.9% | 73% | 89.3% | 97.4% P_{max} |

Fig. 11

for structures subjected to dynamic loads to reduce the magnitudes of permissible design loads by 50 %. The tests on plate-girder webs under repeated loads, (at frequency of loading pulses being 3Hz), have proved that the decisive plastification arises also just under the loaded flange.

3. Load carrying capacity of plate-girder webs

3.1. Derivation of an semi-empirical formula

The failure of a plate-girder web depends on the structural and loading parameters, e. g. the applied load P, loading length l_0 web thickness t_w, web depth b, web breadth a, flexural rigidity J_{fb} and torsional J_{ft}, of the loaded flange, flange thickness t_f, yield stress of material in both the web f_{yw} and the flanges f_{yf}, moduls of elasticity of the web E, initial deflections of the web w_0, interacting stresses, parameters of web stiffening geometry, number and frequency of loading cycles. Making advantage of the dimensional analysis let us define the failure of the web as a relation between members of a complete set of linearly independent dimensionless numbers composed of the above mentioned parameters.

For the construction of these dimensionless numbers we can utilize compatibility requirements with egs. (2) and (5), which gives the basic numbers $(P/f_{yw}t_w^2)$ and (J_{fb}/t_w^4). Then, for the concentrated, (point), load the ultimate collapse force can be estimated in the form

$$P_{ULT} = A \cdot f_{yw}t_w^2 \; (J_{fb}/t_w^4)^B \tag{7}$$

where A, B are constants.

3.2. Influence of other parameters

The influence of the loading length l_0 has been experimentally tested by several researchers. From the tests it follows that for webs of slenderness ratio higher than 40 the beneficial influence of the length of load application can be estimated by a coefficient

$$C_1 = 1 + 0.0025 \; (l_0/t_w)(\ln(J_{fb}/t_w^4))^{-1} \tag{8}$$

The influence of the slenderness ratio is graphically presented in Fig. 12. Obviously the reduction of the web slenderness below $\lambda = 40$ results in a remarkable increase of the load bearing capacity of the plate-girder panel.

The torsional rigidity of the loaded flange J_{ft} cannot be omitted in combination with the loading excentricity, see eq. (6), but for centric loading it affects the load carrying capacity very little (Fig. 13). The same is valid as far as the flange thickness is concerned, (Fig. 14).

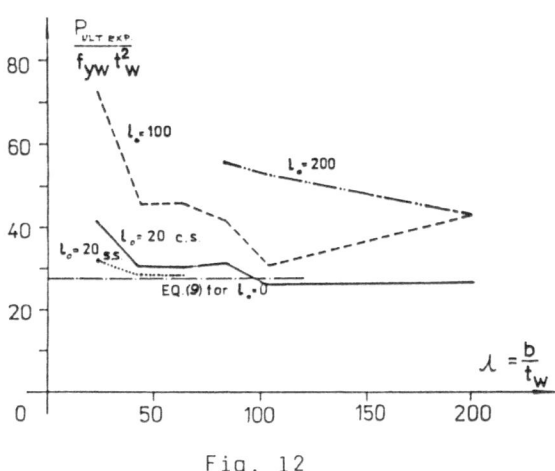

Fig. 12

116

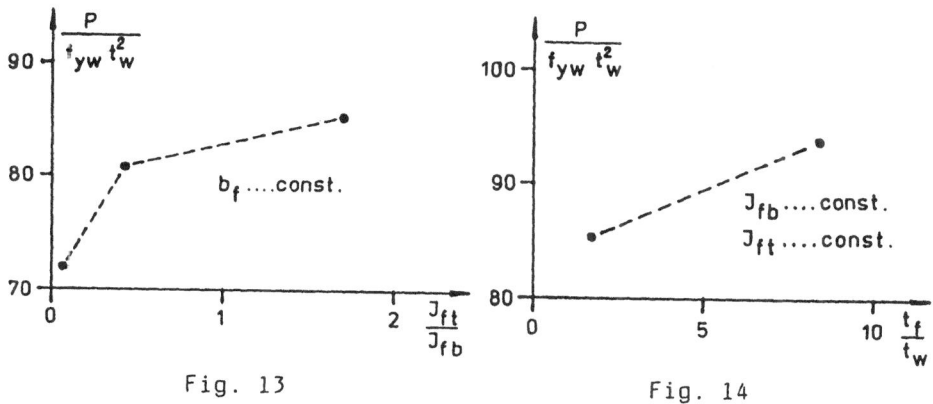

Fig. 13   Fig. 14

The web breadth is of use in combination with rigidity of the loaded flange. According to recent experiments it seems possible to neglect this number at girders where $J_{fb}/a^3 t_w \leq 3.10^{-5}$.

3.3 Tentative formula

Taking into account the above mentioned facts, the author has proposed in 1978 a tentative formula

$$P_u = 19.54 \, f_{yw} t_w^2 \, (J_{fb}/t_w^4)^{0.10} \cdot C_1 \qquad (9)$$

For the estimate of the bearing capacity of thin webs with flanges subjected to partial edge loads. The adequacy of this form tested on 166 experimental values is characterized by the mean value of 1.000, the standard deviation of 0,161 and the coefficient of variation   16.13 %.
According to new experimental results this form should be improve, which the presented method easily enables being sufficiently general and "open".

4. Experimental techniques

The complex behaviour of web-flange systems under discussion requires continuous monitoring of large structural parts.

At our experiments, the web deflections were recorded by means of the shadow moiré method or the stereophotogrammetric method and usually chacked locally by dial gauges. The stereo-photogrammetry is very suitable for measurements of large webs, (larger than 1 m by 1 m), it gives a complete spatial evaluation is very lengthy even in case of analogue devices. The shadow moiré gives a real time contour map of out-of-plane displacements, which is very comfortable. This method is easy applicable to smaler, flat objects (Fig. 15).

The stress distribution in a web was studied by means of strain-gage rosettes and by photoelastic coating method. Due to the above mentioned nonhomogenity of stresses in webs, the photoelastic measurements are complicated and the two usual data - isoclinics and isochromatics parameters - are not sufficient for evaluation and they should be completed by an additional

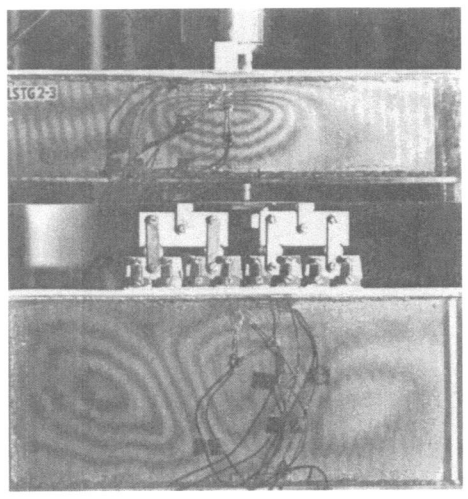

Fig. 15

information.

The strain gages were also used for investigations into the plastification of a web material. To overcome the disadvantage of discrete point measurements, the accoustic emission method has also been applied. It has proved to be satisfactory rather for a membrane yielding than for a surface yielding detection. In any case a correlation of the results measured on test specimens with those from tension tests was hardly possible to evaluate.

Nevertheless, this method can be used for recognition of cracks at dynamic loading tests. But more suitable we have found thermal emission methods. In plastic range the released heat at low frequency dynamic loadings enables to utilize simple thermovision cameras detecting temperature changes of 0.1 K. The stress concentrations at dynamic loads can be measured by means of the thermoelastic effect which is proportional to the sum of principal stresses. This method is very promissible and should be more widely applied in experimental research of thin-walled structures. /5/

5. References

/1/ Drdácký, M.: Flanged Slender Steel Webs under Partial Edge Loading, Chalmers tekniska högskola, Rapport Serie Int. skr. S82.14, Göteborg 1982.

/2/ Drdácký, M.: Limit States of Steel Plate Girder Webs under Patch Loading, Proc. of the 2nd Regional Colloq. on Stability of Steel Structures, Vol. II/1 pp. 49-56, Hungary, September 1986.

/3/ Drdácký, M., Jaroš, P., Weinberg, O.: Experimental Investigations into the Plastification of Thin Webs, Proc. of the 1st Cong. IMEKO TC-15, Plzeň, May 1987.

/4/ Drdácký, M.: On Two Particular Problems of Plate Girder Webs under Partial Edge Loads, Proc. of Int. Colloq. on stability of steel structures, Vol. II, pp. 29-36, Budapest 1990.

/5/ Contactless Measurements of Deformations and Stresses in Structures, (Editor M. Drdácký), Proceedings of the Workshop, Prague November 1990.

BUCKLING ANALYSIS OF TRAPEZOIDALLY CORRUGATED WEB UNDER PATCH LOADING USING SPLINE FINITE STRIP METHOD

Ruoshan LUO   and   Bo EDLUND

Dept. of Structural Engineering
Chalmers University of Technology
S-412 96 GÖTEBORG, SWEDEN

Abstract

Elastic buckling of trapezoidally corrugated webs under patch loading is studied by use of a spline finite strip method. A computer code, STRIPBA, is developed on a mainframe computer and on a Unix workstation. An example girder, which was used for full-scale experiments, is used for application of the computer program and to see if reasonably accurate computational results can be obtained. The influence of geometric parameters and of the location of the patch loading on the bifurcation (buckling) load is analysed. Some preliminary suggestions for a practical design of this kind of web is also given.

1. Introduction

Welded plate girders with a thin-walled trapezoidally corrugated web have been used successfully for nearly two decades in Sweden. These girders are very competitive and the rather stiff web makes them easy to handle and to erect when compared to slender plate girders with a plane web.

In [1] welded girders with corrugated web have been tested under partial edge loading.The girders had rather large depth and a short span. The flanges were rather stiff, so the component plates of the web may be regarded as having clamped support at the upper edge, i.e. the rotation is prevented but vertical translation is permitted. The webs were thin-walled, thickness 2 or 2.5 mm.The location of the load was varied with respect to the corrugation of the web. Fig.1 illustrates the case where the patch load was over the central part of one of the corrugation. The vertical trapezoidal folds give an effective stiffening of the web. It is convenient to define different stages

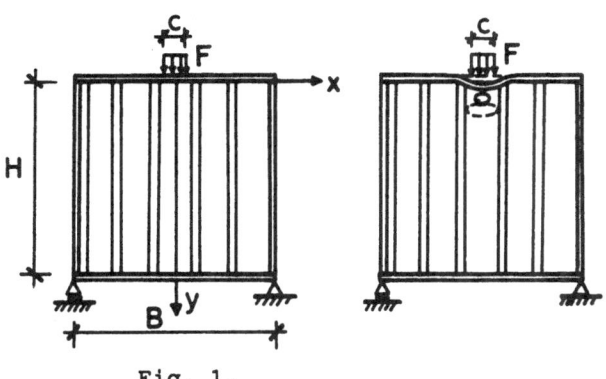

Fig. 1.

in the behaviour of the web when increasing the vertical displacement of the loaded point. In the case shown in Fig.1 the first buckle appears within the flat field below the load. When the load then increases the buckle develops and will at first be limited by the two neighbouring oblique fields which act as stiffeners. Only the formation of the initial buckling pattern is studied in this paper (bifurcation load). In the tests, however, new buckles will be observed in the neighbouring oblique fields at a somewhat higher load. In the next stage the first buckle extends sideways and will break through and deform the first knuckle lines (trapezoidal folds). At the maximum load the flange deforms strongly at the loaded patch and a plastic mechanism is formed in the web, as shown in [1].

In this paper the spline finite strip method is used for the elastic buckling analysis of trapezoidally corrugated webs under patch loading. The finite strip method is found to be more efficient and convenient than the finite element method, when analyzing prismatic thin-walled structures. By using spline functions as displacement functions [2] the method can also easily be used in cases with complicated boundary and loading conditions. In our analysis, the corrugated web is assumed to have perfect initial geometry (i.e. no imperfections) and only the first bifurcation load is sought for. Only isolated webs with prescribed boundary conditions are studied, i.e. the influence of the flanges is disregarded.

2. Spline finite strip method and buckling analysis

The spline finite strip method recently developed by Cheung and his coworkers [2] has been shown to be both accurate and efficient for analysing the buckling of stiffened plates subjected to combined compression and shear, see [3].

2.1 Structural subdivision and knot displacements

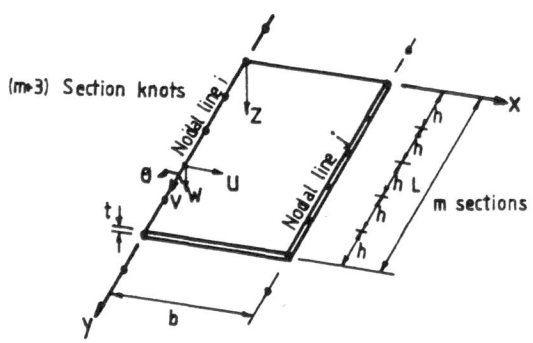

A trapezoidally corrugated web is subdivided transversely into a number of strips using n nodal lines and subdivided longitudinally into m sections using (m+3) section knots. The nodal lines and section knots for one strip are shown in Fig.2. Each section knot has four degrees of freedom corresponding to the two out-of-plane deformations, w and

Fig. 2. A B_3-spline strip.

θ, and the two in-plane displacements u and v.

2.2 B_3-spline function

The displacement function of a strip is expressed as the product of transverse interpolation polynomials and in the longitudinal direction a linear combination of local B_3-splines, Eq.(2).

The displacement is taken as the summation of (m+3) local B_3-spline functions by

$$f(y) = \sum_{i=-1}^{m+1} \alpha_i \, \psi_i \, (y) \tag{1}$$

where $\psi_i(y)$ is a local B_3-spline as shown in Fig.3a and α_i is a coefficient to be determined.

The length of the structure is divided into m equal length sections as shown in Fig.3b.

A standard B_3-spline function is defined by

$$\psi_i(y) = \frac{1}{6h^3} \begin{cases} 0 & y < y_{i-2} \\ (y - y_{i-2})^3 & y_{i-2} \leq y \leq y_{i-1} \\ h^3 + 3h^2(y-y_{i-1}) + 3h(y-y_{i-1})^2 - 3(y-y_{i-1})^3 & y_{i-1} \leq y \leq y_i \\ h^3 + 3h^2(y_{i+1}-y) + 3h(y_{i+1}-y)^2 - 3(y_{i+1}-y)^3 & y_i \leq y \leq y_{i+1} \\ (y_{i+2}-y)^3 & y_{i+1} \leq y \leq y_{i+2} \\ 0 & y_{i+2} < y \end{cases} \tag{2}$$

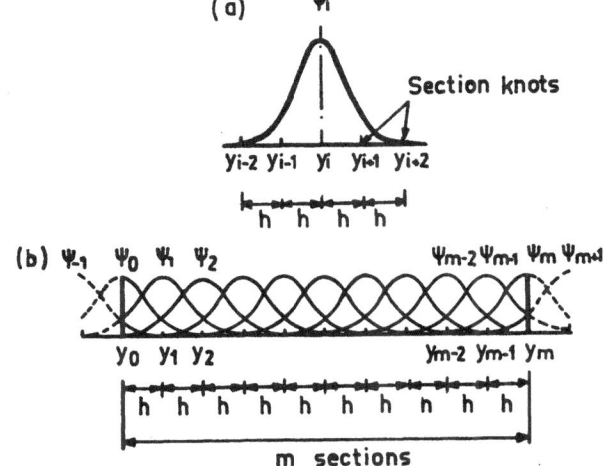

Fig.3(a) Typical B_3-spline function, Eq.2.
(b) Basis of B_3-spline expression.

2.3 Displacement functions and generalized displacement vector
The displacement functions {f} of a strip are expressed as

121

conventional transverse shape function N_i.

For plane stresses

$$\{f\} = \begin{Bmatrix} u \\ v \end{Bmatrix} = \begin{bmatrix} N_1 & 0 & N_2 & 0 \\ 0 & N_1 & 0 & N_2 \end{bmatrix} \begin{bmatrix} [\psi_{ui}] & & & \\ & [\psi_{vi}] & & \\ & & [\psi_{uj}] & \\ & & & [\psi_{vj}] \end{bmatrix} \begin{Bmatrix} \{u_i\} \\ \{v_i\} \\ \{u_j\} \\ \{v_j\} \end{Bmatrix} \qquad (3)$$

For plate bending

$$\{f\} = \{w\} = [N_3 \ N_4 \ N_5 \ N_6] \begin{bmatrix} [\psi_{wi}] & & & \\ & [\psi_{\theta i}] & & \\ & & [\psi_{wj}] & \\ & & & [\psi_{\theta j}] \end{bmatrix} \begin{Bmatrix} \{w_i\} \\ \{\theta_i\} \\ \{w_j\} \\ \{\theta_j\} \end{Bmatrix} \qquad (4)$$

where

$$\begin{aligned}
N_1 &= 1-\bar{x} \\
N_2 &= \bar{x} \\
N_3 &= 1-3\bar{x}^2+2\bar{x}^3 \\
N_4 &= x(1-2\bar{x}+\bar{x}^{2)} \\
N_5 &= 3\bar{x}^2-2\bar{x}^3 \\
N_6 &= x(\bar{x}^2-\bar{x}) \\
\bar{x} &= x/b
\end{aligned} \qquad (5)$$

The terms $[\psi_{ui}]$, $[\psi_{vi}]$, $[\psi_{wi}]$, $[\psi_{\theta i}]$, $[\psi_{uj}]$, $[\psi_{vj}]$, $[\psi_{wj}]$, $[\psi_{\theta j}]$ are row matrices in terms of B_3-splines and $\{u_i\}$, $\{v_i\}$, $\{w_i\}$, $\{\theta_i\}$, $\{u_j\}$, $\{v_j\}$, $\{w_j\}$, $\{\theta_j\}$ are the corresponding displacement parameter vectors for the two adjacent nodal lines i and j respectively.

The displacement functions of a trapezoidally corrugated web can be expressed as a combination of plate bending and in plane (membrane) deformation. Define a generalized displacement vector for a strip with

$$\{\delta\} = [\{u_i\}^T \ \{v_i\}^T \ \{w_i\}^T \ \{\theta_i\}^T \ \{u_j\}^T \ \{v_j\}^T \ \{w_j\}^T \ \{\theta_j\}^T]^T \qquad (6)$$

2.4 Elastic buckling analysis

The strain energy of a strip resulting from buckling deformation is given by

$$u_i = \frac{1}{2} \int_0^L \int_0^b \{\sigma\}^T \{\varepsilon\} \ dx \ dy \qquad (7)$$

where

122

$$\{\sigma\} = [\sigma_x \; \sigma_y \; \tau_{xy} \; M_x \; M_y \; M_{xy}]^T \tag{7a}$$

$$\{\varepsilon\} = [\frac{\partial u}{\partial x} \; \frac{\partial v}{\partial y} \; (\frac{\partial u}{\partial y} + \frac{\partial v}{\partial x}) \; -\frac{\partial^2 w}{\partial x^2} \; -\frac{\partial^2 w}{\partial y^2} \; 2\frac{\partial^2 w}{\partial x \partial y}]^T \tag{7b}$$

The stress matrix $\{\sigma\}$ and strain matrix $\{\varepsilon\}$ can be related to the generalised displacement vector $\{\delta\}$ by

$$\{\varepsilon\} = [B]\{\delta\} \tag{8}$$

$$\{\sigma\} = [D]\{\varepsilon\} = [D][B]\{\delta\} \tag{9}$$

The strain energy U for the structural system is

$$U = \Sigma \, u_i = \Sigma \frac{1}{2}\{\delta\}^T \int_0^L \int_0^b [B]^T [D][B] \; dx \; dy \, \{\delta\} \tag{10}$$

where Σ means the sum of all strips.

The increase in potential energy of the membrane forces resulting from the buckling deformations was derived by [4] as

$$W = -\Sigma \frac{1}{2}\int_0^L \int_0^b \{\sigma_y \; \{(\frac{\partial u}{\partial y})^2 + (\frac{\partial v}{\partial y})^2 + (\frac{\partial w}{\partial y})^2\} + \sigma_x \, (\frac{\partial w}{\partial x})^2 + 2\tau_{xy}\frac{\partial w}{\partial x}\frac{\partial w}{\partial y}\} \, t \, dxdy \tag{11}$$

By applying the principle of minimum potential energy we have

$$\frac{\partial (U+W)}{\partial \{\delta\}} = 0 \tag{12}$$

That is

$$([K] - \lambda\,[G])\,\{\delta\} = \{0\} \tag{13}$$

where

[K] = stiffness matrix
λ = buckling load factor
[G] = stability matrix

This is an eigenvalue problem. By solving for the smallest eigenvalue of (13) we obtain the first bifurcation (buckling) load.

2.5 Computer program

Based on the method described in the previous sections a computer code STRIPBA has been developed on an IBM3090 computer and on an

123

engineering workstation and written in the FORTRAN 77 language. The code is easy to modify, for example to include various imperfections of the trapezoidally corrugated webs, or different loading and boundary conditions.

3. Computational analysis

A series of computational tests (numerical experiments) has been performed on slender trapezoidally corrugated web panels using the method and program presented in section 2. The panels are subjected to partial edge loading as shown in Fig.1 and have hinged supports at all four boundaries (panel width B = 1.0 m, depth H = 1.5 m). The corrugation geometry is defined in Fig.4. The geometry assumed for the analysis has sharp corners instead of the smooth corners in the test girders of ref. [1]. The location of the load was varied with respect to the corrugation of the web. The load was appplied at three different locations (a) directly over the central part of one flat field of the corrugation, (b) directly over the junction in a fold of the corrugation with load distribution length c = $25(1+\sqrt{2})$ mm and (c) over one oblique part of the corrugation (c = $50 \cdot \sqrt{2}$ mm).

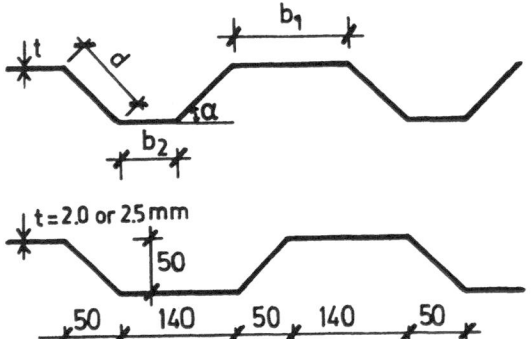

Fig.4. Corrugation geometry.
   (a)   Notations.
   (b)   Dimensions of girders
         tested in [1].

The boundary conditions assumed for the web panels are

1. At the loaded edge (upper edge)

$$\sigma_y = \begin{cases} F/ct & \text{for } -c/2 \leq x \leq c/2 \\ 0 & \text{elsewhere} \end{cases} \qquad (14)$$

$$u = w = M_x = \theta_y = M_z = 0$$

2. At the vertical edges

$$u = v = w = M_y = \theta_x = \theta_z = 0 \qquad (15)$$

3. At the lower boundary

$$\sigma_y = u = w = M_x = \theta_y = M_z = 0 \qquad (16)$$

124

4. Results

When calculating the buckling loads, the webs are divided into 23 strips in the transverse direction and 14 sections in the longitudinal direction. (In fact, we have also made calculations using 20, 21 and 22 strips with 4 to 14 sections. Due to space limitations, those results are not shown here. However, it is noticed that the convergence is fast.) According to the full-scale tests on steel girders, Young's modulus $E = 2.1 \times 10^5$ MPa. Poisson's ratio $v = 0.3$. The load distribution length is $c = 50$ mm, $25(1+\sqrt{2})$ mm, and $50\sqrt{2}$ mm respectively, see Table 1.

Table 1.

Web thick- t mm	Web depth H mm	Load length c mm	Load location	Buckl.stress theoretical σ_{cr}	Experimentally observed[1] σ_{buc}
2.5	1500	50		498	600
2.0	1500	50		319	-
2.5	1500	$25(1+\sqrt{2})$		2392	-
2.0	1500	$25(1+\sqrt{2})$		1536	-
2.5	1500	$50\sqrt{2}$		1247	-
2.0	1500	$50\sqrt{2}$		811	430-525[2]

1) σ_{buc} is the stress at the load level where the linear part of the load-deflection curve ends and nonlinear behaviour starts.
2) For web depth 2000 mm.

The following factors (see definition in Fig.4) which influence the buckling load are studied: 1) angle of the corrugation α, with values of 15°, 30°, 45°, 60°, 75° and 90°; 2) ratio $\gamma=b1/b2$, with values of 1,2,3,4,and 5. Oblique part of the corrugation d is remained constant. Results are shown in Fig.5. Fig.5a shows that when the loading is over the centre of one flat part, the buckling stress is increased when the angle α and the ratio γ increase, although the difference between minimum and maximum buckling stress is quite small. When the load is loaded over a junction of a corrugation and over an oblique part of one corrugation, the buckling stresses (Fig.5b and Fig.5c) decrease when the ratio γ increases while the angle α is 15° and 30°. For the cases when α is 45°,60°,75° and 90°, the buckling stress increases somewhat when the ratio γ increases.

125

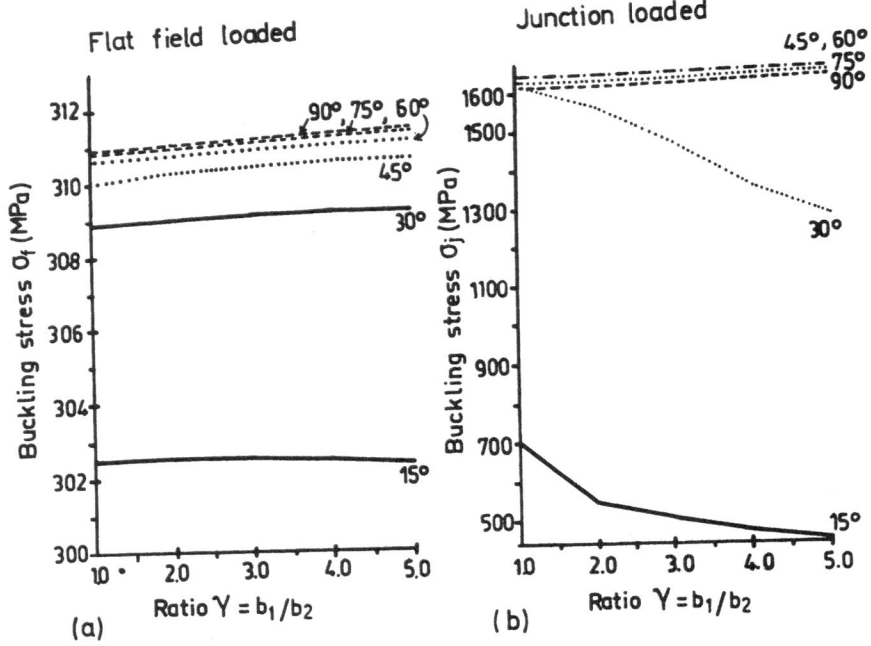

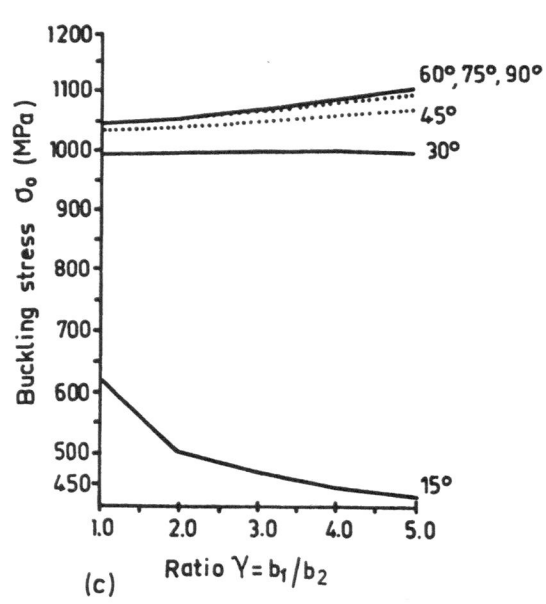

Fig.5. Influence of angle α and ratio γ = b₁/b₂ on elastic
        buckling stress. Figures a, b and c show results for the
        three load locations illustrated in Table 1.

5. Conclusions

The computational results show a similar trend as the results obtained from the full-scale tests [1]. For example Table 1 shows that the web thickness is an important factor in the behaviour of a corrugated web under patch loading, i.e. the buckling loads are proportional to the square of the thickness.

The location of the load with respect to the corrugation geometry is another important factor, which influences the buckling load. When the load is located over the junction in a fold of the corrugation the buckling load is the highest. When the load is located over the flat part, the buckling load is the lowest.

When the angle $\alpha \geq 45°$, the buckling stress increases somewhat when the ratio γ increases in the range from 1 to 5. When the load is located over the centre of one flat part, the geometry with $\alpha = 90°$ gives the highest buckling stress. When the load is over a junction (corner) of a corrugation, $\alpha = 45°$ and $\alpha = 60°$ give the highest buckling stress. When loading over an oblique part of one corrugation, $\alpha = 60°$, $\alpha = 75°$, and $\alpha = 90°$ lead to higher buckling stress than $\alpha = 45°$. When $\alpha = 15°$ the buckling stress decreases as the ratio γ increases. When $\alpha = 30°$ the increase of γ has very small influence on buckling stresses σ_f and σ_o, but on the contrary σ_j will decrease as γ increases.

6. References

1. L. Leiva-Aravena and B. Edlund, Buckling of Trapezoidally Corrugated Webs, ECCS Colloquium on Stability of Plate and Shell Structures. Ghent University, 6-8 April 1987, 107-116.

2. Y.K. Cheung and S.C. Fan, Static analysis of right box girder bridges by spline finite strip method. Proc. Instn. Civ. Engrs., Part 2, 75, 311-323 (1983).

3. S.C. Lan and G.J. Hancock, Buckling of Thin Flat-Walled Structures by a Spline Finite Strip Method. Thin-Walled Structures 4, 269-294 (1986).

4. R.J. Plank and W.H. Wittrick, Buckling under combined loading of thin flat-walled structures by a complex finite strip method. Int. J. Numerical Methods in Engineering 8(2), 323-339 (1974).

BEHAVIOR OF WEBS UNDER ECCENTRIC COMPRESSIVE EDGE LOADS

Mohamed Elgaaly, Professor
Raghuvir K. Salkar, Graduate research assistant

Department of Civil Engineering,
University of Maine,
Orono, Maine 04469,
USA.

Abstract

Webs of rolled and built-up beams can be subjected to local
in-plane and eccentric compressive edge loads. For practical
and/or economic reasons, transverse stiffeners are to be
avoided or minimized. Most of the research to-date has
addressed the web behavior under in-plane loads, and has
been of experimental nature. This paper reports the
experimental and analytical work done at the University of
Maine to study the effect of parameters like t_f/t_w, c/d,
b/d, e/b_f, b_f/t_f on web behavior under eccentric edge loads.
Practically, there is no reduction in the ultimate capacity
of the web due to eccentricity, when the load is applied
through a thick patch plate placed eccentrically with
respect to the plane of the web. However, reductions are
observed when the load is applied through a cylindrical bar,
and these depend mainly on the ratios t_f/t_w and e/b_f.

1. Introduction

Webs of rolled and built-up beams can be subjected to local
in-plane and out-of-plane compressive loads. Examples are,
wheel loads, loads from purlins and roller loads during
construction. For practical and/or economic reasons,
transverse stiffeners are to be minimized or avoided, except
at critical sections. It is, therefore, necessary to check
the unstiffened web under edge compressive loading to ensure
that no localized failure will occur.

During the past 60 years, tests have been performed to study
the web behavior under in-plane compressive edge loads.
These are mostly of the type shown in figure 1; however, the
compression of the web over a support bearing block, was
also investigated. Research to study web behavior under out-
of-plane, henceforth referred to as eccentric, edge
compressive loading was initiated at the University of Maine

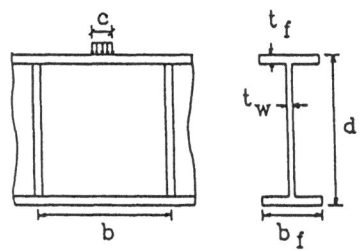

Fig. 1.

by Elgaaly in 1988. Recently, similar research was started at the Czechoslovak Academy of Sciences by Drdacky [1].

The purpose of this paper is to briefly summarize the available experimental and analytical results for the loading type shown in figure 1, and to present the recent and ongoing research at the University of Maine.

2. Webs under in-plane edge compressive loading

Failure in webs under in-plane edge compressive loading could be initiated by either its buckling or yielding; the ultimate failure mode being web crippling. As a rule, failure in stocky webs is initiated by yielding, whereas failure in slender webs is initiated by buckling.

The study of webs subjected to this type of loading has been carried out by many researchers. All test results indicate that the web ultimate capacity $'P_u'$ is almost independent of the web slenderness ratio and the flange width to thickness ratio. It is, however, more or less directly proportional to the square of the web thickness, and is influenced to a lesser extent by the length of the patch load $'c'$, the flange thickness $'t_f'$, and the web material yield stress $'F_y'$. The coexistence of global bending stresses in the beams has also been studied. Very limited work has been done to study the effect of global shear stresses on the ultimate capacity of webs. All this work has been summarized by Elgaaly [2]. The ratio of the length of the patch load to the web depth $'d'$, in most of the tests conducted to-date, is limited to a maximum value of 0.33. Recently, work has been carried out in Japan by Shmizu and others, where the ratio c/d was as much as 0.50 [3].

3. Webs under eccentric edge compressive loading

Eccentricities in edge loads are unavoidable in practice. They are due to fabrication or rolling tolerances, construction loads and/or construction tolerances. For this reason, the strength of webs under eccentric loads needs to be studied.

Research in this area was started by Elgaaly and Nunan in 1988, at the University of Maine. Twenty-two rolled sections of the same cross-sectional properties were tested to study the effect of the eccentricity of load. It was found that there was little reduction in P_u due to eccentricity, when the load was applied through a thick patch plate, placed eccentrically with respect to the web. However, reductions occurred when the load was applied through a cylindrical roller; the reduction being a function of the eccentricity [4]. Further research was carried out by Elgaaly and Sturgis [5,6]. Several rolled and built-up sections were tested. The parameters studied were the ratio t_f/t_w, the eccentricity of the load e and the panel aspect ratio b/d. The conclusions of the above study were that the web ultimate strength increases with an increase in t_f/t_w ratio. The panel aspect ratio did not cause any definite changes in web strength. The above mentioned research established a need for further investigation, to quantify the reduction in P_u due to the eccentricity of the load. Hence, effort in this direction was started by Elgaaly and Salkar in June 1989, with emphasis on experimental as well as analytical (computer) work, and the following reports that work.

Fig. 2.

All the experimental work was carried out in a Baldwin Testing Machine with a capacity of 400 kips. As shown in figure 2, the specimens were supported on roller supports, located under transverse stiffeners. The load was applied on the top flange through a cylindrical bar. All test samples were instrumented with strain rosettes, linear strain gages and displacement transducers.

Table 1 describes the test program (tests S1 to S9) and gives the test P_u values. Moreover, it also gives the analytically predicted P_u values and its comparison with the test P_u values. Ptst refers to the test P_u values, Panl refers to the analytically predicted P_u and R is the ratio of the former to the latter. Tests S7 and S8 were done with longitudinal stiffeners at a clear distance of 0.2d from top flange. Tests S5 and S6 had c/d ratio of 0.6, whereas all other tests had a c/d ratio of 0.2. Tests J1 to J4 were done by Elgaaly and Sturgis [6], and tests T1 to T5 have been reported by Elgaaly in appendix 1 as tests 43, 77, 84, 92 and 118 [2]. All test specimens had an aspect ratio of 1, except in test 118, where it was 2.

The failure in the web was local and crippling occurred within a depth of 0.16d. The yield line(s) were closer to the top of web for thinner flanges and larger eccentricities. The mode of failure was crippling for in-plane as well as eccentrically loaded specimens. As seen from tests J3, S7 and S4, S8 in table 1, the addition of the longitudinal stiffener did not increase the web strength by a great extent.

A linear computer analysis started by Elgaaly and Nunan [4], was continued in more depth by Elgaaly and Sturgis [6]. This analysis, being linear in nature, was useful in studying the web behavior only before its buckling or yielding. Elgaaly and Salkar used a non-linear program developed by Elgaaly, Caccese and Du [7]. It uses a 3-dimensional isoparametric doubly curved shell element, which is a degeneration of the isoparametric hexahedron H20. The analysis combined the effects of large displacements and material nonlinearities. An updated Lagrangian method was used. The inelastic material behavior was modelled based on Prandtl - Reuss flow theory of plasticity and the Von Mises yield criterion.

Initially, the stability of the element and the software was examined with respect to the load increments and the number of nodes in a model. The validity of the finite element analysis was established by comparing the analytically predicted P_u values with the test P_u values, as shown in table 1. Further, a detailed comparison was made between the strains and displacements predicted by analysis and those observed from tests. Figures 3, 4 and 5 give a comparison between the results from test S4 and the corresponding computer analysis. The computer model had a total of 475 nodes and the smallest load increment was 500 pounds. Figure 3 shows the out-of-plane web displacements along its central cross section, near failure. Figures 4 and 5 show the membrane and bending strains at the same location, near failure.

Table 1.

No	c/d	d/t_w	b_f/t_f	t_f/t_w	t_w	e/b_f	F_y	Ptst	Panl	R
J1	.2	208.3	24.3	2.1	.120	0	41	24.2	24.5	.988
J2	.2	208.3	23.8	2.1	.120	1/24	40	22.4	22.0	1.02
J3	.2	208.3	23.8	2.1	.120	1/12	40	19.0	17.5	1.09
J4	.2	208.3	23.8	2.1	.120	1/8	41	13.9	13.0	1.07
S1	.2	213.7	12.1	4.24	.117	0	44	28.4	29.5	.963
S2	.2	208.3	12.1	4.13	.120	1/12	39	25.5	24.0	1.06
S3	.2	208.3	12.1	4.13	.120	1/12	39	26.0	24.0	1.08
S4	.2	213.7	12.1	4.24	.117	1/6	44	25.8	24.0	1.08
S5	.6	213.7	12.1	4.24	.117	1/6	44	37.0	34.5	1.07
S6	.6	213.7	24.0	2.14	.117	1/12	44	26.5	26.0	1.02
S7	.2	213.7	24.0	2.14	.117	1/12	44	19.4		
S8	.2	213.7	12.1	4.24	.117	1/6	44	29.0		
S9	.2	213.7	24.0	2.14	.117	0	44	25.8		
T1	.1	250.0	2.0	12.3	.079	0	35	13.2	14.5	0.91
T2	.1	165.0	4.9	2.78	.143	0	41	34.6	33.0	1.05
T3	.05	168.0	5.1	2.80	.141	0	37	28.0	27.0	1.04
T4	.2	250.0	16.7	3.00	.079	0	35	11.6	12.0	0.97
T5	.07	127.0	12.8	2.09	.118	0	36	23.3	22.2	1.06

Figure 6 shows the results of the analysis done to study the effect of the parameters t_f/t_w, e/b_f, b_f/t_f and c/d. All cases considered had a slenderness ratio of 149.33 and t_w equals 3/16 of an inch. As seen from table 1 and figure 6, the analysis was limited to slender webs only, with d/t_w around 120 or more.

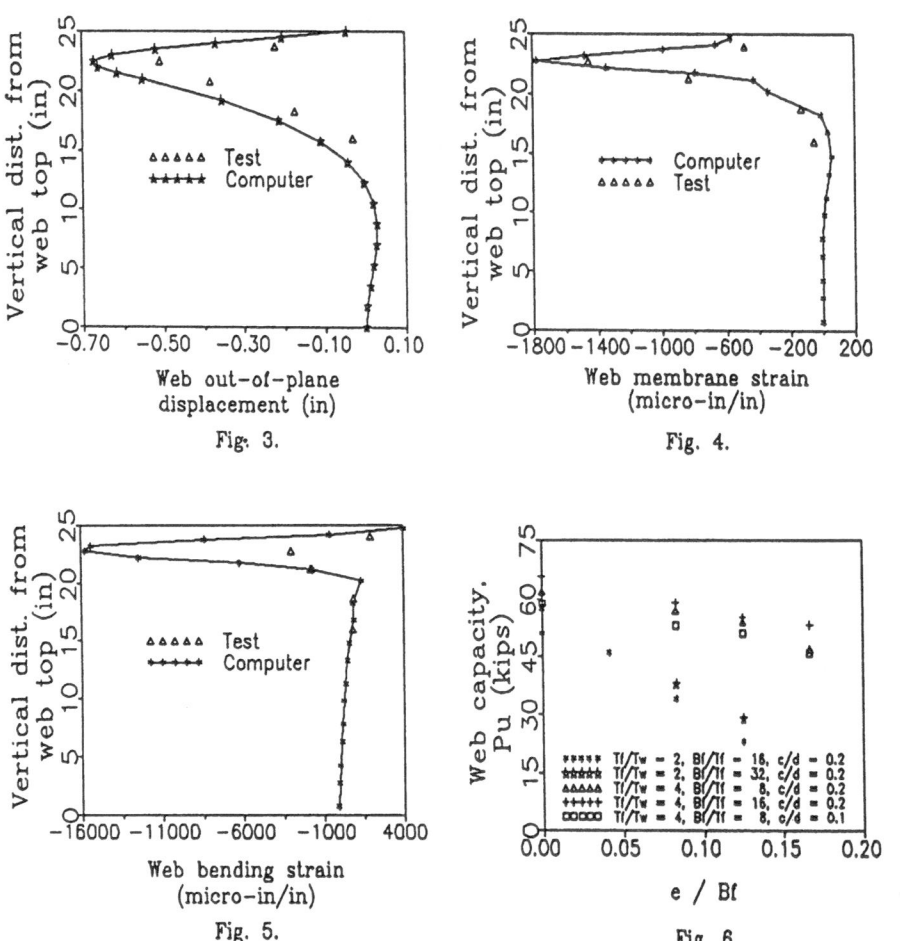

Fig. 3.

Fig. 4.

Fig. 5.

Fig. 6.

4. Concluding remarks

1. Eccentric loading can occur in practice, and can reduce the web ultimate strength by an amount that depends on various factors.

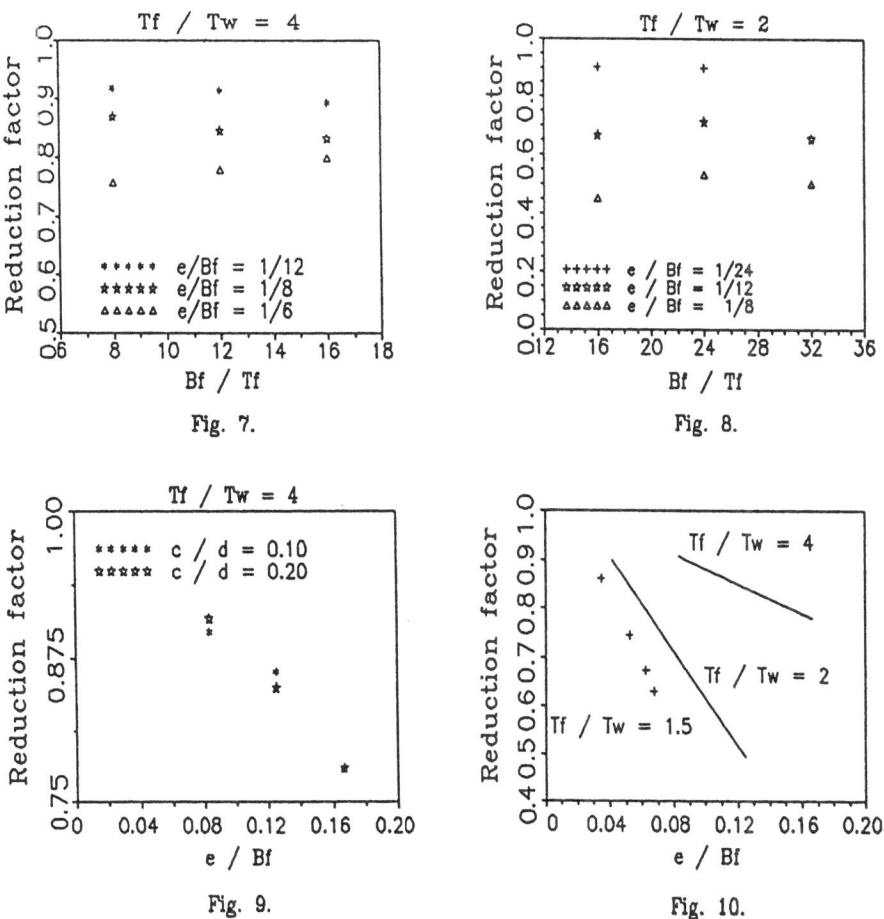

Fig. 7.

Fig. 8.

Fig. 9.

Fig. 10.

2. The research by the authors considered the parameters t_f/t_w, e/b_f, b_f/t_f, c/d and F_y. It showed that the parameters that control the reduction are t_f/t_w and e/b_f; the factors b_f/t_f and c/d have little effect, and this can be seen from figures 7, 8 and 9. Reduction factor appearing in these figures and figure 10 is the ratio of P_u with some eccentricity to that with no eccentricity of loading.

3. For all practical purposes, the reduction in P_u for a given beam is directly proportional to e/b_f ratio. As a rule of thumb, a ratio of e/b_f equal to 1/8 reduces P_u by about 50 % for beams with t_f/t_w between 1 and 2. For beams with t_f/t_w equal to or more than 4, a ratio of e/b_f as high as 1/6 causes a reduction as low as only 20 % . Figure 10 shows curves for reduction in P_u where nominal t_f/t_w ratio is equal to 2 and 4; these being a result of the analytical

work. The scattered points in the figure are a result of the testing work by Elgaaly and Nunan [4], and are for a nominal t_f/t_w ratio equal to 1.5.
4. Currently, the effect of the ratio c/d with values greater than 0.33 is under investigation.

Conversion to S.I. units

1 ksi = 6.89 Mpa
1 in = 25.4 mm
1 kip = 4.45 KN

References

1. Drdacky, M., On two particular problems of plate girder webs under partial edge loads, Personal communication.
2. Elgaaly, M., Web design under compressive edge loads, AISC Engineering Journal, 153-171, Fourth Quarter, 1983.
3. Shigeru Shimizu, Hisanori Yabana & Shunya Yoshida, A collapse model for patch-loaded web plates, J. Construct. Steel Research 13, 61-73, 1989.
4. Elgaaly, M., Nunan, W., Behavior of rolled section web under eccentric edge compressive loads, Journal of Structural Engineering, 1561-1578, Vol. 115, No.7, July 89.
5. Elgaaly, M., Sturgis, J., Nunan, W., Stability of plates under eccentric edge loads, Proceedings of the American Society of Civil Engineers, Structural Congress, San Francisco, California, May 1989.
6. Sturgis, J., Web behavior under eccentric edge loading", Master's Thesis May 1989, Univ. of Maine, Orono, USA.
7. Elgaaly, M., Caccese, V., Du, C., Post-buckling strength of plates: Finite elements vs empirical models, Proceedings of the 4th International Colloquium, Structural Stability Research Council, 123 - 134, 1989.

LOCAL BUCKLING OF STEEL BRIDGE GIRDER WEBS DURING LAUNCHING

TORSTEN HÖGLUND

The Royal Institute of Technology
Division of Steel Structures
S-100 44 Stockholm
Sweden

Abstract

When launching steel-concrete bridges two or more rollers are used at the support. In this paper some tests on girders with two concentrated loads are presented. Comparison with existing formulas showes that two loads can be considered as one load distributed over the distance between the rollers.

1. Tests

The most common method to build a composite steel-concrete bridge is to first launch the steel girders and then cast the bridge deck. The casting could be rationalized if the concrete slab could be casted before the launching. The major problem with this method is the large concentrated support reactions that will occur during launching and the risk of buckling of the web.

The purpose of this paper is to illustrate the influence of launching with one or two rollers and to investigate if the design methods for one concentrated load could be extended to two concentrated loads.

Five experiments have been performed, four with two concentrated loads and one with one concentrated load. The space between the loads and the ability for the flanges to rotate was varied. All tests where made on the same welded plate girder according to figure 1.

The tests are summerized in table 1.

Test no. 1 and 5 where made on one half of the girder the other tests 2, 3 and 4 where made on the other half.

To make the loading similar to the equipment commonly used for launching, two steel cylinders with a diameter of 80 mm where used at the loading points. When calculating the theoretical ultimate load these rollers where supposed to have a 5 mm loading length c.

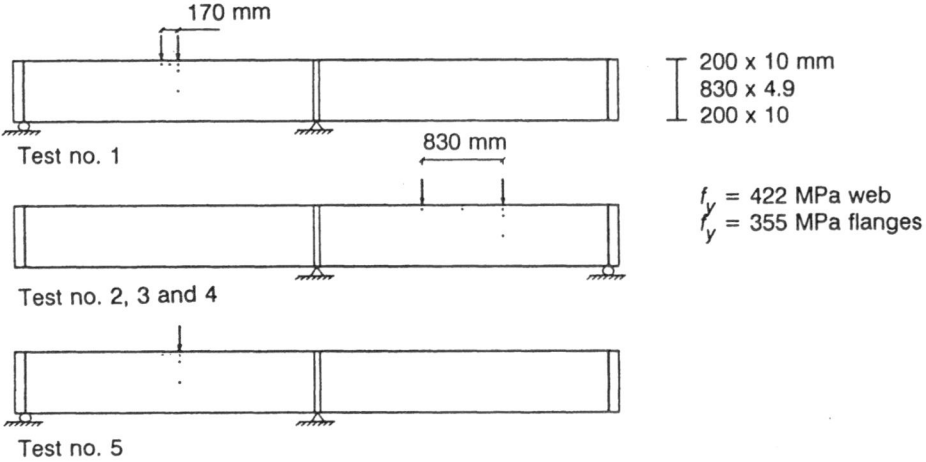

Fig. 1. Test girder and tests

Table 1. Experimental and theoretical loads

Test no	No.of loads	Load dist. a	Max. load F_{max}	Failure load F_u	Theor. load F_{theor}	$\dfrac{F_{max}}{F_{theor}}$	$\dfrac{F_u}{F_{theor}}$	Re-mark
1	2	170	250		285	0.88		
2	2	830	324		285	1.14		
3	2	830	339		285	1.19		a)
4	2	830	345	345	285	1.21	1.21	b)
5	1	-	224	224	208	1.08	1.08	

a) The rotation of the unloaded flange was prevented
b) The rotation of both flanges was prevented

The deflections of the web as well as the strains where measured at five locations marked with dots in figure 1. The vertical deflection of both the top and the bottom flanges where also measured in the same five points. In this paper only the web deflection under one of the loads is reported, se figure 2.

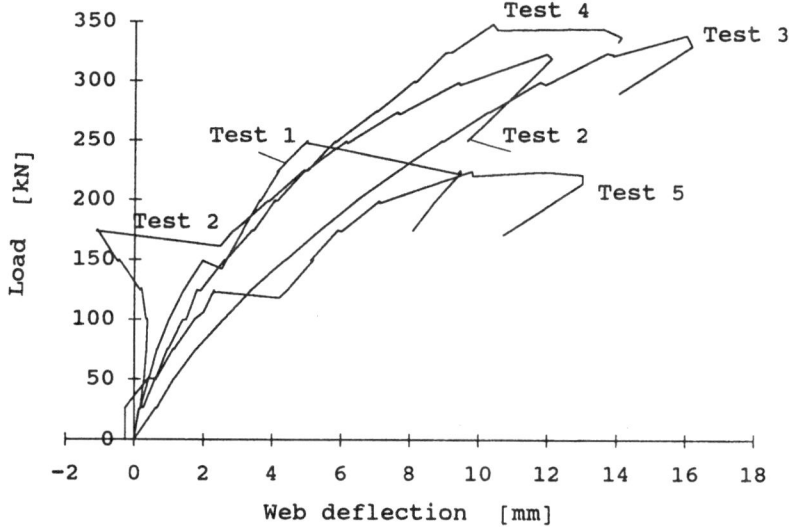

Fig. 2. Load - web deflection curves.

2. Test results

The web aspect ratio a_w/h_w was quite large, aboute 3.8. Initially there were three buckles in each web panel. At test no. 1, 2 and 5 these three buckles where formed into one, accompanied by a loudly bang. In test no. 2 there was a change in the direction of the deflection under the concentrated loads. The deflection configuration at maximum load was primarely one elastic buckle along the whole web panel except for test·no. 5, with only one concentrated load, where the typical local collapse mechanism, with yield lines in the web, arose. See figure 3.

The loaded top flange was to some extent torsional restrained by the rollers.

To simulate the restraint by the concrete slab, the unloaded bottom flange of test nr 3 was torsionally restrained by five long beams supported at the ends against the floor. This restraint changed the buckling configuration to some extent, compared to test no. 2, but had a minor effect (5%) on the maximum load.

Additional torsional restraint of the loaded flange on both side of the load points was introduced in test no. 4. The load could then be increased just a little (2%) compared to test no. 3.

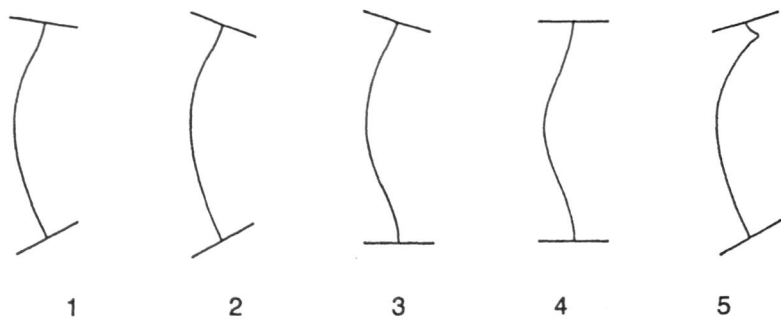

Fig. 3. Cross sectional deflection close to the load.

Test no. 5 with one concentrated load was made for comparison. The web deformation started with a global deflection surface along the web panel. At about half the ultimate load the deflection surface was divided into three half waves and at about 90% of the ultimate load the typical web crippling deformation appeared close to the load. See figure 3.5.

3. Comparision with design formulas

During the past 20 years a large number of model tests with one concentrated load has been performed by several research workers. Based on these tests several design formulas have been proposed.

Bergfeldt [1] has proposed the formula

$$F_u = 0.70 \; t_w^2 \; \sqrt{E \, f_y} \; \alpha_1 \; \alpha_2 \; \alpha_3 \tag{1}$$

where

$$\alpha_1 = 1 + \frac{40 \, c \, t_w}{c_0 \, b_w} \tag{2}$$

$$c_0 = \sqrt[4]{\frac{4 \, b_w \, l_f}{t_w}}$$

$$\alpha_2 = 9.2 \cdot \sqrt[8]{\frac{l_f}{t_w^4}} \tag{3}$$

138

$$\alpha_3 = 1.25 - 0.5 \frac{M}{f_y W} \tag{4}$$

According to the formula (2) the influence of the loaded length c seems to be of miner importance if

$$c > c_0 = \frac{c_0 \, b_w}{100 \, t_w}$$

where c_0 is 58 mm for the test reported in this paper. If a load from two rollers with spacing c is supposed to have the same action as one load with the loaded length c then all tests with two rollers will have the same theoretical ultimate load 285 kN according to formula (1). If the spacing between the rollers is very large it is to be expected that the loads don't interact on each other. According to the tests this spacing is larger than the web depth, probably larger than two times the web depth.

In practice it is not possible to have launching equipment with so large distance between the rollers. Therefore larger spacing than c_0 (= 96 mm in the test) is of less meaning.

If there is only one roller, the theoretical web crippling load is 208 kN. Two rollers increases the ultimate load to 285 kN (37% increase). If the flange is thin compared to the web, then the increase is smaller.

The test results are compared with the theoretical values in table 1.

In many cases two rollers do not give much larger ultimate load than one roller. To avoid local yielding two or more rollers are recommended, nevertheless.

[1]   Bergfeldt, A., Patch loading on a slender web. Chalmers Technical Institute, Steel and Timber Structures, Publ. S 79:1, Göteborg 1978/79.

[2]   Mannberg, O., & Schagerström, L., Livbuckling vid lansering av samverkansbroar. Diploma work. Div. of Steel Structures, Royal Institute of Technology, Stockholm 1989.

ELASTIC AND PLASTIC STATES OF METALLIC STRUCTURES UNDER BENDING WITH ALLOWANCE FOR LOCAL STRESSES

A.KOLESOV, B.LAMPSI, V.GUSEV
Chair of Metal Constructions, Gorky Civil Engineering Institute
65, Krasnoflotskaya str., Gorky, 603000, USSR.

A method is suggested to perform practical analysis of thin-walled rods with singly- and doubly-connected cross-section under local loads at the stages of elastic and elasto-plastic work. Numerical examples for theoretical solutions and experimental studies are offered. A new constructive solution is proposed for a crane girder in which upper chord is connected with the web by two arc-like plates.

Introduction.

The present lecture deals with theoretical and experimental results concerning the application of the local stress theory [1] developed by prof. B.Lampsi and his students as a means to improve reliability of thin-walled metal structures subject to local loads and having structural concentrators.
The theory is based on the solution of boundary-value problems of the elasticity theory for a long strip as a primary component of a thin-walled rod with singly- or doubly-connected cross-section. The latter can be easily divided into constituent strips by making appropriate cuts (Fig.1). Knowledge of the stressed state of the strip due to loads at its edges and, if necessary, due to prescribed displacements of the latter allows one to unite the strips into a rod imposing interaction forces in the locations of cuts which provide connectivity condition.
This theory permits us to estimate the stressed state of a rod with singly- and doubly-connected cross-section to the accuracy corresponding to the elasticity theory concepts. This is particularly important to consistently implement the ideas of the method of limiting states when analysing building constructions both at elastic and elasto-plastic stages of work.

1. Elastic stressed state.

If, for example, we consider an I-beam then after separating the web from the chords total stressed state of the beam due

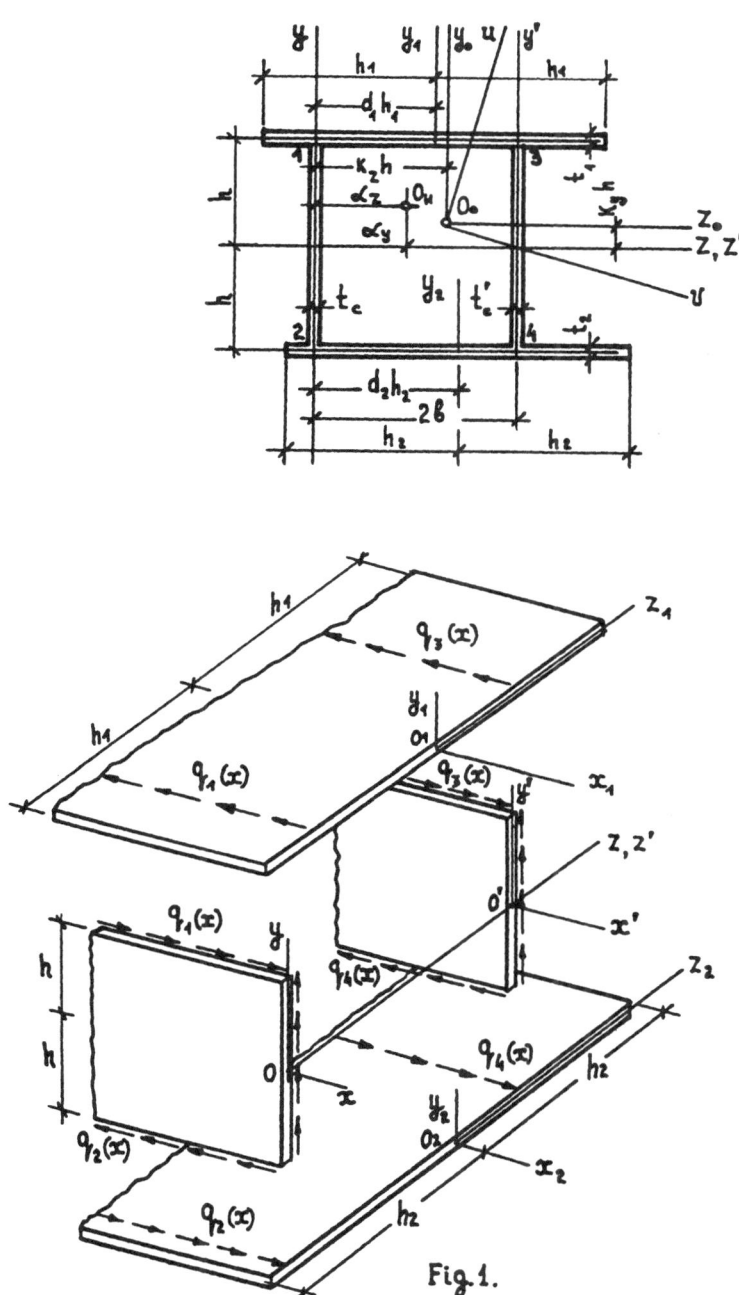

Fig. 1.

141

to local loads is composed of four components:

- elementary stresses σ_{x0} , τ_{xy0} ,

$$\sigma_{x0} = \alpha_0 \left(\xi, \mu\right)\frac{F}{t \cdot h} , \quad \tau_{xy0} = \gamma_0 \left(\xi, \mu\right)\frac{F}{t \cdot h}$$

- local stresses due to normal pressure $p(x)$ against the edge

$$\sigma_{xp} = \alpha_p \left(\xi, \mu, x\right)\frac{F}{t \cdot h} , \quad \sigma_{yp} = \beta_p \left(\xi, \mu, x\right)\frac{F}{t \cdot h}$$

$$\tau_{xyp} = \gamma_p \left(\xi, \mu, x\right)\frac{F}{t \cdot h} ;$$

- local stresses due to interaction force jumps $q_{0i}(x)$ at the

edges $\sigma_{xq} = \alpha_q \left(\xi, \mu\right)\frac{q_{0i}}{t} , \quad \sigma_{yq} = \beta_q \left(\xi, \mu\right)\frac{q_{0i}}{t} ,$

$$\tau_{xyq} = \gamma_q \left(\xi, \mu\right)\frac{q_{0i}}{t} ;$$

- local stresses due to a surge of tangent interaction forces

$q_i(x)$ $\sigma_{x\Delta} = \Delta\alpha\frac{q_{0i}}{t} , \quad \sigma_{y\Delta} = \Delta\beta\frac{q_{0i}}{t} , \quad \tau_{xy\Delta} = \Delta\gamma\frac{q_{0i}}{t}$

Here α , β , γ are influence functions corresponding
to the type of loading (interaction forces) as in $[1]$, $\xi = x/h$,
$\mu = y/h$ are relative coordinates, $\chi = \lambda_0/2h$. In the case of
doubly-connected cross-section (for example, runway chord of a
crane-footing girder) using the method $[2]$ of passing to an equivalent I-beam with the corresponding load one can calculate
stresses with the help of the above formulae as well.

Table 1 offers stresses in a calculation point of the runway
chord web under plane bending due to four cranes with respect
to two types of crane-footing girders (CFG):

type 1 - top frame member of an assembling shop with 120m bay.
Runway chord cross-section:web-25x1978mm, chord-16x3400mm. Travelling cranes with rated load capacity amounting to 15 t;

type 2- the same top frame member with bay 150m. Runway chord
cross-section:web-16x1972mm, chord- 14x3400mm. Travelling cranes with rated load capacity amounting to 20 t.

Table 2 contains results of comparative strength analysis for
runway chords of CFG of four types:

type 1 and 2 are described above,

type 3 is CFG with bay 9 m. Runway chord cross-section is: web-
6x550mm, chord- 6x850mm,

type 4 is CFG with bay 36 m. Runway chord cross-section is: web-
20x2706 mm, chord - 32x2680 mm.

Table 1

CFG type	elementary USSR SNiP*	elementary suggested theory	local USSR SNiP	local suggested theory	Reduced stresses USSR SNiP	Reduced stresses suggest. theory
1	60 2.37	60 0.2	-9.3 -37.1 11.1	-26.2 -37.1 0	79.8	61.4
	51.5 2.08	51.5 0.02	-9.3 -37.1 11.1	-26.2 -37.1 0	72.4	54.4
2	168.7 3.74	168.7 0.09	-14 -56.0 16.8	-36.1 -56.0 0	193.9	167.7
	83.6 3.52	83.6 0.09	-14 -56.0 16.8	-36.1 -56.0 0	114.5	89.7
3	-17.7 4.07	-1.77 5.91	-28.9 -115.8 34.7	-51.9 -115 0	121.2	101.5

σ_{xo} τ_{xyo} σ_{xc} τ_{xy} $\sigma_{loc,x}$ $\sigma_{loc,y}$ $\tau_{loc,xy}$ $\sigma_{loc,x}$ $\sigma_{loc,y}$ $\sigma_{loc,xy}$ $\tau_{loc,xy}$

Note: in upper lines for type 1 and 2 systems values under total load are given, in bottom lines values are given under constant and crane loads.
* Code of recommended practice acting in the USSR.

Table 2

CFG type	elementary USSR SNiP	elementary suggested theory	local USSR SNiP	local suggested theory	Reduced stresses USSR SNiP	Reduced stresses suggested theory
1	102.2 2.36	102.2 1.02	-8.87 -35.5 10.6	-24.15 -35.5 0	117.4	100.6
2	162.4 3.79	162.4 1.48	-13.4 -53.7 16.2	-33.6 -53.7 0	185.0	162.5
3	-12.12 6.71	-12.12 5.42	-23.6 -94.4 28.3	-36.7 -94.4 0	102.4	82.3
4	-32.5 2.02	-32.5 1.09	-11.7 -46.8 14.0	-22.1 -46.8 0	53.4	51.1

Data entering the tables demonstrate that a more precise calculation provides, in all the cases, drop of reduced stresses within the limits of 15-27 %.

2. Allowance for plastic deformations.

The theory of local stresses is extended to an elasto-plastic
stage of work when applying the Prandtl diagram. The procedure to
solve such a problem is reduced to obtaining the expressions for
the plane problem stresses satisfying the Mises-Hencky yield condition for any stage of plastic evolution. The variable Poisson
ratio grows with plastic strains. Stress distribution is obtained
when solving the problem of transverse bending of a beam under
local loads.

Stresses for plastic hinge state are used to define reduced bending moment and transverse force under local loads whose values
are compared with those for pure bending.

Load capacity of single beams under local loads and of continuous beams with a central prop is estimated from this viewpoint.
Table 3 contains theoretical results $\Psi_c = M_{st}/M_o$, while Table 4 is for those of the experiment.

Table 3.

$\eta = \frac{\ell}{h}$	x_1	z_o	μ_1	φ_c	Ψ as per conditions		
					1	2	3
			$\beta = 0.2164$;	$x = 0.375$			
3	0.500	0.07	−0.9452	0.4661	0.4357	0.3026	0.6669
4	-"-	0.06	−0.6916	0.5619	0.5724	0.4562	0.8074
5	-"-	0.04	−0.2810	0.5857	0.5514	0.4379	0.8416
6	-"-	0.02	−0.1600	0.6373	0.6294	0.4611	0.9158
			$\beta = 0.0927$;	$x = 0.500$			
3	0.500	0.08	−0.8994	0.3268	0.3358	0.2213	0.4891
4	-"-	0.08	−0.9165	0.3648	0.3183	0.2017	0.5468
5	-"-	0.06	−0.9392	0.4233	0.3292	0.2142	0.6344
6	-"-	0.06	−0.9482	0.4590	0.3389	0.2262	0.6880

Analysis of the results obtained indicates that:
- the first section plastifying along the whole hight does not
coincide with that under load. This can be explained by the lack
of tangent stresses in the loaded section;
- in short beams with a strong chord ($x \geqslant 0.5$) plastic hinge
locking in a web is observed near bottom edge; in long beams
($\eta > 6$) plastic hinge in a web is locked near the neutral

144

axis.

The abovementioned phenomena imply significant effect of local stresses in the case of relatively strong chords;
- values of factors depend on the method for chord effect estimation.

Table 4

Beam	η	F_{st}^{ex} kN	$\psi^e =$ $= M_{st}^e / M_{st}^o$	Theoretical values				
				ψ_1	ψ_2	ψ_3	ψ_4	ψ_5
B-1	3	725	0.578	0.490 / −12.2	0.486 / −15.9	0.812 / +40.5	0.803 / +64.0	0.547 / −1.97
B-2	4	682	0.700	0.695 / +0.7	0.706 / +0.8	0.846 / +20.8	0.896 / +27.0	0.739 / +5.51
B-3	5	614	0.787	0.780 / −0.82	0.753 / −4.32	0.866 / +10.0	0.952 / +22.0	0.857 / +7.6
B-4	6	620	0.978	0.966 / −1.27	0.785 / −19.7	0.888 / −9.22	0.991 / +2.57	0.946 / −3.26

Table 5 includes data on the load capacity of continuous beams with a central prop using the procedure described above for simply supported beams ($\psi_2 = M_{st},2 / M_o$).

Table 5

η	$\psi_2 \cdot 10^4$ for χ_1 equal to			
	0.125	0.25	0.375	0.5
3	7799	8217	8358	8922
4	8527	8590	8661	9013
5	8909	8925	8958	9216
6	9211	9439	9536	9627

Analysis of the last results indicated that:
- predicted load capacity of short beams ($\eta < 6$) is lower than that calculated in pure bending; for $\eta \geqslant$ 6, on the contrary, it becomes higher than for pure bending;
- increase of the stiffening support rib area makes load capacity higher by within 17-26%.

145

3. Limited elasto-plastic deformations under repeated-alternating loads.

Solutions offered in Sections 1 and 2 of this lecture are used in the problem of defining plastic strains in a web of an I-beam under repeated-alternating loading with local loads. To this end an algorithm was developed (which was realized in a thesis by N.Yu.Tryanina, Chair of Metal Constructions of Gorky Civil Engineering Institute) based on the use of V.V.Moskvitin theorem and N.L.Chernov criterion of limited plastic component of deformation. Tests conducted for standard crane girders under light and medium operating modes with application of the above mentioned algorithm showed that plastic deformation growth in these girders does not exceed 0.002. Experiments carried out for 4 beams subjected to seven cycles " loading-unloading" under limited plastic deformations demonstrated that if plastic strain of the web was restricted to 0.2% stabilization of the strain could be observed as the cycle number was increased. Predicted and experimental loads corresponding to the criterion of limited plastic deformations under eight cycles in experimental beams are shown in Table 6.

Table 6

Loads	Beams (experimental) with relative span $\eta = \ell/h$			
	6	5	4	3
F_t kN	638	676	731	687
F_{ex} kN	650	650	750	680
Δ %	-2	+4	+2.5	+1
$f_{max}/1$	1/370	1/370	1/300	1/320

This permits one to design crane girders for light and medium operating modes taking limited plastic strains into account with resulting steel saving up to 6-8%.

4. A way to reduce effects of local stresses.

A new constructive design is suggested for a crane girder in which upper chord is connected with a web through two arc-like plates. The upper edge of the web is planed and tightly fits

the flange being not welded, however. The web is welded to
curved sheets at a distance from the upper chord. Arc-like
plates must have, in the spots where the web fits the flange,
small portions in parallel with the elements to which the
former are welded to rule out significant influence of the
plate elastic strains on the strength of the welds.

When small loads are applied to the beam the pressure from
the upper chord is transferred to the web through curved plates by way of the welds. As the load grows the arc-like plates are somewhat deformed until the temporary clearance between
the web and the upper flange is taken up and, taking part of
the load, transfer it to the web by way of the welds. The main
portion of the pressure, however, is taken up by a planed up-
per edge of the web where local normal stresses $\sigma_{loc, x}$ and $\sigma_{loc, y}$
and bending linear stresses σ_x emerge.

Tangent stresses, welding stresses and those due to rotation
arise in a web within the region of the welds and below it, i.e.
in places where local stresses $\sigma_{loc, x}$ have insignificant value while local stresses $\sigma_{loc, y}$ and bending stresses σ_x have
reduced value. In the welds directly tangent stresses τ_{xy} and
$\tau_{loc, xy}$, welding stresses and those due to the upper chord
rotation are generated. Local stresses $\sigma_{loc, y}$ within the weld
region are reduced already as the main portion of the pressure is transferred through the planed edge of the web.
Therefore, the stressed state in the weld region will be mark-
edly changed towards reducing of the stresses both quantita-
tively and in their composition. All this can result in increased life of the crane girders of a new constructive design.

REFERENCES
1. B.B.Lampsi. Thin-walled metal load-carrying structures under
   local loads. M.: Stroiizdat. 1979. 270p. (In Russian)
2. B.B.Lampsi. Strength of thin-walled metal structures. M.:
   Stroiizdat. 1987. 280p. (In Russian)

ULTIMATE LOAD BEHAVIOUR OF LONGITUDINALLY STIFFENED STEEL
WEBS SUBJECT TO PARTIAL EDGE LOADING

I.KUTMANOVA[1],M.ŠKALOUD[2], K.JANUŠ[1] and O.LÖWITOVÁ[1]

[1] Czech Technical University
    Building Research Institute
    Šolínova 7
    166 08 Prague 6
    Czechoslovakia

[2] Czechoslovak Academy of Sciences
    Institute of Theoretical and Applied Mechanics
    Vyšehradská 49
    128 49 Prague 2
    Czechoslovakia

Abstract

The objective of the contribution is to describe the main results and conclusions of a several years' experimental investigation into the ultimate limit state of steel plate girders whose thin webs were loaded by a partial edge load. Altogether 184 experimental girders were tested, with the following geometrical characteristics of the test girders being varied from girder to girder: (i) the depth-to-thickness ratio of the webs, (ii) the flange size, (iii) the position of the longitudinal rib, (iv) its dimensions and (v) the length of load. Special care was given to a study of (a) the stiffeners location and size and (b) the dimensions of the loaded flange upon the progression of plastification, the failure mechanisms and ultimate strengths of the girders tested.

1. Introductory Remarks

One of the stability problems that attracted plenty of attention over the last years was that of the ultimate load behaviour of thin steel webs subject to partial edge loading. Several years ago, the second author, jointly with V. Křístek and P. Novák, carried out an extensive investigation into the problem (see [1] and [2]). These studies, representing the first stage of our research on the performance of the plate girders under the action of a patch load, were concerned mainly with the ultimate limit state of webs without longitudinal stiffeners.

However, the webs of deep plate girders are often stiffened by longitudinal ribs. At least, this is frequent practice with webs loaded by combined shear and bending; and sometimes the use of longitudinal stiffeners is necessary also for other reasons: for instance, to reduce web deflection or even initial "dishing". Then, of course, the question arises whether the presence of longitudinal stiffening can also favourably influence the behaviour of a web if this is loaded by a patch load, or by such a combination of loading in which the effect of patch loading predominates.

Little information was available in regard to this problem; therefore, the authors decided to contribute to its solution. Desiring to look profoundly into the progression of plastification in the web and the whole girder, and into its effect upon the ultimate limit state, and bearing in mind that a theoretical treatment of such a problem would be very complex and time-consuming, they chose the experimental way of investigation.

The first stage of the research dealt with the performance of steel webs under the action of a static patch load. In conclusion, the authors established formulae for (i) the optimum rigidity of the longitudinal rib and (ii) the predicted ultimate loads of webs subject to partial edge loading and stiffened by a longitudinal rib.

2. Test Girders

The first stage of the authors' experimental investigation, focused on static loading, comprised 184 tests, sub-divided into six stages, the following quantities being varied in them:

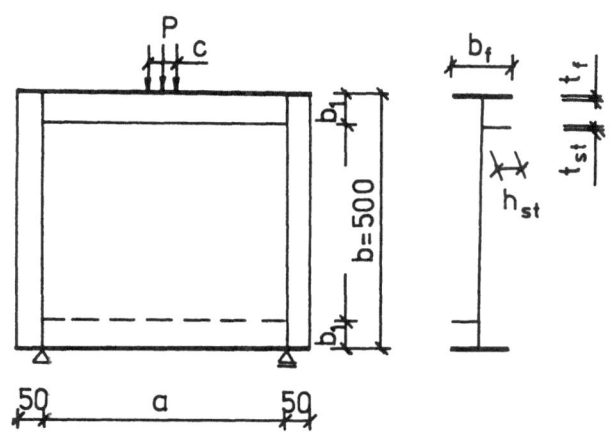

Fig. 1.

- the distance of the longitudinal stiffener from the loaded flange,
- the size (and flexural rigidity) of the longitudinal stiffener,
- the character (single- or double-sided) of the longitudinal stiffener,
- the depth-to-thickness ratio of the web,
- the aspect ratio of the web,
- the size (and flexural rigidity) of the loaded flange, and
- the length of load.

149

Table 1.   The general details of the test girders can be seen in Fig.1, their main characteristics are listed in Table 1.

Test series	Number of tests	Web thickness (mm)	Constant quantities	Variable quantities
1	26	2	Load length $c=a/10$ Flange rigidity Single-sided longitudinal stiffener	Aspect ratio of the web; $=a/b$ ($=1$, $=2$) Distance b_1 of the longitudinal stiffener from the loaded flange $(b_1=0.1-0.5b)$ Longitudinal stiffener rigidity
1	24	6	$c=a/10$ $\alpha=1.245$ Flange rigidity Double-sided longitudinal stiffener	Distance b_1 of the longitudinal stiffener from the loaded flange $(b_1=0.15-0.4b)$ Longitudinal stiffener rigidity
2	42	2,4,6	$c=a/10$ $\alpha=1$ $b_1=0.2b$ Single-sided longitudinal stiffener	Longitudinal stiffener rigidity Flange rigidity
3	36	2,4,6	$c=a/10$ $\alpha=1$ $b_1=0.2b$ Single-sided longitudinal stiffener	Longitudinal stiffener rigidity Flange rigidity
4	24	4	$c=a/10$ $\alpha=1$ $b_1=0.2b$ Single-sided longitudinal stiffener	Longitudinal stiffener rigidity Flange rigidity
5	16	4	$\alpha=1$ No longitudinal stiffener	$c=0.1a$, $0.2a$, $0.3a$ Flange rigidity
6	16	4	$c=a/10$ $\alpha=1$	$b_1=0-0.2b$ Longitudinal stiffener rigidity Flange rigidity

3. Test Set-Up and Experimental Apparatus

The test set-up is seen in Fig. 2. The measurements were concentrated mainly on
(i)   strains at a number of characteristic places of the web and longitudinal stiffener,
(ii)   web buckling, and
(iii) stiffener deflection (perpendicularly to the web sheet).

Besides that, also flange deflection (in the direction of the web sheet plane) and potential rotation of the free edge of the flat stiffener, which - if it did occur - would represent another component of stiffener deformation, were also detected.

Strains were measured by means of electric resistance strain gauges C 120 positioned at various places of interest for our study of the progression of plasticization in the test girder, and mounted on the both surfaces of the web sheet and of the flat stiffening element. The measurements were facilitated by

150

Fig. 2.

using a measurement unit Peekel controlled by a computer PDP 11/40.

The buckled pattern of the web was measured by means of a special piece of apparatus, which transforms the measured quantity, i.e. deflection, into an electric signal and then plots it via a plotter. Also part of web initial curvatures were detected with the aid of this device.

The other initial curvatures and all post-failure plastic residues in the webs were measured using the stereophotogrammetric method, which already proved to be of advantage for the purpose considered in the previous experiments conducted by the authors.

The deflections of the upper and lower flanges, and those of the longitudinal rib, were detected via pickups linked again with the measurement unit Peekel.

It goes without saying that also due attention was paid to cutting off large enough pieces from the material of all girders components, from which coupons for tensile tests were fabricated, to be used later on in order to determine the actual material characteristics of the element.

4. Main Results of the Tests

4.1. Ultimate loads and failure mechanisms of test girders

The relationship between the average values of the experimental ultimate load, P_{ult}^{exp}, of the test girders and the distance, b_1, of the stiffener from the loaded flange is shown in Fig. 3.

All test girders collapsed by the formation of a failure mechanism which consisted of (i) a segmental line plastic hinge in the web sheet and (ii) three point plastic hinges in the loaded flange. The segmental line plastic hinge in the web formed in the vicinity of the loaded flange, the size of this flange considerably influencing the configuration of the line hinge. The larger the size of the flange, the longer and

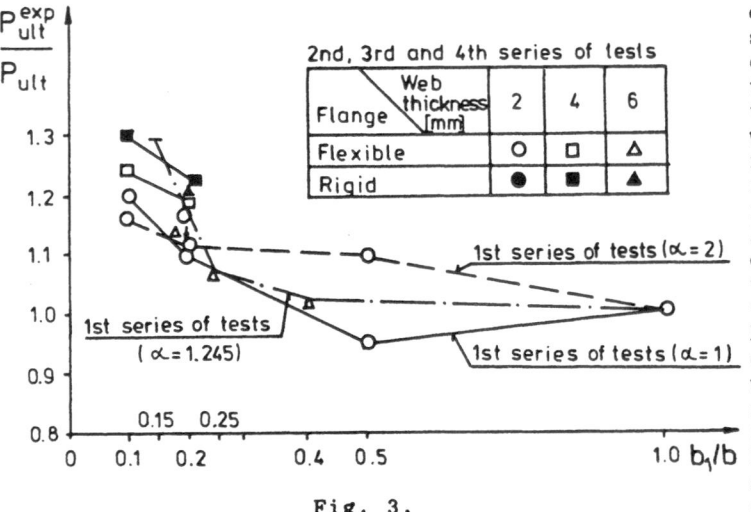

deeper the segment. The effect of the longitudinal rib was then reflected in shortening the length and depth of the arc of the line hinge, if compared with the situation in a web having the same dimensions but not being reinforced by a longitudinal rib.

Fig. 3.

The plastic hinges in the loaded flange occurred (i) at mid-span, i.e. under the partial edge load, and (ii) at those flange sections from which the segmental line plastic hinge in the web sheet emanated. The distance of the outer hinges from the central one was influenced by the flange size; i.e. the larger the size of the flange, the greater was the distance of the outer hinges.

M. Drdácký established in [3] the following criterion for the formation of outer plastic hinges:

$$I_f/a^3 t_w < 3.10^{-5} \tag{1}$$

where I_f is the moment of inertia of the loaded flange with respect to its centroidal axis perpendicular to the web plane, a is the web width and t_w the thickness of the web sheet.

This criterion has the following meaning: When condition (1) is satisfied, the outer flange hinges form between the adjacent transverse stiffeners, i.e. within the web panel under consideration.

Then, as flexible we regard those flanges in which all three plastic hinges do form between the two adjacent transverse stiffeners; if the opposite is the case, the flange is regarded as rigid. In this context, the upper flange of all test girders of the first series can be classified as flexible. In the following series, the test girders had flanges of two thicknesses; the thinner flange can then be regarded as flexible and the thicker as rigid.

152

4.2. Onset of yielding in test girders

By way of analysis of the measured strains (and related stresses), the writers found the loads that corresponded to the onset of plasticization, whether this be surface (load $P_{y,w}^{su}$) or membrane (load $P_{y,w}^{me}$) plastification, in the individual test girders, and also quantities $P_{2\varepsilon}^{su}$, $P_{3\varepsilon}^{su}$, $P_{2\varepsilon}^{me}$ and $P_{3\varepsilon}^{me}$, which are related to unitary deformations ε attaining values equal to twice or three times the strains corresponding to the onset of surface or membrane plastification. The main results are given in Fig. 4.

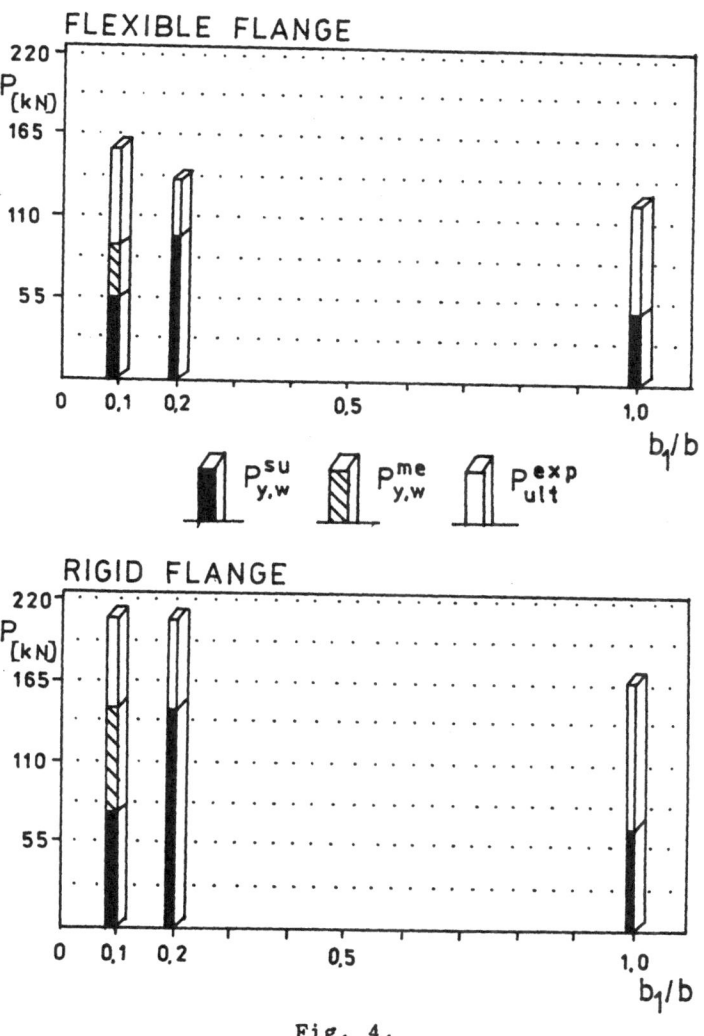

Fig. 4.

4.3. Effect of the size of the loaded flange

An examination of the results obtained indicates that the effect of the size of the loaded flange upon the load-bearing capacities of the webs (and of the whole plate girders) is very significant. It considerably influences the dispersion of the load into the web, its buckling and even the collapse mechanism of the plate girder.

4.4. Effect of the position and size of the longitudinal stiffener

An analysis of the results reveals that a longitudinal rib can substantially improve the behaviour of a plate girder web subject to partial edge loading only when the rib is located in the vicinity of the loaded flange, i.e. if $b_1 < b/4$. The conclusion follows from the observation that, in the case of a web under the action of a discrete edge load the buckling is pronouced merely in the region immediately adjacent to the applied patch load, see Fig. 5 (showing the plastic residue of an unstiffened test girder). If the longitudinal stiffener is positioned off this region, it is not able to influence the web buckling and, consequently, the ultimate load behaviour of the girder.

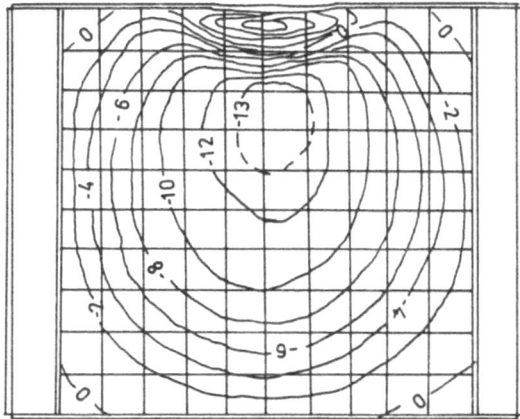

TG 0 : PLASTIC RESIDUE

Fig. 5.

As for the effect of the size (and rigidity) of the rib, this was pronounced only in the range of flexible ribs, where the curve plotting the relationship between ultimate load P_{ult}^{exp} and stiffener rigidity γ_l exhibited a conspicuous rising tendency. In the domain of larger stiffeners, the curve was flat, so that the growth of ultimate strength with a further increase in stiffener size and rigidity was very slow (see Fig. 6). The average increase in the ultimate loads of the girders tested with a longitudinal stiffener was of 24.2 % for $b_1 = 0.1b$ and 15.7 % for $b_1 = 0.2b$, if compared with the ultimate loads of girders without longitudinal stiffeners.

On the basis of the experimental results obtained, the authors established the following formula for the optimum moment of inertia, $I_{st,r}^{opt}$, of a rigid (i.e. not deflecting in the course of web buckling and, therefore, able to provide the buckling sheet with rigid support until the ultimate strength of the girder has been exhausted) longitudinal stiffener positioned at $b_1 \leq 0.2b$:

154

$$I_{st,r}^{opt} = 0.1 bt_w^3 \; \mathcal{H} \, \mu^*. \tag{2}$$

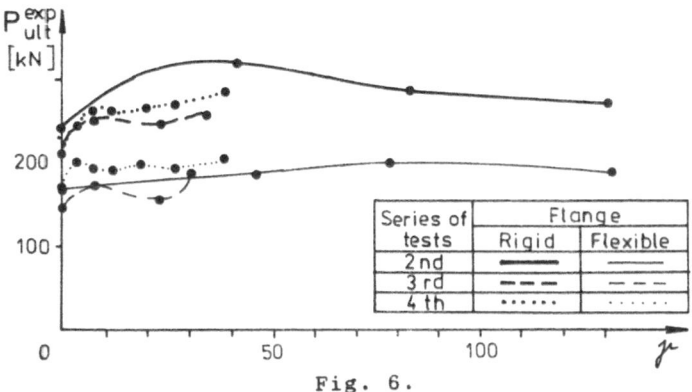

Fig. 6.

Series of tests	Flange	
	Rigid	Flexible
2nd	——	——
3rd	— — —	— — —
4th	······	······

There I_{st} is the moment of inertia of the longitudinal stiffener (calculated for a cross-section consisting of (i) the stiffener proper and (ii) an effective portion of the web sheet having a width of $2 \times 15 t_w \; 240/R_{y,w}$) with respect to its centroidal axis parallel with the web plane), μ^* the optimum rigidity resulting from linear buckling theory, and $\mathcal{H}$ a coefficient according to Table 2, which respects the effect of initial imperfections and of post-buckled behaviour.

Table 2.

λ	≤ 90	≥ 230	Between the values of $\lambda = 90$ and
$\mathcal{H}$	2.25	4.0	$\lambda = 230$ can be linearly interpolated.

However, when examining the curves "Ultimate load versus stiffener rigidity", it was found that for larger rigidities these curves were very flat. This led to the conclusion that a major reduction of stiffener rigidity would lead to only an insignificant reduction of ultimate load. This conclusion was further confirmed by the authors evaluating the relation of the experimental ultimate loads to the predicted ones for those of the test girders whose webs were stiffened by ribs the moment of inertia of which was less than $I_{st,r}^{opt}$ according to formula (2).

Therefore, the writers came to the conclusion that, at least as far as strength considerations were concerned, a less strict formula would do for the calculation of the optimum stiffener rigidity, viz.

$$I_{st}^{opt} = 0.5 I_{st,r}^{opt}. \tag{3}$$

And this is also the formula the authors recommend to be used in design.

It should be noted, however, at this juncture that, unlike Eq.(2), formula (3) is not able to ensure for the stiffener to be rigid during the whole "life" of the plate girder, so that the stiffener then more or less deflects with the buckling

155

sheet and even the buckled pattern of the web sheet is more pronounced.

4.5. Effect of the longitudinal stiffener on web deflection

One of the objectives of the tests was to study how the presence of a longitudinal stiffener can help to reduce the deflection of webs under partial edge loading if compared with the buckled patterns of unstiffened webs.

An analysis of the data obtained indicates that, in the case of the writers' experiments, this reduction was pronounced and amounted on an average to 42%.

4.6. Effect of load length

To look into this effect, 16 tests were carried out, in which the length of load, c, was larger then in the first series of tests (where $c/a=0.1$), viz. (i) $c/a=0.2$ and (ii) $c/a=0.3$. An analysis of the results obtained revealed that an increase in ultimate load by 5 % is reached when the load length c is enlarged from 50 to 100 mm (i.e., from $c/a=0.1$ to $c/a=0.2$) and by 10 % when c is increased from 50 to 150 mm (i.e., from $c/a=0.1$ to 0.3).

4.7. Effect of the aspect ratio of the web and of the character of the longitudinal stiffener

Neither the aspect ratio of the web, nor the character (single- or double-sided) of the stiffener influenced in a significant way the ultimate load behaviour of the webs of the test girders.

6. References

1. Škaloud, M. - Novák, P.: Post-buckled behaviour of webs under partial edge loading. Rozpravy (Transaction of) ČSAV, řada tech. věd (Series of Techn. Sciences), No. 3, Vol. 85, Academia, Prague (1975).

2. Škaloud, M. - Křístek, V.: Stability problems of steel box girder bridges. Rozpravy (Transaction of) ČSAV, řada tech. věd (Series of Techn. Sciences), No. 1, Vol. 91, Academia, Prague (1981).

3. Drdácký, M.: Steel webs with flanges subjected to partial edge loading. Chalmers University of Technology, Göteborg (1982).

4. Škaloud, M. - Kárníková, I.: Experimental research on the limit state of the plate elements of steel bridges. Rozpravy (Transaction of) ČSAV, řada tech. věd (Series of Techn. Sciences), No. 1, Vol.95, Academia, Prague (1985).

5. Januš, K. - Kárníková, I. - Škaloud, M.: Experimental investigation into the ultimate load behaviour of longitudinally stiffened steel webs under partial edge loading. Acta technica ČSAV, No. 2 (1988).

6. Kárníková, I. - Škaloud, M.: Experimental research on the ultimate load behaviour of steel plated structures. Journal of Constructional Steel Research, Vol. 12, No. 1 (1989).

7. Kutmanová-Kárníková, I. - Škaloud, M. - Januš, K.: Ultimate load behaviour of longitudinally stiffened steel plate girders subject to (i) stationary or (ii) variable repeated patch loading. International Colloquium on Structural Stability, Beijing (1989).

8. Januš, K. - Kutmanová-Kárníková, I. - Škaloud, M.: Design of longitudinally stiffened thin webs under patch loading. Colloquium on Stability of Steel Structures, Budapest (1990).

"BREATHING" OF LONGITUDINALLY STIFFENED STEEL WEBS
SUBJECT
TO REPEATED PARTIAL EDGE LOADING

I. KUTMANOVÁ[1], M. ŠKALOUD[2] and K. JANUŠ[1]

1) Czech Technical University
   Building Research Institute
   Šolínova 7
   166 08 Prague 6
   Czechoslovakia

2) Czechoslovak Academy of Sciences
   Institute of Theoretical and Applied Mechanics
   Vyšehradská 49
   128 49 Prague 2
   Czechoslovakia

Abstract

The aim of the first part of the Prague experimental research on web "breathing" was to look into the effect of the repeated character of patch loading on the ultimate limit state of steel plate girder webs fitted with longitudinal ribs. 80 experimental panels were tested to date, the flange size and the position and dimensions of the longitudinal ribs being varied. During the tests, the "breathing" of the girder webs under the action of a cyclic patch load, the initiation and propagation of cracks in the most heavily stressed areas of the webs, and their influence on the "erosion" of the plastic failure mechanisms of the test girders were carefully studied. Final analysis of the data obtained made it possible for the authors to establish the limit fatigue loads, P_{fat}, of longitudinally stiffened steel plate girders whose thin webs are subject to a repeated partial edge load.

1. Introduction

The first stage of the authors' experimental research on web "breathing" was focused on the behaviour of webs under the action of repeated patch loading. The test girders had the same dimensions, and were fabricated from the same material, as those used in the constant patch loading tests (see [5]), this being so as to enable the authors to compare the results (ultimate loads, onset-of-yeilding loads, etc.) of both respective experimental series.

80 test girders subject to repeated patch loading have been tested to date, their webs being fitted with a longitudinal stiffener positioned at (i) one-tenth and (ii) one-fifth of

the web depth. For the sake of comparison, several girders had
no longitudinal stiffener.

2. Test Set-Up and Experimental Apparatus

The variable repeated loading was materialized by means of a
1000 kN AMSLER pulsator, the frequency of loading cycles being
of 3.75 Hz.

The following quantities were registered during the tests:
(i)   the initial imperfections of the experimental girders,
(ii)   the values of deflections and strains at a number of
        selected places,
(iii)   the initiation and propagation of cracks, and
(iv)   the acoustic emission in the web in the neighbourhood of
        the applied load.

The load P cycled between (i) zero and (ii) a value P_{max},
which in turn was varied between α) the statical ultimate load
P_{ult}^{exp} and β) the onset-of-surface yielding load, detected also
in the related statical test. Thus, under the above loading,
the webs of the girders tested behaved in the elasto-plastic
range and, consequently, their performance was expected to be
governed by low-cycle fatigue. Therefore, the basic number of
loading cycles was chosen so as to be equal to 5×10^4.

In the case of girders where after 5×10^4 cycles no failure
(whether through initiation of cracks or through excessive
plastic buckling of the girder web) occurred, the experiment
was continued under a higher load level. If, however, a crack
appeared in a certain loading cycle, the experiment went on,
under the same (i.e. unchanged) load level, as long as the
load-carrying capacity of the test girder was exhausted.

The general details of the test girders, divided into three
series, are shown in Fig. 1, with thick solid lines indicating
rigid flanges and longitudinal ribs and thin lines indicating
flexible ones.

3. Main Test Results

During the analysis of the measurements obtained, the
experimental panels of all series were divided into groups:
first, according to the position of the longitudinal stiffener
($b_1=b/10$, $b_1=b/5$, $b_1=0$), and then in each of the groups
according to flange rigidity (i.e., flexible flange - rigid
flange). This reflected the authors' experience from the
previous constant loading tests, which had demonstrated that
stiffener location and flange rigidity were the two factors
that influenced the behaviour of webs in the most significant
way.

3.1. Failure mechanisms of the test girders

One of the most important questions to which the experiments
were expected to give a reply was to what extent the failure

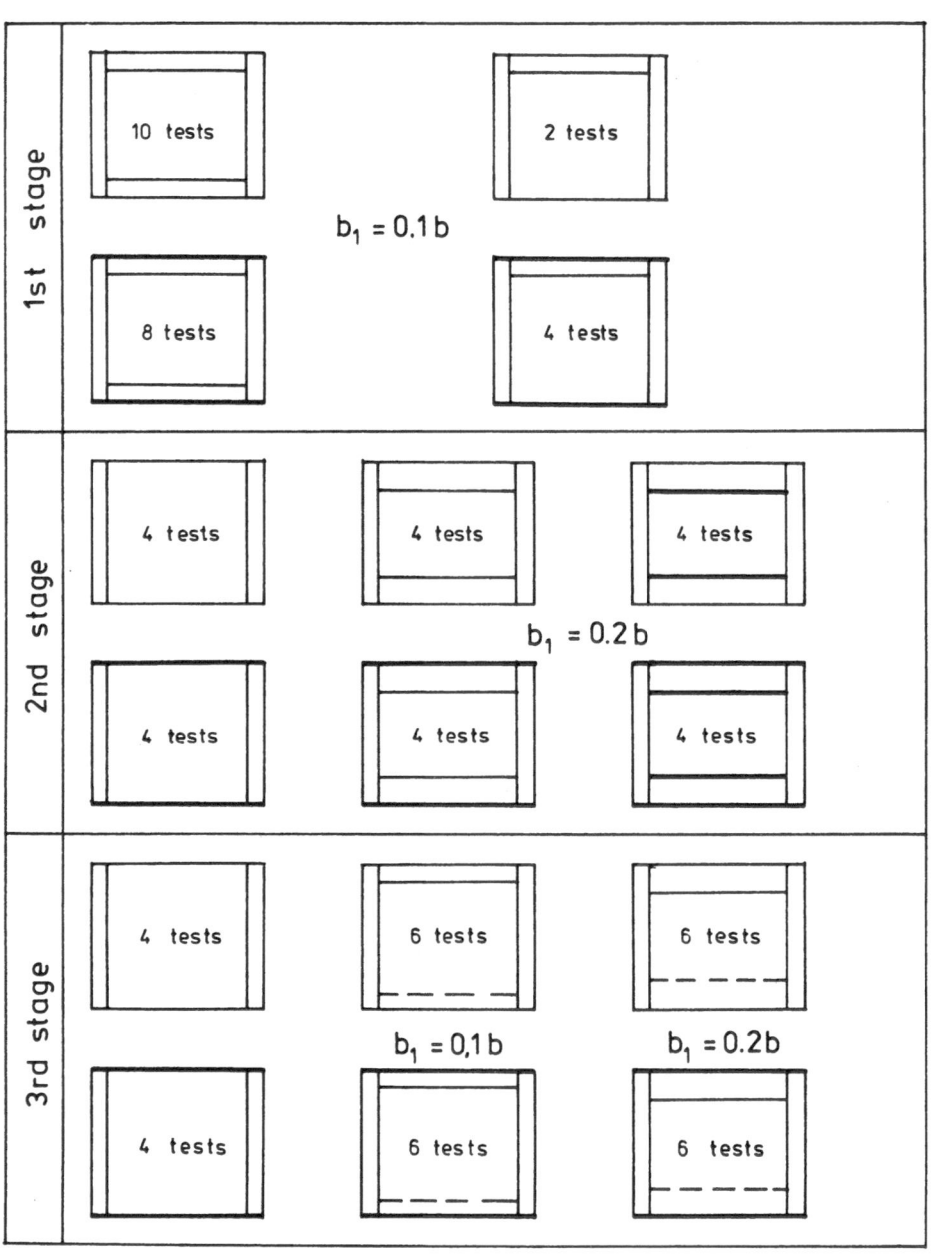

Fig. 1.

mechanisms of girders subject to repeated loading was different from that of girders under the action of constant load. An examination of the data obtained revealed that in both cases the failure mechanism consisted of a set of three plastic hinges in the loaded flange and of a segmental line plastic hinge in the adjacent zone of the web sheet, which in the case of flexible flanges emanated from the two outer flange hinges (then the length of the chord of the segmental line hinge was about one third of the web length) and in the case of thick flanges (when no flange hinges developed) from the web corners (then, of course, the length of the chord of the segmental line hinge equaled the web length). In addition, during the repeated loading tests, cracks appeared in the web sheet, either in the zone of the (subsequent) segmental line plastic hinge or close to the weld connecting the web sheet to the loaded flange. In most cases, the cracks occurred in the course of the last loading step. In some cases, they developed as late as the failure of the girder, and in others they did not appear at all.

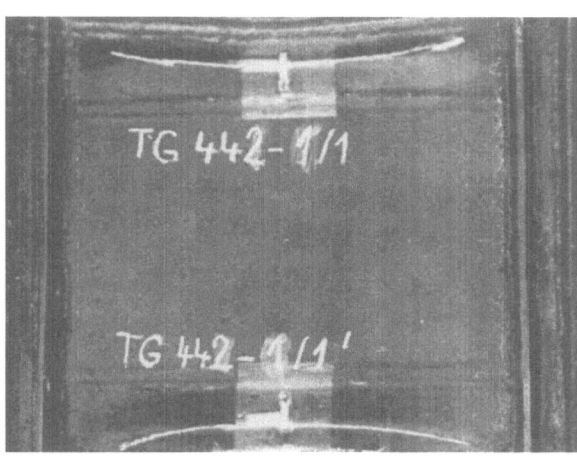

Fig. 2.

To illustrate this phenomenon, a photograph of test girder TG 442 -1/1(having a thick flange and $b_1/b=0.2$), which was first tested under a constant load and then - in an upside down position - under a repeated one, is shown in Fig. 2. The loading of the girder proceeded in four steps. Pending the first three ones (i.e., $P_{1max} = 0.8P_{ult}^{exp}$, $P_{2max} = 0.85P_{ult}^{exp}$ and $P_{3max} = 0.9P_{ult}^{exp}$, P_{ult}^{exp} being the ultimate load obtained in the constant loading experiments), no cracks were detected after 5×10^4 loading cycles. As late as the fourth loading step, under a load $P_{4max} = 0.95P_{ult}^{exp}$, a crack appeared in the vicinity of the weld (connecting the web sheet with loaded flange) after 3.76×10^4 cycles. This crack was first detected on the front side of the web and a little later, after 4.89×10^4 cycles, also on the rear side. The collapse mechanism developed much later, viz. as late as the 12.65×10^4-th loading cycle of this fourth loading step.

161

3.2. Initiation and propagation of cracks

The initiation and propagation of cracks, a phenomenon being very characteristic of the performance of webs under repeated loading, was very carefully studied in the course of all tests. In all experiments, this was done visually with the aid of a magnifying glass, but in the third series of tests, in the case of 12 test girders, the acoustic emission method was employed (via a dosimeter SEDO 2.1) to verify visual observations. It was concluded that, in comparison with the visual observations, the acoustic emission approach gave more accurate results; for example, it was able to detect the initiation of cracks by 4500-9000 cycles (i.e., by 20-40 minutes) earlier. This means that, within the regime of 5×10^4 repeated cycles, the difference was of 9-18 %.

Also other pieces of information were used to verify the initiation of cracks. For instance, it was observed that this initiation was preceded by a sudden change in vertical strains in the respective zone of the web and, at least with some test girders, also by a fast enlargement of web deflections.

Two general conclusions may be drawn from the above evidence: (i) the propagation of cracks was not uniform and frequently proceeded with more or less long interruptions, (ii) the appearance of a crack by far did not herald the end of the useful life of the girder concerned. The girder was thereafter able to sustain a good many more loading cycles before its load-bearing capacity was exhausted.

The plastic residue in a test girder (i.e., in girder TG 542 -1/1') with a crack in the web is seen in Fig. 3.

Fig. 3.

3.3. Ultimate limit state of webs subject to a repeated partial edge load

All test results are summarized in bar charts in Fig. 4. They are plotted in terms of (i) stiffener position and (ii) flange size (the left-hand bar being related to girders with thin

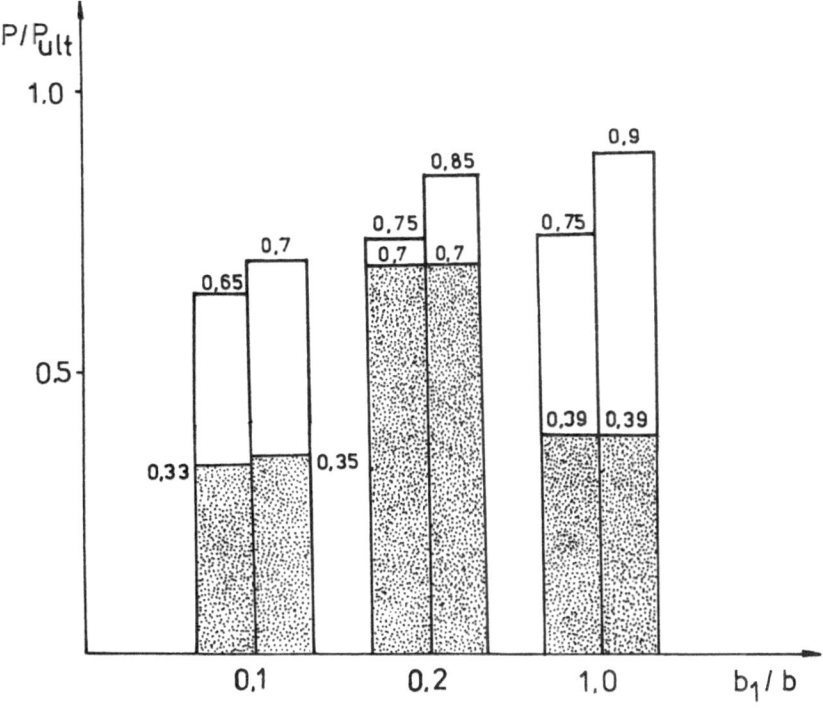

Fig. 4.

flange and the right-hand one to girders with a thick
flange).

The following quantities are given there:
(a) the onset-of-surface-yielding loads, P^{su}, (see the black
    portions of the bars), and
(b) the low-cycle fatigue limit loads, P_{fat}, determined as the
    maximum load values under which no appearance of cracks
    (or of other kind of failure) was detected yet (white
    portions of the bars).
Both aforesaid quantities are related to P_{ult}^{exp}, i.e., to the
ultimate loads obtained in the foregoing constant loading
tests.

An examination of the results obtained shows that even in the
case of webs "breathing" under the action of a repeated
partial edge load a certain degree of plastification in the
webs does not jeopardize the safety of the girders.

4. References

1. Kutmanová, I. - Januš, K. - Löwitová, O. - Škaloud, M.: Mezní stav štíhlých ocelových stěn zatížených opakovaným tlakem na části jedné hrany. Celostátní konference "Moderné ocelové konštrukcie", Bratislava (1989).

2. Januš, K. - Kutmanová, I. - Škaloud, M.: Návrh stěn ocelových nosníku namáhaných místním tlakem od pohyblivého zatížení. Inženýrské stavby č.3 (1989).

3. Kutmanová-Kárníková, I. - Škaloud, M. - Januš, K.: Ultimate Load Behaviour of Longitudinally Stiffened Steel Plate Girders Subject to (i) Stationary or (ii) Variable Repeated Patch Loading. Fourth International Colloquium on Structural Stability, Beijing (1989).

4. Kutmanová-Kárníková, I. - Škaloud, M. - Januš, K.: "Breathing" of Thin Webs under Variable Repeated Patch Loading. International Colloquium on Stability of Stcel Structures, Budapest (1990).

5. Kutmanová, I. - Škaloud, M. - Januš, K. - Löwitová, O.: Ultimate Load Behaviour of Longitudinally Stiffened Steel Webs Subject to Partial Edge Loading. Symposium IUTAM Prague '90 "Contact Loading and Local Effects in Thin-Walled Plated and Shell Structures", Prague (1990).

BUCKLING OF THIN-WALLED BEAMS UNDER CONCENTRATED TRANSVERSE LOADING

C.M. MENKEN and G.M. van ERP[1]

Eindhoven University of Technology,
Fundamentals of Mechanical Engineering Division,
P.O.Box 513,
5600 MB Eindhoven,
The Netherlands.

Abstract

The transversely loaded thin-walled beam under a non-uniform bending moment
forms an example of the detrimental influence that a local effect may have on
the overall behaviour. The local effect is the plate buckling in the region
of maximum bending moment. The overall behaviour is the lateral-torsional
buckling of the beam as a whole. If the local buckling load is smaller than
the overall buckling load, the overall mode may be triggered in the
post-buckling region even if the individual buckling loads are well
separated. This interaction problem is being investigated both
experimentally, and by analysis by means of a simple discrete model as well
as by means of a spline finite strip computer program. A special feature of
this program is the use of splines instead of Fourier series, so that
strongly non-periodic buckling modes, such as occur in the afore mentioned
problem, can be handled.
Preliminary results show a qualitative agreement between the experiments and
the discrete model. Moreover, the numerical model describes the shape of the
non-periodic local buckle very well.

1. Introduction

A beam loaded in bending may, when a certain critical load is reached, show
the phenomenon of so-called lateral-torsional buckling. This type of overall
buckling is characterised by a combination of lateral bending and twisting.
In particular beams having a low lateral stiffness in comparison with the
stiffness in the plane of loading are susceptible to this type of buckling.
If the beam comprises thin-walled (in our case, plate-like) parts, local
buckling may also occur, which means bending of plate elements out of plane
while the junctions between the components of the beam, e.g. between web and
flange, remain straight. (Note that in the concept "local buckling" the
adjective local has a different meaning than in the concept "local effects".)
The local buckling mode is a short wave mode. If the critical load pertaining
to local buckling is smaller than the critical load pertaining to overall
buckling, the occurence of local buckling may immediately trigger overall
buckling, notwithstanding the fact that the overall buckling load has not yet
been reached. This phenomenon, associated with buckling mode interaction, was
investigated analytically and/or numerically by Koiter [1], Pignataro [2] and
Ali [3] among others. However, these investigations were predominantly
confined to structural members under uniform compression, such as stiffened
plates, shells and thin-walled columns. In all these cases the local buckle

---

[1]Now with: School of Engineering, University College of Southern
Queensland, Australia.

has a simple sinusoidal character. Ali [3] and some other authors also
considered uniform bending. Comparable studies of bending caused by
transverse loads are unknown to the authors. That is why we included this
aspect in our study of interactive buckling of thin-walled prismatic members.
This ongoing study comprises experiments and the analysis of both a discrete
model and numerical simulation by means of a spline finite strip computer
program.

2. Some experimental observations

In the experiments, a simply supported prismatic T-beam was loaded in bending
by a concentrated load at midspan (Fig.1) in such a way that the flange was
in compression. The beam was built up from a thin flange carefully machined
from aluminium sheet metal and glued to a relatively stiff web (Fig.2).

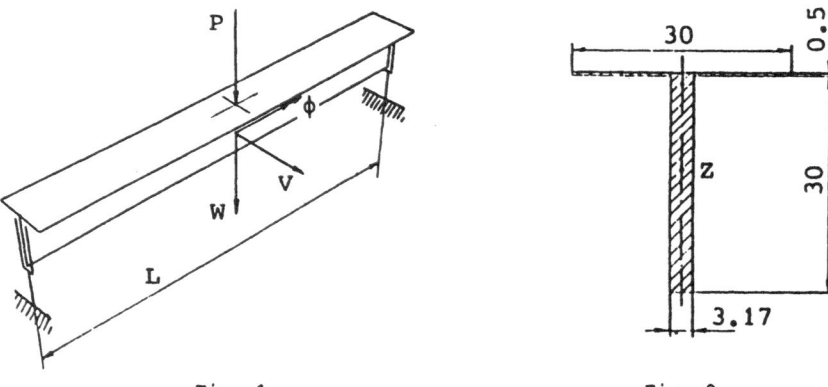

Fig. 1   Fig. 2

In our opinion, this was a practical way of providing the flange with a
uniform thickness. It was verified experimentally that the glue had no
influence on the bending stiffness. The design of the test rig was based on
the earlier buckling experiments of Cherry [4]. More details will be
published elsewhere [5]. The beam was loaded by applying a simple dead-load.

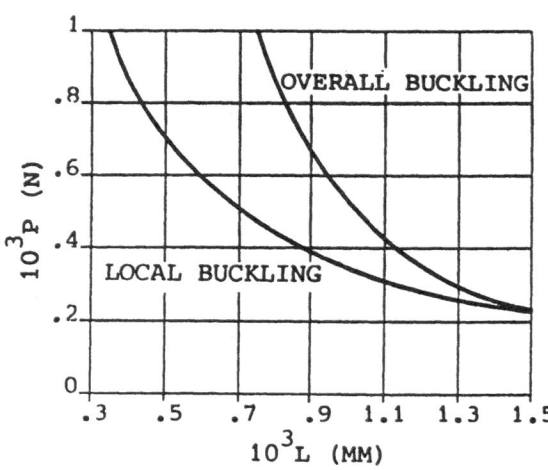

Fig. 3.

As a consequence, descending
equilibrium paths, as might
occur at nearly coincident
buckling loads, could not be
observed during these
experiments. The independent
variables were the magnitude
of the load, P, and the
length, L, of the beam. By
changing the length of the
beam, the distance between the
overall critical load and the
local critical load could be
varied, as shown in Fig.3. To
draw this figure, it was
assumed that the critical
stress for flange buckling
equalled the critical stress
for a periodic buckle.

166

At each length, two experiments were performed consecutively:
- the first experiment, where overall buckling was prevented by means of a hold-up, so that only local buckling could occur;
- the second experiment, where both local and overall buckling could occur.

The variables measured were:
- the shape of the flange in the neighbourhood of the site of maximum prebuckling stresses (the maximum deflection will hereafter be called the local buckling amplitude).
- the overall buckling components; these were the lateral displacement, v, of the centre of gravity of the midspan cross-section, together with the rotation, φ, of the relatively stiff web (deflection v will hereafter be called the overall buckling amplitude).

The shape of the local buckle was measured by mounting a number of retro reflective markers on the free edge of the flange. The positions of these markers were measured by means of a video tracking system (Hentschel GMBH, Hannover).

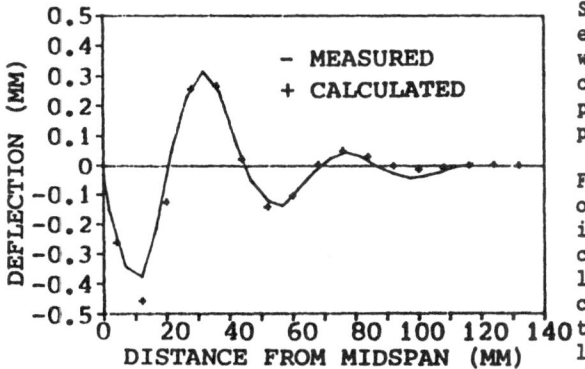

Fig. 4.

Since in these exploratory experiments only 2-D software was available, only one video camera was used with the axis placed perpendicular to the plane of pre-buckling bending.

Fig. 4 shows a measured shape of a local buckle, illustrating that, under concentrated transverse loading, local buckling is confined to a small region and thus is non-periodic and localized.

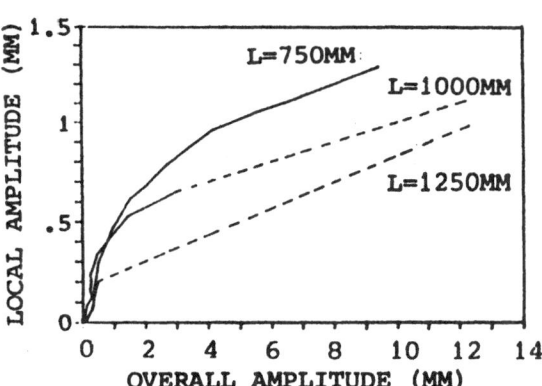

Fig. 5.

Fig. 5 shows some experimental relations between the local buckling amplitude and the overall buckling amplitude. All these experiments were performed in the range where the local buckling load was smaller than the overall buckling load. In all these cases, the curves show in the origin a tangent to the vertical axis, confirming that the buckling process is initiated by the local buckling, but showing that this is immediately followed by an increasing amount of overall buckling.

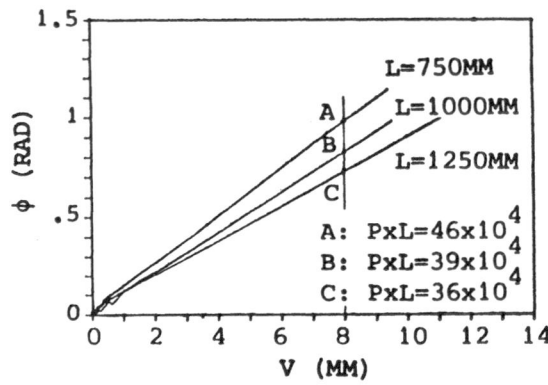

Fig. 6 shows the relationship between the two displacements involved in the overall buckling mode. Strictly speaking, the buckling mode only refers to vanishing displacements at the critical load. This figure shows, however, that the overall mode is roughly maintained even though overall buckling is relegated to the post-buckling range and the displacements are finite.

Fig. 6.

3. The simple discrete model

Since it is a well-known fact that many buckling features can be simulated qualitatively by means of simple discrete models, we analysed such a model to learn more about this type of buckling. The model is described in [5] and [6]. The results are based on a potential energy approach with the potential energy truncated after the quartic terms in the displacements. This is the only limitation; the solutions are not restricted to (nearly) coincident buckling loads. The overall lateral-torsional buckling is characterized by a rotation about the axis Q_1 and a lateral bending Q_3. Each flange half is assigned an independent local buckling amplitude. Accordingly, Q_5 and Q_6 characterize the local buckling of the individual flange halves. As both Q_1 and Q_3 appear in the overall buckling mode, the rotation Q_1 can be eliminated from the potential energy V (Q_1, Q_3, Q_5, Q_6) with the relation

$$Q_1 = - \frac{PL}{S_t} Q_3, \tag{1}$$

where L is the length of the beam in the discrete model and S_t represents the torsional stiffness. This linear relation holds at any load P and corresponds to the overall buckling mode if P equals the critical load for overall buckling P_0. After elimination of Q_1, the potential energy expression becomes

$$V(Q_3, Q_5, Q_6) = \frac{1}{2} A_{33} Q_3^2 + \frac{1}{2} A_{55} Q_5^2 + \frac{1}{2} A_{66} Q_6^2 + \frac{1}{2} A_{355} Q_3 Q_5^2 +$$

$$\frac{1}{2} A_{366} Q_3 Q_6^2 + \frac{1}{24} A_{5555} Q_5^4 + \frac{1}{4} A_{5566} Q_5^2 Q_6^2 + \frac{1}{24} A_{6666} Q_6^4 \tag{2}$$

with $A_{55} = A_{66}$ and $A_{355} = - A_{366}$. There exist two coupled equilibrium solutions with either $Q_5 = 0$ or $Q_6 = 0$. We choose $Q_6 = 0$, which reduces the potential energy (2) to

$$V(Q_3, Q_5) = \frac{1}{2} A_{33} Q_3^2 + \frac{1}{2} A_{55} Q_5^2 + \frac{1}{2} A_{355} Q_3 Q_5^2 + \frac{1}{24} A_{5555} Q_5^4 \tag{3}$$

and this is precisely the same expression as the one for the so-called parabolic umbilic catastrophe, which is a characteristic for an important class of interactive buckling problems. The cubic term leads to mode interaction and destabilization. From the requirement $\partial V / \partial Q_3 = 0$ we obtain:

$$Q_3 = - \frac{A_{355}}{2A_{33}} Q_5^2,$$ (4)

where the buckling coefficient A_{33} is proportional to the difference $(P-P_0)$ between the prevailing load P and the critical load for overall buckling, P_0.

4. The numerical model

Although the finite element method is regarded as the most powerful and versatile tool for structural analysis, a finite strip program was developed because it can save computer time [7]. In most existing finite strip methods, the displacement field is described by harmonic functions in axial direction and simple polynomials in transverse direction. Thus, this method can be regarded as a combination of the Ritz method in axial direction and the finite element method in transverse direction. Thin plate theory requires C^1 continuity of the deflections. The saving of computer time follows from the fact that this C^1 continuity is automatically satisfied by harmonic functions. This approach, however, suffers from one or more of the following limitations:
- localized non-periodic buckling modes, as shown in Fig.4, are difficult to describe;
- in most cases, a more wavy solution than the true one is produced, because of the excitation of the higher terms in the harmonic series [7];
- the boundary conditions are limited to simply supported ends.

Splines, unlike harmonic functions, do not suffer from these limitations and therefore are more suitable for analysing prismatic members in bending and/or shear. The spline finite strip method was originally applied to geometrically linear analysis by Cheung et al [7]. Lau et al [8] were the first who applied the method to buckling problems. Van Erp [9] extended the method to the post-buckling region by combining Koiter's asymptotic post buckling analysis [10] with the spline finite strip method.

One of the features of Koiter's perturbation approach is, that subsequent to the solution of the critical loads and the pertinent buckling modes u_i from the quadratic potential energy expression, the initial post-buckling behaviour is determined by again using the potential energy expression, but now supplemented with the cubic and quartic terms.

According to Koiter's theory, in the case of M coincident or nearly coincident buckling loads, the post-buckling displacement field η can be written as a linear combination of the pertinent buckling modes:

$$\eta = \sum_{i=1}^{M} a_i u_i + v,$$ (5)

where a_i is the amount of (normalised) buckling mode u_i and v is an additional displacement field orthogonal to the buckling modes u_i. From the structure of the potential energy expression, it follows that this additional field is of the type

$$v(\lambda) = a_i a_j u_{ij}(\lambda)$$ (6)

169

where u_{ij} are second order displacement fields which can be obtained by making stationary the supplemented potential energy expression for a fixed value of the amplitudes a_i and the load parameter λ. Expression (6) corresponds qualitatively with the discrete model (4). Both relations show that the additional field triggered in the post-buckling region is quadratic in the amplitude(s) of the buckling mode(s) pertaining to the lowest buckling load(s).

Notwithstanding the fact that the initial post-buckling behaviour can be analysed with the combination of Koiter's asymptotic theory and the spline finite strip method, we will now concentrate on the analyses of the local buckle.
The buckling part of the spline finite strip program is also described in [11]. A problem in the numerical analysis is that even in our case of localized (non-periodic) local buckling, there exist several local buckling eigenvalues very close together. As a consequence, a large proportion of the computer time is required for determining the eigenvalues and eigenvectors.

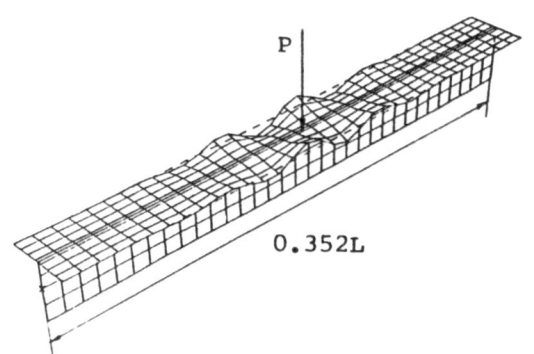

P

0.352L

Fig. 7

Fig.7 shows a calculated local buckle that is antimetric with respect to the middle of the beam. The pertinent eigenvalue is 521.2N, whereas the symmetric mode had an eigenvalue of 520.9N. If the (normalized) calculated buckling mode of the perfect structure, after adaptation of the amplitude, is compared with the shape of the local buckle in the actual structure, a good correspondence can be seen (Fig.4).

Table I shows some measured wavelengths, together with the wavelength obtained from the numerical model and the wavelength pertaining to a periodic buckling mode. It can be seen that there are no great differences.

TABLE 1

Length [mm]	Wavelength [mm]		
	Measured	Num.Model	Periodic
750	45	44	45.07
1000	44	--	45.07
1250	45	--	45.07

5. Additional discussion

The parabolic relation between the overall buckling amplitude and the local buckling amplitude, as predicted by Koiter's asymptotic theory for (nearly) coincident buckling loads (6), is also found in the discrete model (4) and the experiments (Fig.5), which are not restrained to (nearly) coincident buckling loads. Moreover, the experiments extend over finite displacements. The linear relation between the displacements appearing in the overall buckling mode, as found with the discrete model (1), is also approximately found in the experiments (Fig.6) and shows qualitatively the same dependency of the product PL. In the experiments, this linear behaviour extends again over finite displacements.

From these observations we expect that for the type of structures considered here, Koiter's theory, although being of an asymptotic character, may also be applicable for finite post-buckling deflections.

6. Conclusions

- Localized local buckling of a beam may trigger overall lateral torsional buckling, even if the distinct buckling loads are well separated.
- There is a good qualitative correspondence between the experiments and a simple discrete model
- The results presented in this paper show that the spline finite strip method is a good one to analyse localized local buckling.

7. References

[1] W.T. Koiter. General theory of mode interaction in stiffened plate and shell structures. Delft University of Technology, Report WTHD-91 (1976).

[2] M. Pignataro, A. Luongo and N. Rizzi. On the effect of the local-overall interaction of the postbuckling of uniformly compressed channels. Thin-Walled Structures 3, 1470-1486 (1985).

[3] M.A. Ali and S. Sridharan. A versatile model for interactive buckling of columns and beam-columns. Int. J. Solids Structures, Vol. 24, No. 5, 481-496 (1988).

[4] S. Cherry. The stability of beams with buckled compression flanges. Struct. Eng. 38(9), 277-285 (1960).

[5] C.M. Menken, W.J. Groot and G.A.J. Stallenberg. Interactive buckling of beams in bending. Paper intended for publication in Thin-Walled Structures.

[6] C.M. Menken and G.A.J. Stallenberg. Interactive buckling of beams in bending. Proceedings of the Twelfth Canadian Congress of Applied Mechanics, May 28 to June 2, Carlton University, Ottawa, Ontario (1989).

[7] Y.K. Cheung, S.C. Fan and C.Q. Wu. Spline finite strip in structural analysis. Proceedings of the International Conference on Finite Element Methods, Shanghai, 704-709 (1982).

[8] S.C.W. Lau and G.J. Hancock. Buckling of thin flat-walled structures by a spline finite strip method. Thin-walled structures 4, 269-294 (1986).

[9] G.M. van Erp. Advanced buckling analysis of beams with arbitrary cross sections. Ph.D.Thesis, Eindhoven University of Technology, the Netherlands (1989).

[10] W.T. Koiter. Over de stabiliteit van het elastisch evenwicht. (in Dutch). Ph.D.Thesis, University of Delft (1945) Engl. transl. NASA TTF 10, 833 (1967) and AFFDL, TR 70-25 (1970).

[11] G.M. van Erp and C.M. Menken. The spline finite strip method in the buckling analysis of thin-walled structures. Paper accepted for publication in Communications in Applied Numerical Methods.

Acknowledgement

This research is supported by the Technology Foundation (STW). The authors gratefully acknowledge the contributions of Mr. W.J. Groot and Mr. J.M.A. de Vries to this paper. We are also indebted to Dr. E. Riks for his comments.

TESTS OF BUCKLING OF PANELS SUBJECTED TO IN-PLANE PATCH LOADING

RAOUL Joël : D.T.P.E. Engineer, Bridges Division, Service d'Etudes techniques des Routes et Autoroutes (S.E.T.R.A.), Bagneux, France

SCHALLER Isabelle : T.P.E. Engineer, Stuctures Division, Laboratoire central des Ponts et Chaussées (L.C.P.C.), Paris, France

THEILLOUT Jean-Noël : Dr. Engineer, Structures Division, Laboratoire Central des Ponts et Chaussées (L.C.P.C.), Paris, France

ABSTRACT

The experiments described in this paper are the continuation of these which were presented at the colloquium of Stability of Plate Structures in Ghent in 1987. We analyse here the results obtained from three tests caried out on full-scale panels of identical dimensions : length 1800 mm, depth 1350 mm, thickness 6 mm.

The first test is identical to the last test presented in the paper "Tests of Buckling of Panels subjected to in-plane Patch Loading, GALEA, GODART, RADOUANT, RAOUL, in the proceedings of the Ghent Colloquium" but the load is applied by 2 rolling supports instead of 4.

The two other tests are intended to investigate the influence of an error in the layout of the 2 rolling supports.:

- for the second test, the rolling supports have an excentricity of 30 mm out of the plane of the web. Such an excentricity is possible on site when launching a bridge.

- In the third test, an imposed rotation is applied to the flange under the rolling supports. The rotation is about 3°. Such a rotation is consistent with these observed on site.

To monitor the post-buckling behaviour of the panels, the tests are carried out by controlling the vertical displacement of the upper flange under the rolling supports.

The out of plane deflection is measured and the strain condition in the panel is obtained with strain gauges on both sides on the web.

1. INTRODUCTION

Most steel or composite bridges are today launched into place on rolling supports. This poses the problem of the stability of the webs of the metallic girders, simultaneously deflected and highly loaded in-plane by the reaction of the rolling supports.

This problem is all the more crucial in that the launched spans are sometimes very long and that, for reasons of cost, consideration is being given to pushing some decks with part of the slab already poured.

A first series of five tests has already been conducted at the LCPC, working with CTICM and SETRA, to test the influence of the concomitant bending moment and of the presence and position of a longitudinal stiffener. The results of these tests were presented at the "Stability of Plate and Shell Structures" Colloquium at Ghent in 1987 [1].

Another series of three tests on panels of the same size is now in progress at the LCPC to test other parameters:

- the influence of the number of rollers;

- the influence of an error in the positioning of the rolling support.

Two tests have already been conducted and the third is in preparation.

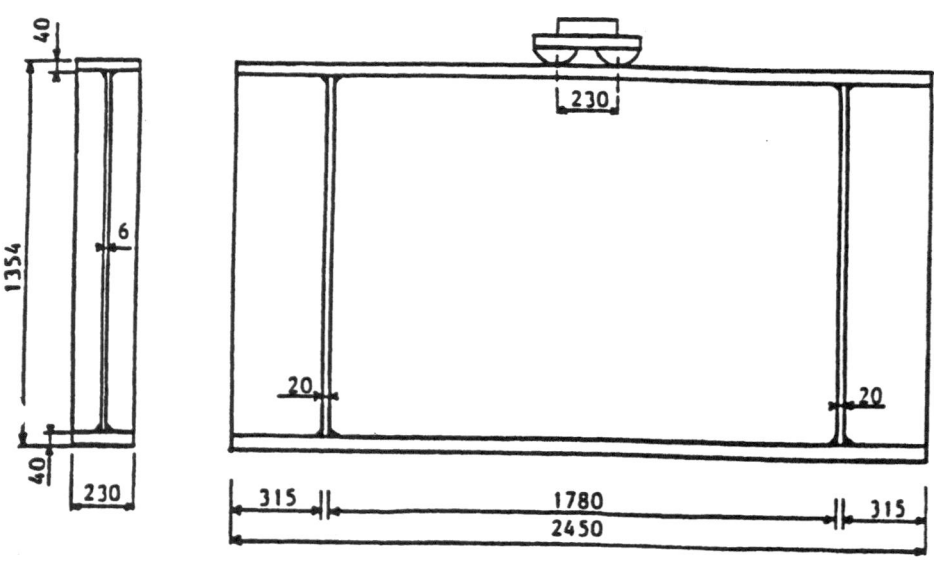

Figure 1

2. DESCRIPTION OF PANELS

The panels tested are 1/2-scale reproductions of the girder panels in common use in metallic construction. As shown by figure 1, the panels are 1780 mm long and 1274 mm high, giving an aspect ratio of 1.4. The web thickness is 6 mm, giving a slenderness ratio of 212.

The steel used in these panels is E 24-3 steel. For each of them, the tensile yield strengths of the web and flanges (σ_w and σ_f) were measured; they are indicated in the table of figure 2.

Test	σ_w(MPa)	σ_f(Mpa)	w_0(mm)
1	362	286	13,9
2	357	301	3,0
3	*	*	5,6

* not yet known

Figure 2

The initial deformations of the webs differ from panel to panel : the production conditions without heat redressment make it difficult to control these initial deformations. The table accordingly also gives for the panels the maximum distance of initial deformation w_0 with respect to a median reference plane.

3. DESCRIPTION OF TESTS

3.1. Application of load

The panels are placed on simple supports under the vertical stiffeners.

In these three tests, the load is applied by a mechanical part simulating the action of a support with two rollers 230 mm apart, rather than four rollers as in the previous series of tests (see figure 3).

The load was centered on the web plane in the first test and 30 mm off centre in the second. For the third test, a rotation of approximately 3° will be imposed on the upper flange of the panel, just below the load. Figure 4 shows the loading arrangements for each of the tests.

175

The first test serves to determine the influence of the number of rollers, while tests 2 and 3 determine the effect of an error in the positioning of the rolling support in the field.

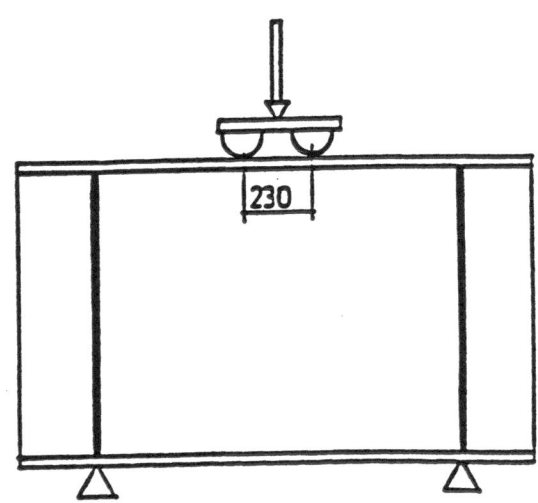

Figure 3

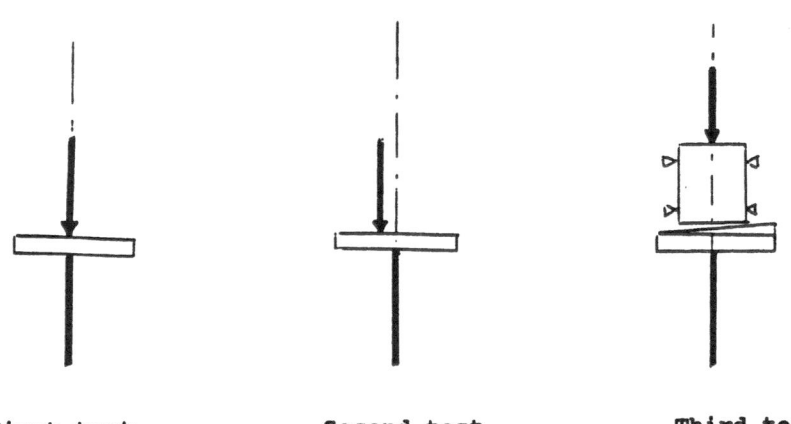

First test   Second test   Third test

Figure 4

3.2. Description of instrumentation

The instrumentation placed on each panel is shown in figure 5:

- the evolution of deformation of the web is measured using 8 displace-ment transducers;

- 24 rosettes of 3 strain gauges at 45° are placed on either side of the web to track stress conditions during loading;

- 16 strain gauges are glued to the flanges and vertical stiffeners to monitor certain local effects.

In addition to this instrumentation, two displacement transducers, one on the upper flange and one on the lower flange, are used to monitor changes in deflection.

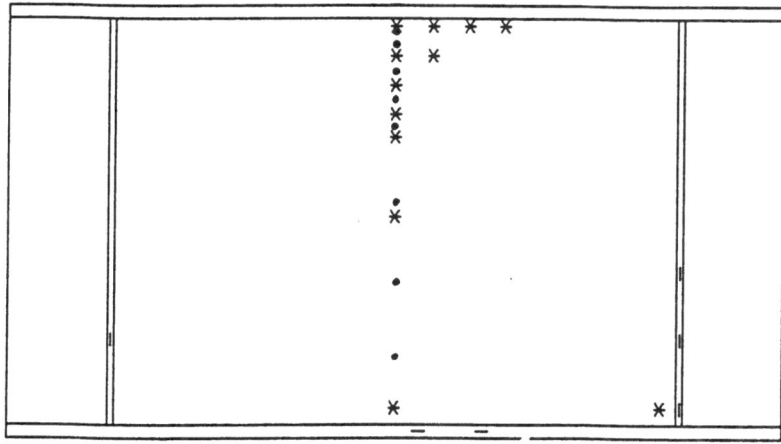

       * strain gage rosette at 45
       — strain gage
       . displacement transducer

Figure 5

3.3. Test procedures

The tests are conducted in the same way. The panels are loaded without plateaus, with periodic unloading.

The test is controlled by imposing a constant rate of deflection of the upper flange (and not force), to make it possible to observe the behaviour of the panel after the failure load.

An acquisition unit regularly records all measurements for later analysis. This is accompanied by practically real-time tracking of some measurements on a monitor to check that the

test is proceeding properly and be able to take action if
necessary.

4. RESULTS OF TESTS 1 AND 2

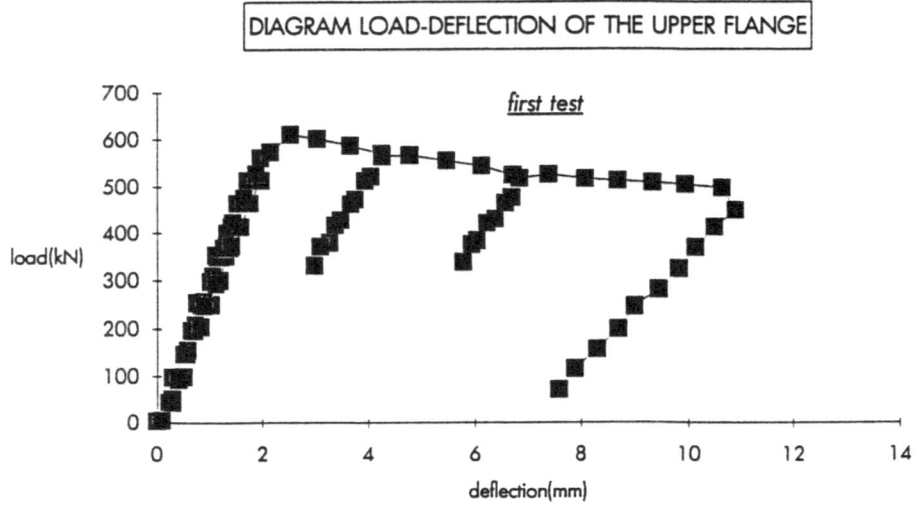

Figure 6

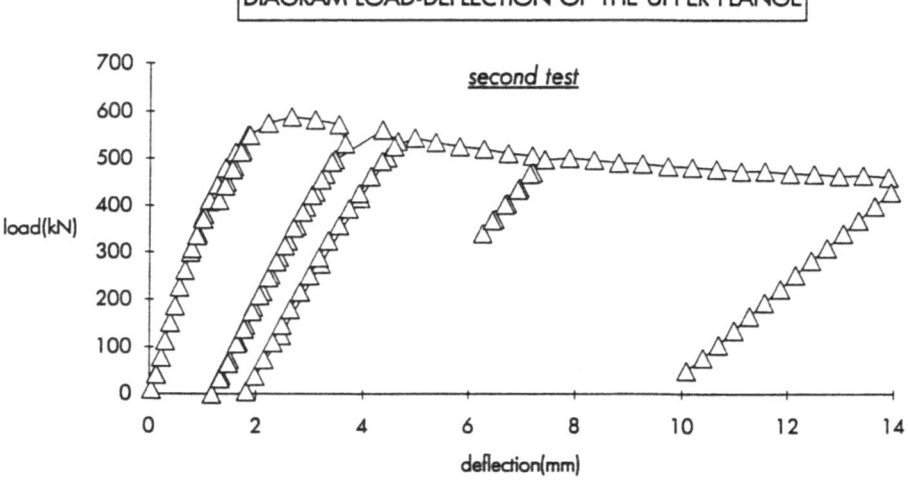

Figure 7

4.1. Ultimate loads and evolution of deflections

The failure loads observed in tests 1 and 2 are 610 kN and 590 kN, respectively.
Figures 6 and 7 show, for the two tests, the evolution of deflection of the upper flange versus load. These curves are characteristic of the tests, and it can be seen that the behaviour of the two panels does not different significantly, even after the ultimate load is reached. Up to about 90 % of the failure load deflection increases linearly with load.

The deflection of the lower rib of the panel remains very small and linear with respect to load up to the end of testing.

4.2. Deformation of web of panels

Figures 8 and 9 show the deformation of the web with respect to its centreline at different load values. The point of maximum deformation is close to the upper flange at final failure of the panels.

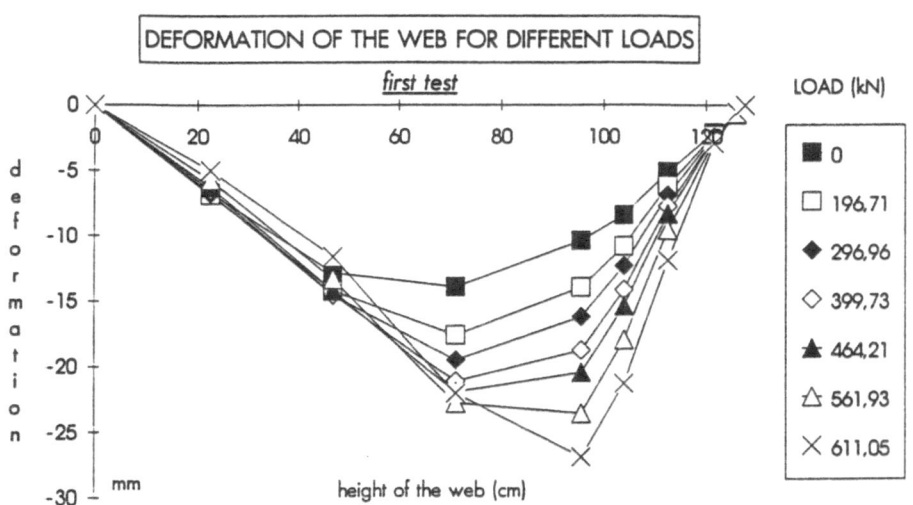

Figure 8

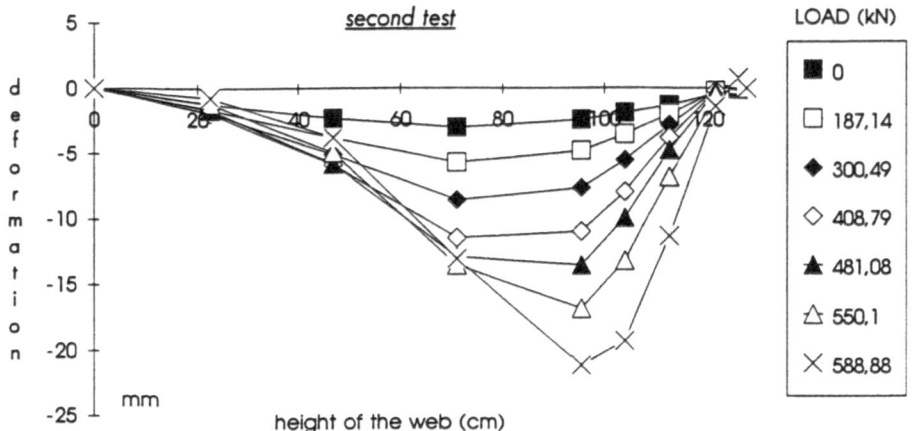

Figure 9

The initial deformation, while very large in the case of test 1, does not seem to have any influence on either the final deformation or the ultimate load, as can be seen by comparison with the second test, in which the initial deformation of the web was much smaller.

4.3. Plasticizing of panels

The zones of plasticizing of the panels webs are practically the same in both tests, as shown by figures 10 and 11. The first plastic points appear in both tests towards 400 kN and are located under the load near the web-upper flange weld.

5. CONCLUSIONS

A comparison of test 1 with a test previously conducted at the LCPC in which the load was transmitted via four points rather than two seems to show that the number of rollers in the support does not have a significant influence on the behaviour of the web of a girder, even if the zones plasticized during the test are somewhat different. The failure load in test 1 is only 6 % less than that obtained with four rollers, which was 650 kN.

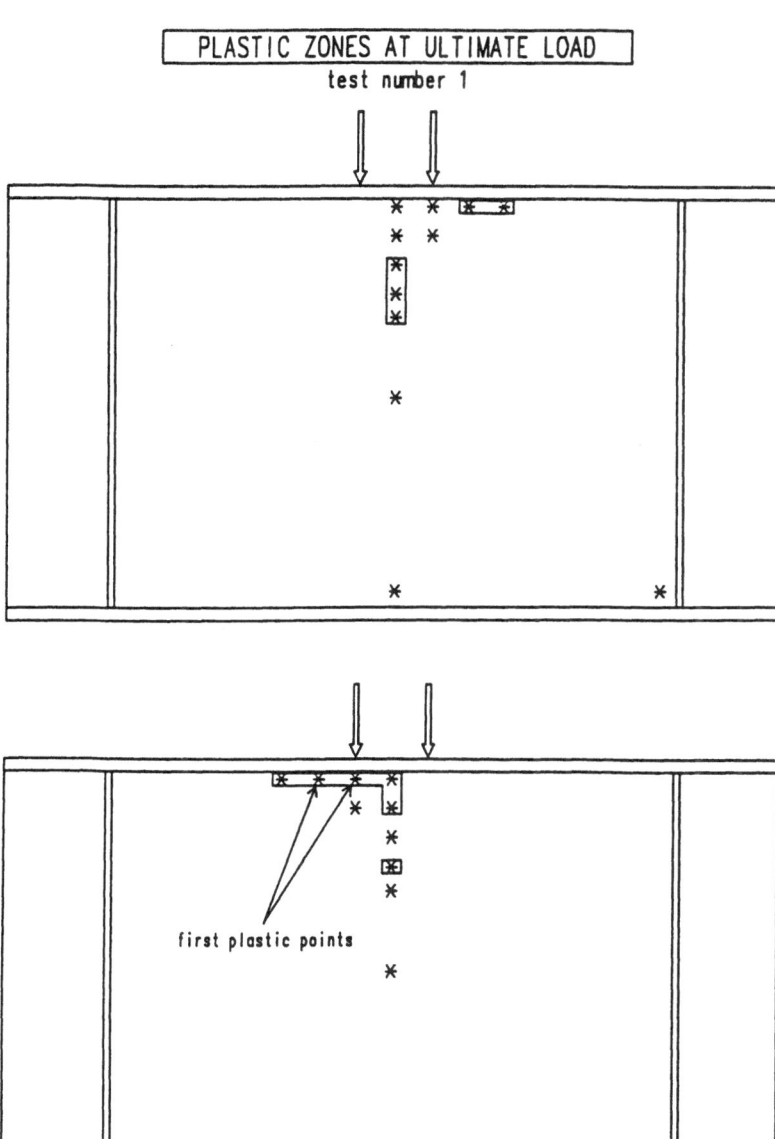

Figure 10

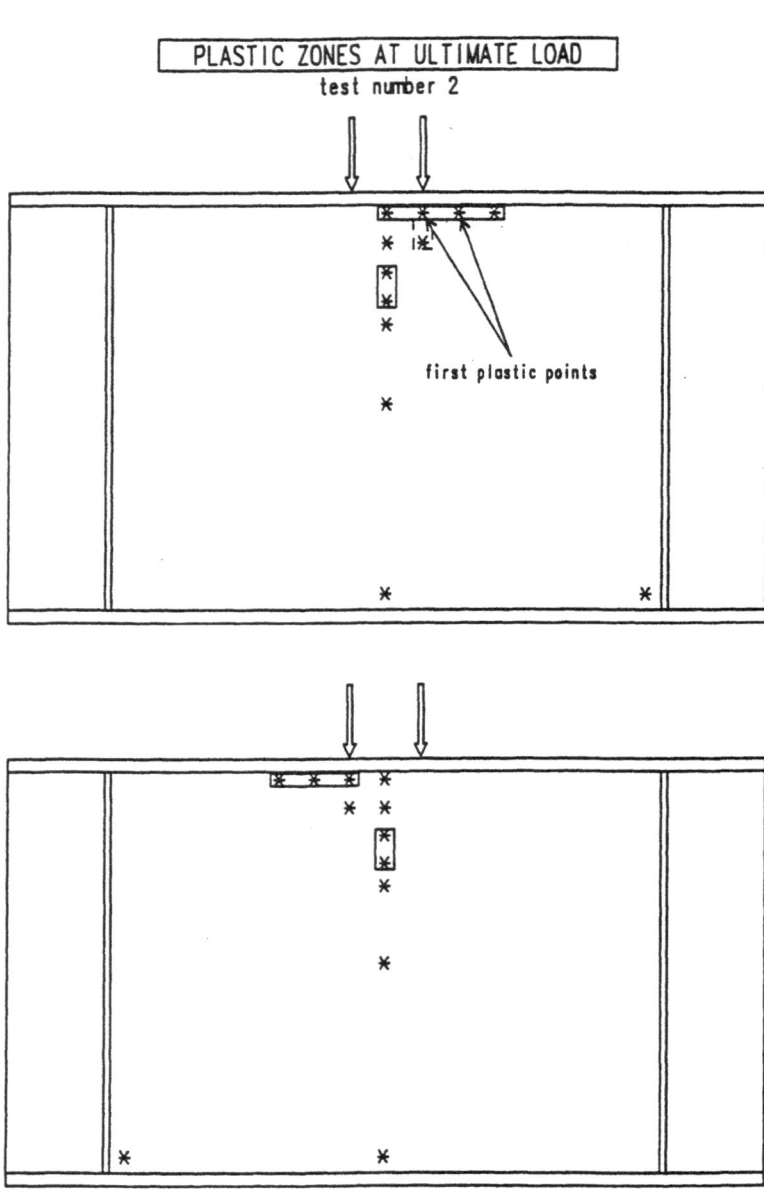

Figure 11

Moreover, comparing tests 1 and 2, we find that an eccentricity of the load, simulating incorrect positioning of the rolling support, is not a primordial factor in the buckling behaviour of the panel (which is practically the same in both tests).

We have now only to conduct the final tests to know for sure whether poor positioning of the rolling support (within reasonable limits, of course) can affect the buckling behaviour of launched metallic girders.

REFERENCE

[1] GALEA Y., GODART B., RADOUANT I., RAOUL J.: "Tests of buckling of panels subjected to in-plane patch loading", Stability of plate and shell structures, International Colloquium, Ghent, Belgium, 6-8 April 1987.

BEHAVIOUR OF STIFFENED WEB PLATES SUBJECTED TO THE PATCH LOAD

S. SHIMIZU[1], S. HORII[2] and S. YOSHIDA[1]

[1]   Department of Civil Engineering,
       Shinshu University
       500 Wakasato, Nagano, Nagano 380, Japan

[2]   Yokogawa Bridge Works,Ltd.,
       4-44 Shibaura 4-Chome, Minato-Ku, Tokyo 108, Japan

Abstract
This paper gives results of numerical analysis on the web plates subjected to a patch load having a horizontal stiffener. The Dynamic Relaxation Method is used for the numerical method. Patch length and the support condition of upper edges of web plates are adopted as parameters, and relatively larger patch length is used. Results obtained through the analysis are compared to those of authors' previous study which deals with patch-loaded plates having no stiffener. Through the analysis, behaviour of patch-loaded web plates up to collapse and layouts of yield lines are cleared. The contribution of stiffener is also found; in most cases a stiffener is effective to increase the strength of plates, however, in certain case, a stiffener decrease the strength.

1. Introduction
A web plate on a launching shoe is often modelized as a patch-loaded plate. To predict the ultimate strength of a patch loaded web plate, Roberts et al. proposed a procedure using the idealized collapse mechanism [1]. In this procedure, the accuracy of the predicted ultimate strength of the plate depends on the definition of the idealized collapse mechanism. According to the authors' experimental study [2], the collapse behaviour which did not confirm with the idealized mechanism given in [1] is observed for relatively larger patch length. So, a procedure presented in [1] is extended by authors in Ref.[3] for the relatively larger patch length.

On the other hand, to study the behaviour of plates subject to the patch-load with relatively larger patch length, authors made an elasto-plastic large deflection analysis on the unstiffened patch loaded web plate [4] by using the Dynamic Relaxation Method (DRM). Through [4], contributions of the patch-length and the stiffness of flanges to the collapse mode are cleared.

In a general way, when a web plate has insufficient strength, it is stiffened by a horizontal stiffener which is similar to one arranged against the in-plane bending.

The contribution of a horizontal stiffener to the ultimate strength of patch-loaded plates is mentioned in the report of TWG 8.3 of ECCS [5] in which a concentrated load or a patch-load having small patch length is considered.

In this paper, as a sequel to authors' previous study [4], the results of the analysis on the patch-loaded web plates having a horizontal stiffener are given. As similar to [4], the patch-length and the stiffness of the upper flange are adopted as parameters, and the ultimate strength, collapse mode and so on are studied. Results of the analysis are compared to those described in [4].

2. Analized Models

Table 1. Model Names and Parameters

Model Names	Upper Edge Conditions	Patch Length $\beta=c/a$
FS1		0.5
FS2	Fixed	0.25
FS3		0.125
ES1	Elastically Supported	0.5
ES2		0.25
ES3		0.125

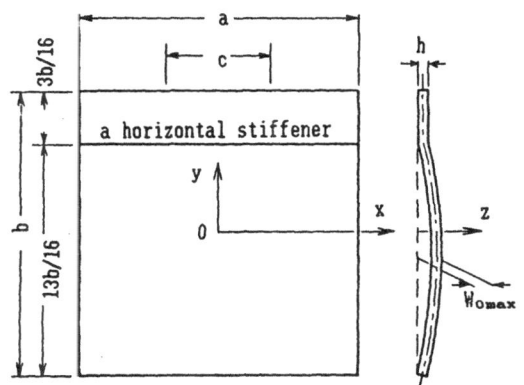

Fig.1 A Typical Web Plate

Fig.1 shows a typical web plate used in this study. This model corresponds to part of a plate girder, and is a web plate subtending two vertical (transverse) stiffeners; its width a, whole depth b and web thickness h. The length c in the figure denotes the patch-length. This model is same with one in [4] except the existence of a horizontal stiffener.

In JSHB [6], no specification is given on the stiffener arrangement of a web plate subjected to the patch-load. Thus, design rules on the horizontal stiffener against the in-plane bending specified in JSHB is often applied to decide the arrangement of a stiffener under a patch load. JSHB specifies the location of a horizontal stiffener as about 0.2b from the upper flange, so in this study the stiffener is assumed to be at 3b/16 from the upper edge of the web plate. In the current analysis, a stiffener is considered merely to restrict the out-of-plane web deflection and to have no torsional or bending stiffness itself.

185

The web plate is assumed to have the cosine-shaped out-of-plane initial deflection of its maximum value w_0 max of w_0 max=0.1h in the larger (lower) panel. The material properties assumed as the yield stress σ_y=235 MPa, Young's modulus E=206 GPa and the Poisson's ratio ν=0.3. The web plate is discretized into 16×16 meshes for its in-plane direction and into 10 layers for its thickness. In the analysis, the patch load is replaced as the displacement of $-V_a$ for y direction at the upper edge as same as in [4]. This is corresponding that the load is applied through a very stiff launching shoe.

The dimensionless patch length (defined as β=c/a) is adopted as a parameter. Aspect ratio (α=a/b) and width-thickness ratio (a/h) are assumed to be 1.0 and 100.0 respectively.

Two types of boundary conditions (FS and ES) are adopted to take the contribution of stiffness of the loaded edge into account. Models of boundary condition type FS is fixed at its upper edge. This is corresponding to the very stiff upper flange. Type ES has elastically supported upper edge corresponding to the torsionally flexible flange. The remained edges of each model are considered to be simply supported.

The parameter β and the support conditions are listed in **Table 1** with model names.

3. Behaviour of Models

3-1 Models with a Fixed Upper Edge

Fig.2 shows load out-of-plane web deflection curves (P-δ curves) of models FS1-FS3 which have fixed upper edges. The vertical axis of this figure is the dimensionless load parameter σ/σ_y and the horizontal axis is the dimensionless out-of-plane deflection w/h at the center of the larger (lower) panel. In **Fig.2**, plots of yield lines of each model defined by a procedure described in [4] are also illustrated with out-of-plane web deflections. It is found that the model FS3 which is subjected to the smallest patch length has three yield lines and remained two models have two yield lines. P-δ curves in the figure seem to show that a smaller patch length gives the larger maximum load. However it should be noted that the vertical axis of this figure denotes the dimensionless load parameter σ/σ_y and not the practical load applying the plates. To discuss the load-carrying capacity of the models, the dimensionless load parameter should be translated into the practical load. On the translated load, discussion will be made in the next section.

Behaviour up to the final stage of each model is as follows:
Model FS1 Small out-of-plane deflection appeared in the upper (and smaller) panel, and no yield line is developed up to the maximum load is reached. On this model, two yield lines are developed almost simultaneously at the stage shown by a black point with a character "a" on the P-δ curve; one of them are

186

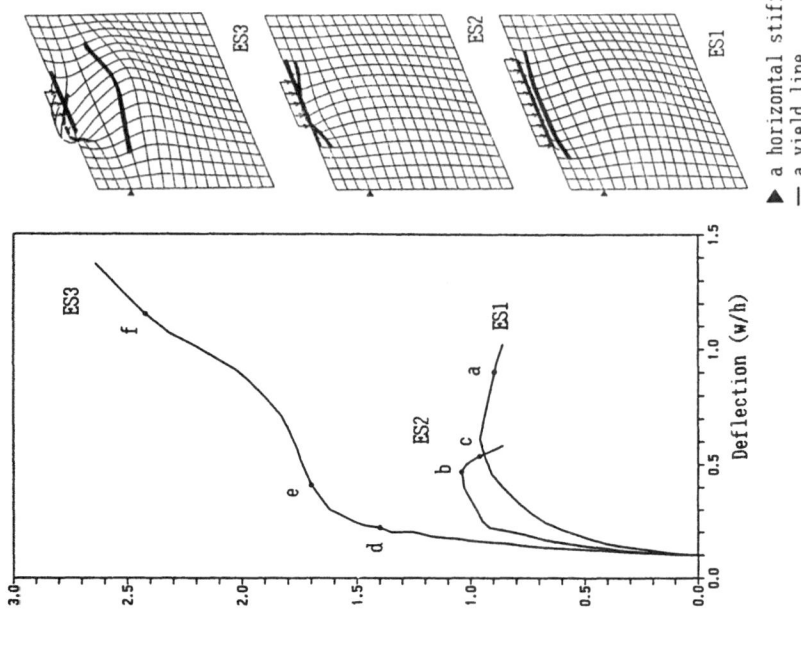

Fig.3 · Load-Deflection Curves (ES)

▲ a horizontal stiffener
— a yield line

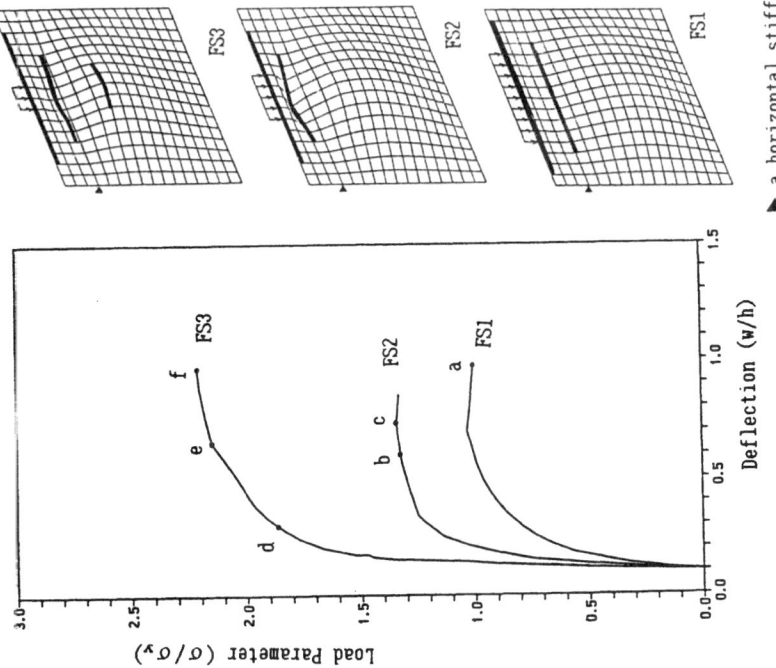

Fig.2 Load-Deflection Curves (FS)

▲ a horizontal stiffener
— a yield line

appeared along the horizontal stiffener, and another one along
the upper edge. Although the curvature for the out-of-plane
direction along the upper edge of this model is not so large as
illustrated in Fig.2, the numerical results show that the upper
zone along the edge in the panel is fully yielded by bending.
So the upper yield line is regarded to be developed.

Model FS2   The first yield line is appeared at the stage
shown the black point "b" in the curve along the upper edge,
and the second yield line is developed above the stiffener when
the maximum load is reached (stage of "c"). On this model, out-
of-plane deflection of the upper panel is begun to increase
rapidly after the P-δ curve reaches to the plateau, and it is
found that the web plate is folded at the panel. In the lower
panel, the out-of-plane deflection is increased gradually,
however, no yield line is appeared in the panel.

Model FS3   On this model, behaviour up to the maximum load
is reached cannot be traced, and P-δ curve of this model is
interrupted prior the maximum load. The first yield line is
developed along the upper edge of the plate at the stage of the
black point "d". On this model, the second yield line is
appeared above the stiffener at the stage "e" as similar to the
case of model FS2, however, the load is still increased after
this stage. The out-of-plane deflection of the lower panel
increased gradually after the stage "e", and the yielded zone
begun to be developed in the panel. At the stage "f", the third
yield line is developed in the lower panel. The deflection of
the upper panel of this model is smaller than one of model FS2.

3-2 Models with an Elastically Supported Upper Edge
     P-δ curves and layouts of yield lines of models ES1-ES3
are illustrated in Fig.3. Model ES3 with smallest patch length
has three yield lines and models ES1 and ES2 has two yield
lines.

     Behaviours of these models are as follows;
Model ES1   As similar to the model FS1, small out-of-plane
deflection is appeared in the upper panel. Prior the maximum
load is reached, yielded zones are begun to developed along the
upper edge and above the stiffener, however, the yield lines
are not completed up to the maximum load. Beyond the maximum
load, two yield lines are completed almost simultaneously at
the stage shown by a black point "a" on the P-δ curve. One of
the yield line is developed along the upper edge, and another
line between the upper edge and the stiffener.

Model ES2   On this model, unlike to the model FS2, the
first yield line is developed between the upper edge and the
stiffener when the maximum load is reached (stage "b"). Beyond
the maximum load, the second yield line is developed along the
upper edge. That is the order of developing of two yield lines
of this model is reverse to one of model FS2. This model has
large out-of-plane deflection at its upper panel which is
increased rapidly after the curve reaches to the plateau.
However, as shown by P-δ curve, this model has the smallest
deflection in its lower panel.

Model ES3   The first yield line has developed along the
upper edge at the stage of black point "d" where the P-o curve
is almost vertical, and the second yield line is developed

between the upper edge and the stiffener at stage "e". Beyond this stage, yielded zone is begun to appear along the stiffener, and this zone is extended downward with the load increment. At the stage "f", the third yield line is developed below the stiffener. As similar to model FS3, the behaviour of this model cannot be traced up to the maximum load is reached.

4. The Maximum Loads

The results are compared to those described in Ref.4 which deals with the behaviour of patch loaded web plates having no horizontal stiffener.

In **Table 2** and **Table 3**, the maximum load and the number of yield lines of each model are summarized. Corresponding data of models with no horizontal stiffener (F1-F3,E1-E3) are also shown in the tables. Figures in () denote the ratios of maximum loads of models with a stiffener to those of corresponding models without a stiffener. As mentioned in section 3, the P-max of models FS3 and ES3 shown in these tables are not the maximum load because the analysis cannot be continued up to the maximum load is reached on these two models.

Table 2. P-max and No. of Yield
Lines(Models F and FS)

	Model Names	P-max [MN]	Number of Yield lines
β=0.5	F1	1.152(1.00)	2
	FS1	1.207(1.05)	2
β=0.25	F2	0.746(1.00)	2
	FS2	0.787(1.05)	2
β=0.125	F3	0.514(1.00)	2
	FS3	0.649(1.26)*	3

* Maximum load is not reached.

Table 3. P-max and No. of Yield
Lines(Models E and ES)

	Model Names	P-max [MN]	Number of Yield lines
β=0.5	E1	1.023(1.00)	2
	ES1	1.129(1.10)	2
β=0.25	E2	0.658(1.00)	2
	ES2	0.606(0.92)	2
β=0.125	E3	0.707(1.00)	3
	ES3	0.782(1.11)*	3

* Maximum load is not reached.

These two tables indicate that a larger patch length β gives generally a larger P-max, and that the models in which three yield lines are developed have somewhat larger P-max.

From **Table 2**, it is clear that a stiffener gives larger P-max than one of with no stiffener. **Table 2** shows that the models FS1 and FS2 have their P-max larger by 5% than those of F1 and F2, and the model FS3 has 26% larger than F3, so it seems that a stiffener is more effective to a model with smaller patch length or to a model in which three yield lines are developed.

In **Table 3**, it is found that model ES1 has the P-max larger by 10% than one of model E1, and the model ES3 has larger by over 11% than E3, and the stiffener is also effective to increase to the P-max of these models.

However, **Table 3** indicates that the P-max of the model ES2

189

is smaller by 8% than one of the model E2, i.e. a stiffener seems to be harmful on this model. The exact reasons of this result are not known to authors, but the following descriptions may be made by using Fig.4. In this figure the shapes of plate sections at the final stage of loadings of models ES1 and ES2 and corresponding models E1 and E2 are illustrated.

Fig.4(a) shows that the model E1 has the out-of-plane deflection shape with its peak at a location approximately 0.3 h from the upper edge. In the model ES1, a stiffener makes the deflection smaller effectively. This was similar to the sets of models of F1/FS1 and F2/FS2.

On the other hand, the model E2 has the deflection with its peak at 1.2 h from the upper edge (**Fig.4(b)**). Unlike to other models, the out-of-plane deflection of the lower panel of this model is decreased after the maximum load is reached [4]. On the model ES2 which has a stiffener, although the deflection of the lower panel is relatively smaller than one of other models (**Fig.3**), the deflection of the upper panel is very large, and the plate is folded in this panel. In addition, as mentioned in Section 3, only this model has its first yield line developed on the upper panel where the plate is folded. Thus, in this model, it can be said that the crippling of the upper panel is promoted by the stiffener.

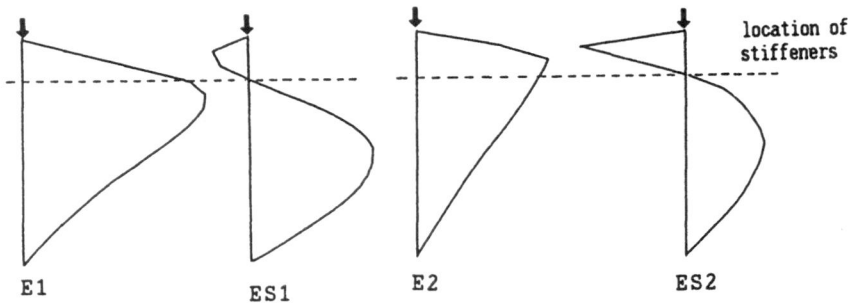

(a) models E1 & ES1   (b) models E2 & ES2

Fig.4   Sections of Web Plates

In [5], it is mentioned that the longitudinal stiffener is most effective when it is located adjacent to the upper flange. However, in JSHB or the other design rules, the specification on the horizontal stiffeners for the patch loaded plates are not given. So, in designing such plates, design rules on the horizontal stiffeners for the plates subjected to the in-plane bending are commonly applied. Above results indicate that such application is sometimes not suitable to design a plate under a condition which is not specified in the design rules.

The analysis in this study is made on web panels with the

extreme conditions that the panel is subjected only to the patch load and no in-plane bending. In a practical web plate which subject not only the patch load but also the in-plane bending, a horizontal stiffener against the in-plane bending shall be more effective. However, results shown in this paper indicate that attention should be paid to the effectiveness on the stiffeners.

5. Conclusion
In this study, an analysis is made on the patch loaded web plates having a horizontal stiffener. Results of the analysis are compared to those of no stiffener. Through the analysis, followings are found;
1) A smaller patch length gives the failure mode of web plates with three yield lines, and a larger patch length gives two yield lines.
2) Models with three yield lines have relatively larger P-max.
3) A horizontal stiffener attached to the web plate generally increase the P-max. However, in the certain case, the stiffener may act to be harmful.

The authors appreciate the sponsorship of the Public Trust JSCE Scientific Exchange Foundation. A part of this study is also sponsored by Grant-in-Aid for Scientific Research from the Ministry of Education in Japan. The numerical analysis in this study is made by using Shinshu University Computer Center and Computer Center, University of Tokyo.

References
1. T.M.Roberts and K.C.Rockey, A mechanism solution for predicting the collapse load of slender plate girders when subjected to in-plane patch loading, Proc. ICE,2(2),155-175(1977)

2. S.Shimizu, S.Yoshida and H.Okuhara, An experimental study on patch-loaded web plates, Proc. of ECCS Colloquium on Stability of Plate and Shell Structures, 85-94(1987)

3. S.Shimizu, H.Yabana and S.Yoshida, A new collapse model for patch-loaded web plates, J. Construct. Steel Research, 13,61-73(1989)

4. S.Shimizu, S.Horii and S.Yoshida, The collapse mechanisms of patch loaded web plates, J. Construct. Steel Research, 14,321-337(1989)

5. ECCS TWG 8.3, Behaviour and design of steel plated structures, Edited by Dubas,P. and Gehri,E., Applied Statics and Steel Structures, ETH Hönggerberg, Zurich(1986)

6. Japan Road Association, Specifications for Highway Bridges (JSHB), Maruzen, Tokyo(1980)

PARAMETRIC STUDY ON PLATE GIRDERS SUBJECTED TO PATCH LOADING

Spinassas I., Raoul J., Virlogeux M.

Service d'Etudes Techniques des Routes et Autoroutes
46, avenue Aristide Briand BP 100
92223 Bagneux Cedex
France

Abstract

The aim of this paper is to present the results of a parametric study dealing with the verification of webs of bridge plate girders during the launching. The study is based on the use of finite element program (SYSTUS) which takes into account material and geometric non linearities. The validity of the program has been verified by a comparison with experimental tests. At the end of this study a simple formula for the ultimate load is proposed which includes the main parameters of the problem.

1. Introduction

Currently composite bridges are very competitive in France and the construction of bridges with large spans becomes very usual. They are erected by launching and in certain cases the plate girders are launched with a concrete deck slab. The reactions of the launching rollers on the plate girder can be very large. The launching requires a special checking of the web under such loading to avoid the failure. This type of loading is more known as "patch loading" and it has been studied during the past 30 years. Empirical and theoretical studies [1][2] have been done and formulae have been proposed although we have not yet a satisfactory response to the behaviour of the plate girder under patch loading. This is due to the many parameters of the problem, which make the cost of experimental studies very important.
Recently the finite element method is widely used for the investigation of such type of structures and we can obtain more informations about its behaviour influenced by many parameters like the problem studied here.
The main purpose of the present study is to clarify the behaviour of a web panel subjected to in plane patch loading and the influence of the main parameters on the failure load.

2. Modelisation

The study is based on the use of finite element program
(SYSTUS Version 230). The element used is a shell three-node
element with six degrees of freedom per node. To obtain this
element the simple approach is used which superimposes a
plate bending element and a plane-stress membrane element.
The membrane element uses a linear polynomial to approximate
the displacement fields and the bending element uses a
cubical. This last one element was based on the Kirchhoff
plate theory. The Code SYSTUS uses the classical theory of
plasticity with the von Mises yield condition and
Prandtl-Reuss flow rule. The stress is evaluated at nine
stations through the thickness. For the geometric
non-linearity, the total Lagrangian formulation was chosen
which uses the 2nd Piola-Kirchhoff stresses and Green-Lagrange
strains.
The modelised girder has been tested at the Laboratoire
Central des Ponts et Chaussées (L.C.P.C.) in 1986 [3]. The
general details of the test specimen are given in figure 1.
The web has an aspect ratio of 1.4 and a slenderness of
212. The girder was simply supported at its ends and loaded
at its centre by four loads applying through a half-cylinder,
which was similar to launching rollers.

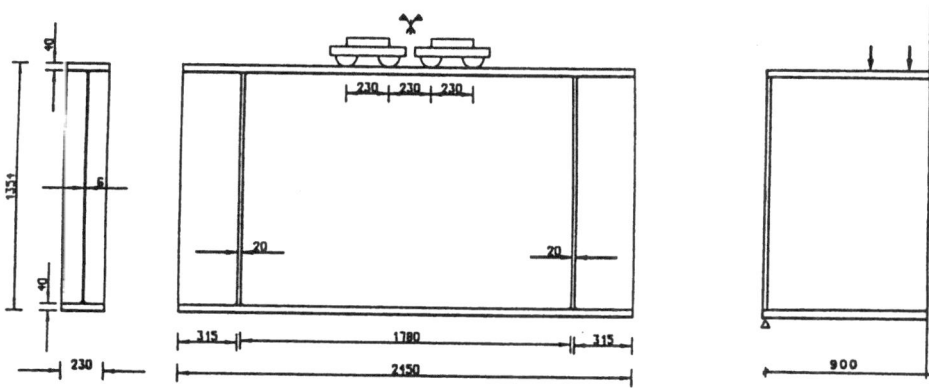

Figure 1 : Dimensions of the test panel

We have modelised the above described test specimen using
the triangular shell elements with a finer mesh under the
loads for a better precision of the web behaviour (figure
2). We have introduced an initial out of plane web deflection
whose the shape is circular with an amplitude w_o equal to
5 mm. This is reasonably representative of practical
imperfections. Four nodal loads were applied on the upper
flange to simulate the edge load. By symmetry reasons, we
have modelised the half panel.

193

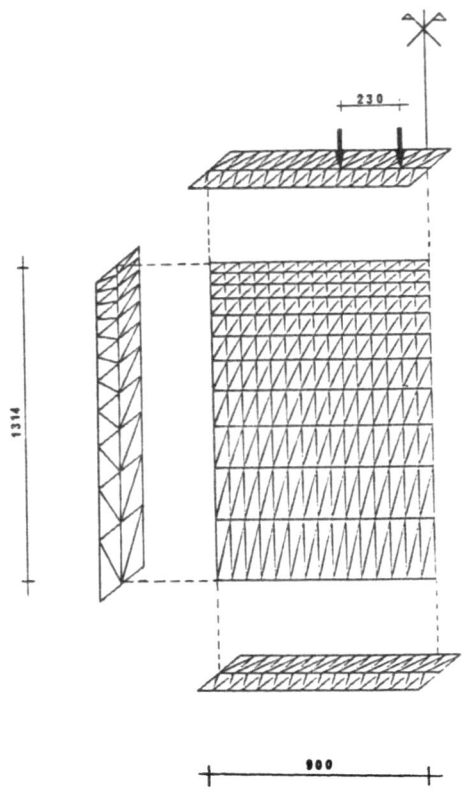

Figure 2 : Finite element model

3. Results of analysis

The element is divided in three parts corresponding to the three Gauss points. When the maximum resistance is reached at a point, the corresponding part becomes black (fig.3). The first plastic zone appears at the top of the panel just under the flange due to the effect of normal forces (AF); with the amplification of the out of plane deflection a second plastic zone appears later at a distance h/4 of the upper flange under the effect of the bending moment in the web (DE). The two yield lines are linked with a third inclined yield line (AD). The width of these three plastic zones increases with the load. This failure mechanism is the same as the one observed in the test at the LCPC.The measured ultimate load of the girder tested was equal to 650 kN and the load obtained by the finite element analysis was equal to 680 KN (i.e.+ 4,6 %).Out-of-plane deflections at the vertical axis of the panel calculated and measured are shown

194

in figure 4 under incremental loads. The calculated deflections wére smaller than the corresponding measured deflections.

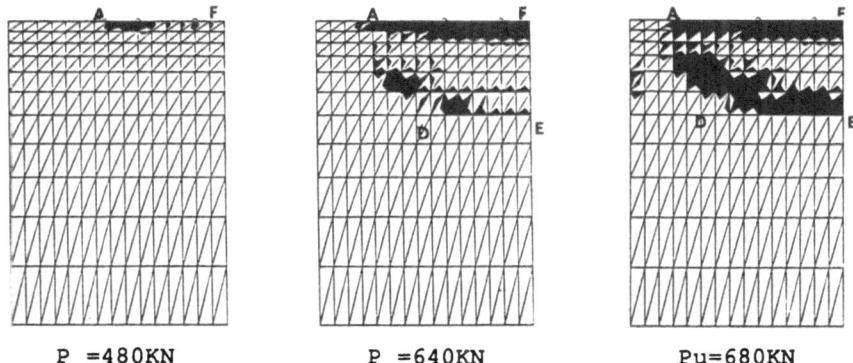

| P =480KN | P =640KN | Pu=680KN |

Figure 3 : Evolution of the plastification

We think that could be due to the residual stresses that have been neglected, and which make that the plastification comes sooner. Nevertheless the calculated deflection shape across the central section of the web is similar at this observed during the test. As it can be noted the point corresponding to the maximum deflection moves from mid height to the upper part of the web. The distance to the upper flange is reduced to approximately h/2 to h/4

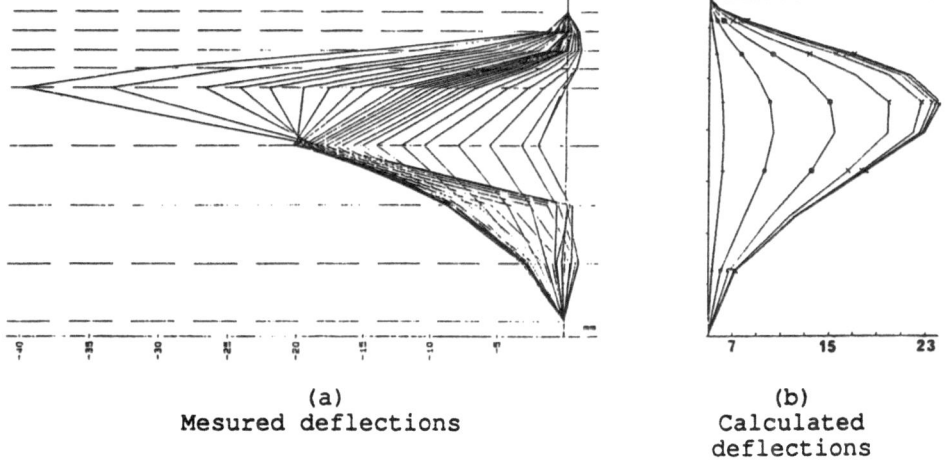

(a)
Mesured deflections

(b)
Calculated deflections

Figure 4 : Deformation of the web

195

4. Parametric study

With the model which has been used for the comparison with the experimental result, a parametric study has been carried out to clarify the influence of the main parameters on the ultimate behaviour of the beam on the launching rollers. The main parameters (fig.5) were :

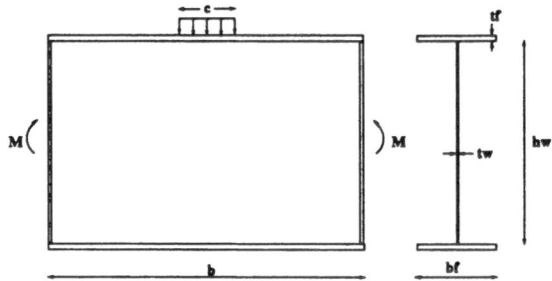

Figure 5 : Notation of the parameters

- the dimensions of the loaded flange (t_f, b_f)
- the web thickness (t_w)
- the bending moments (σ_b)
- the yield stress of the web (σ_w)
- the aspect ratio of the panel between vertical stiffeners (α)
- the initial out-of-plane deflection w_o
- the length of the loading patch (c)

4.1. Dimensions of the flange

i. flange thickness
We have analysed the influence of the loaded flange thickness upon the ultimate load for a panel. Figure 6 evidences that we always have three lines of plastification; when the loaded flange becomes very thick the whole panel is yielded almost throughout its length. The plastified zone at a distance from the upper flange equal to h/4 is larger for t_f=60mm than for t_f=20mm. This is due to the fact that the vertical compressive stresses in the web is more widely distributed with a thick flange than with a thinner one; in this last case the plastification is more localised under the loads. The principal stress distributions given by figure 7 confirm this analysis. The more effective distribution permitted by an increase of the flange thickness improves the ultimate resistance.

196

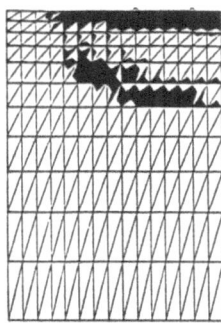 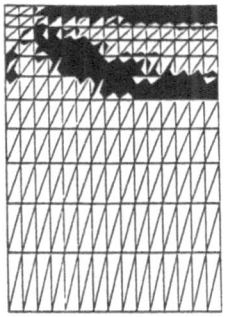

Pu=628KN   Pu=718KN
tw=6mm   tw=6mm
tf=20mm   tf=60mm

Figure 6 : Influence of the flange thickness on the plastification of the web

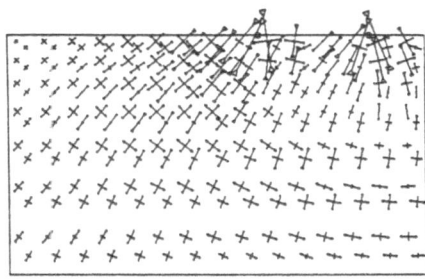 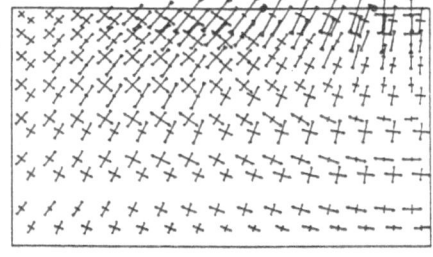

Pu=628KN   Pu=718KN
tw=6mm   tw=6mm
tf=20mm   tf=60mm

Figure 7 : Influence of the flange thickness on the principal stresses distribution in the upper half part of the web

To quantify we have plotted the ultimate loads (Pu) as a function of the flange thickness for several values of the web thickness (Fig.8a) and for different values of aspect ratio (Fig. 8b). These figures show that the increase of the flange thickness is less and less efficient as the web thickness decreases.

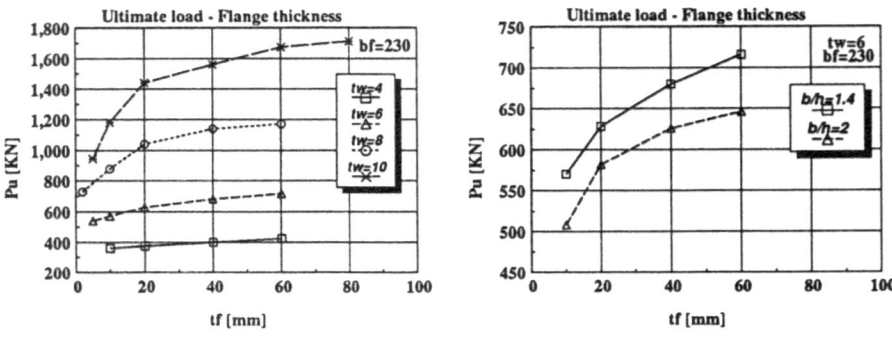

Figure 8 : Variation of ultimate load with the flange thickness

ii. Width of flange

In figure 9 we can see that an increase of the width of the flange from 230 mm to 300 mm does not seem to have a significant influence on the ultimate load.

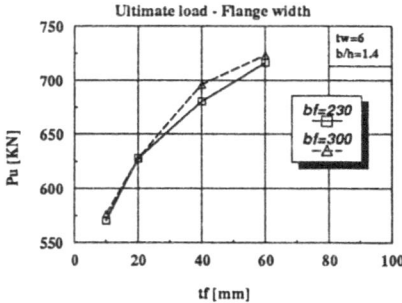

Figure 9 : Variation of ultimate load with the width of the flange

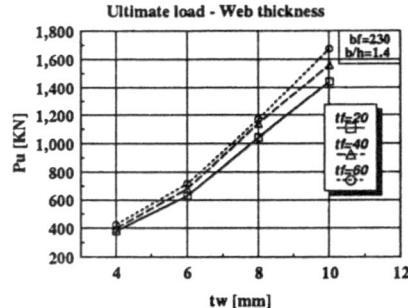

Figure 10 : Variation of the ultimate load with web thickness

4.2. Thickness of web

This parameter seems to influence significantly the ultimate load (figure 10), but we can not find many experimental results concerning this parameter and mainly web thickness at 6 mm and more. Nevertheless for a given thickness of the flange, the increase of the web thickness concentrates the plastification in the region of the load (Fig. 11). The orientation of the principal stresses confirms that with an increase of the web thickness the stresses are concentrated under the load (Fig. 12).

198

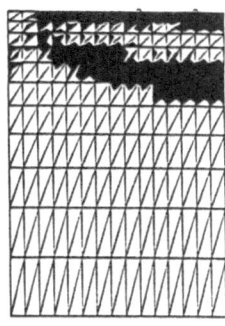 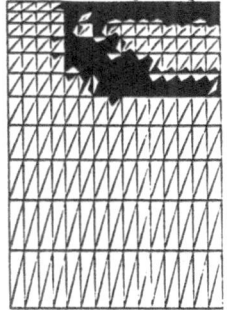

Pu=398KN
tw=4mm
tf=40mm

Pu=1560KN
tw=10mm
tf=40mm

Figure 11 : Influence of the web thickness on the
plastification of the web

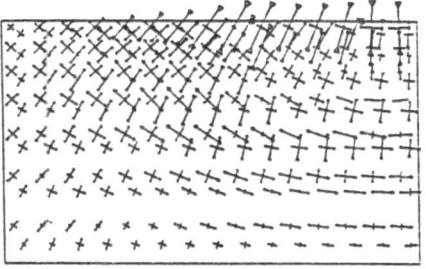 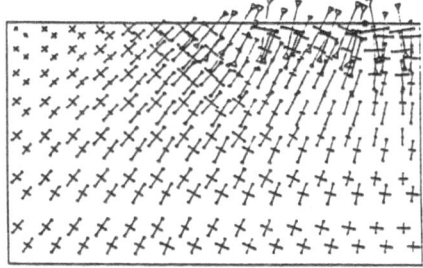

Pu=398KN
tw=4mm
tf=40mm

Pu=1560KN
tw=10mm
tf=40mm

Figure 12 : Influence of the web thickness on the principal
stresses distribution in the upper half part of the web

4.3. Bending moments

We have computed the ultimate load when the panel is subjected to a bending moment and we have investigated its influence. Bergfelt [1] has proposed a reduction factor equal to

$$\left[1-\left(\frac{\sigma_b}{\sigma_w}\right)^2\right]^{0.125}$$

and Roberts [2] has proposed a larger factor given by

$$\left[1-\left(\frac{\sigma_b}{\sigma_w}\right)^2\right]^{0.5}$$

More recently DUBAS and TSCHAMPER [7] proposed another factor based on their tests equal to

$$\left[1.1-\left(\frac{\sigma_b}{\sigma_w}\right)^2\right]^{0.5}$$

In the figure 13 we have compared our computed values to the three proposed factors and we observed a good agreement with the factor proposed by ROBERTS.

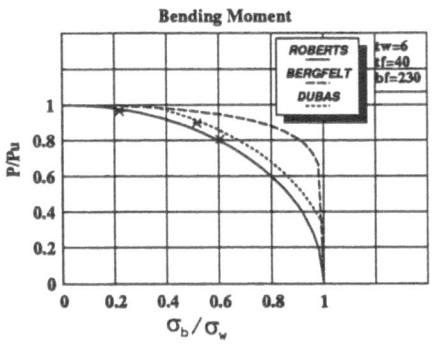

Figure 13 : Influence of the bending moment on the ultimate load

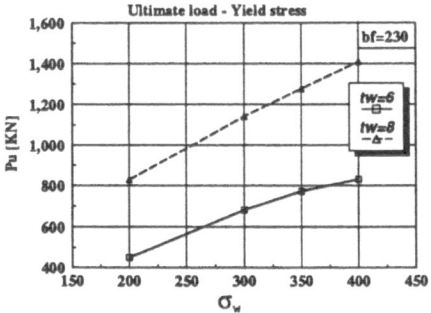

Figure 14 :Variation of the ultimate load with the web yield stress

4.4. Yield stress of web

Figure 14 shows how the ultimate load varies with the web yield stress for two given values of the web thickness (6 mm and 8 mm). In the commonly proposed formulae the ultimate load varies by the square root with the web yield stress. Our analysis seems to lead to a less conservative exponent.

4.5. Aspect ratio of the panel

The influence of the aspect ratio on the ultimate load has been investigated. First for a panel with a given web thickness (tw = 6mm) with different values of the loaded flange thickness and then for a panel with a given loaded flange thickness with different values of the web thickness. From the results of the analysis (figure 15) it can be observed that the increase of the aspect ratio reduces the ultimate load. In our model, due to the position of the supports the increase of the aspect ratio produces a greater bending moment in the panel; more generally a small aspect ratio favorises a better transmission of the patch loading to the stiffeners, but the influence of the aspect ratio seems rather limited.

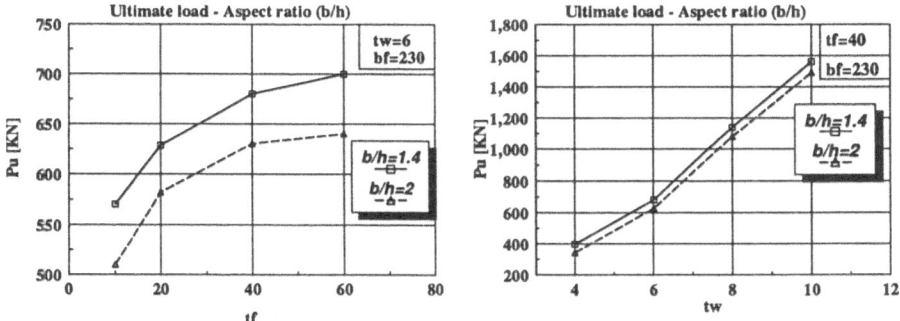

Figure 15 : Influence of the aspect ratio on the ultimate load

4.6. Initial out of plane deflection

The initial out of plane deflection w_o reduces the ultimate load of the panel. As we have seen, the panel failure occurs with a large lateral deflection of the web. The existence of an initial out of plane deflection initiates the lateral deflection and produces an earlier failure. However, the reduction of the ultimate load seems limited (Table 1)

bending moment	tf [mm]	tw [mm]	bf [mm]	h [mm]	wo [mm]	Pu [KN]
no	40	6	230	1274	3	688
no	40	6	230	1274	5	680
no	40	6	230	1274	9	656
no	60	6	300	1274	2	780
no	60	6	300	1274	5	723
yes	40	6	230	1274	5	604
yes	40	6	230	1274	13	572

Table 1

201

4.7. Length of loading patch

Our analysis (figure 16) obviously shows that the ultimate resistance of the panel is increased when the loads are more widely distributed.

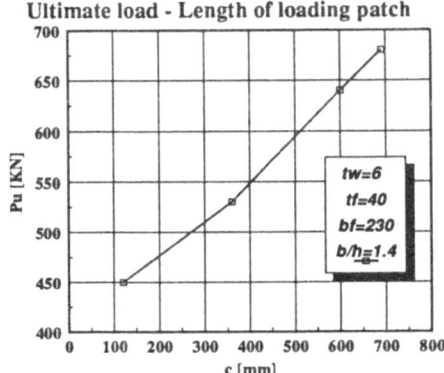

Figure 16 : Variation of the ultimate load with the length of the loading patch

5. Proposed formula

The important difference that we have observed between the experimental results and the different existing formulae leads us to build a new formula based on our parametric analysis. We propose :

$$P_u = 1.3 I_f^{0.06} \ t_w^{1.53} \ c^{0.23} \ \sigma_w^{0.7} \ E^{0.3} \tag{1}$$

We have compared the ultimate loads evaluated from this formula in comparison with 115 tests included in the Background Document for EUROCODE 3. The statistical comparison gives a standard deviation (s) equal to 0.140 and a correlation coefficient (ρ) equal to 0.983. We consider this result as extremely satisfactory.

6. Conclusion

With the help of the finite element method, considering steel plastification and geometric non linearity, we have analysed the influence of the main parameters of the patch loading problem. The results are presented as a form of different curves giving the influence of each parameter on the ultimate load. Finally a simple formula is proposed, which summarise the effects of the different parameters and which seems in a good agreement with the experimental results.

7. References

1. ECCS/TGW8.3 Behaviour and design of steel plated structures, Edited by P.DUBAS and E.GEHRI ECCS no44 1986, Section 4.6, 121-126.

2. ROBERTS T.M., Slender plate girders subjected to edge loading, Proc. Instn Civ. Engrs., Part 2, 1981, 805-819

3. GALEA Y., GODART B., RADOUANT I., Test of buckling of panels subjected to inplane patch loading, Stability of plate and shell structures, Intern Colloquium, Belgium, 6-8 April 1987, Dubas P., Vandepitte D., 1987, 65-71

4. SYSTUS, Manuel Théorique, Framatome, Paris 1989

5. SEDLACEK G., UNGERMANN D., Background document (5.06) on the evaluation of test results on web crippling in EC3, April 1989.

6. SPINASSAS I., RAOUL J., Comportement d'une âme métallique soumise à une charge appliquée dans son plan, Annales des Ponts et Chaussées, no 51, 3e trimeste 1989, 3-9.

7. DUBAS P., TSCHAMPER H., Interaktion von Einzellast und Biegung an Trägern hoher Stegschlankheit, Eidgenössisches Verkehrs und Energiewirtschaftsdepartement Bundesamt für Strassenbau, no 177 Zürich, Mai 1989.

LOAD-INTRODUCTION RESISTANCE OF COLUMN WEBS IN STRONG AXIS BEAM-TO-COLUMN JOINTS

R. MAQUOI and J.P. JASPART
University of Liège,
Quai Banning, 6
4000   Liège
Belgium

Abstract

Present paper is aimed at presenting the results of the study performed at the University of Liège on the deformability and the resistance of column web panels in strong axis ·beam-to-column joints, when subjected to transverse loads carried over by the beam(s).
Formulae for the assessment of the ultimate resistance and stability (buckling and crippling) of the column webs are proposed; they are based on the conclusions of a parametrical study which is briefly described.
Present paper is restricted to the beam-to-column joints with H or I hot-rolled sections.

1. Joint deformability components

The two following sources of deformability of a strong axis beam-to-column joint have to be distinguished :

a) the deformation of the connection associated to the deformation of the connection elements (end plate, angles, bolts,...), to the slip, to the column flange deformation and to the local deformation of the column web in the tension and compression zones;
b) the deformation of the column web under shear associated mostly to the common presence of forces, equal and opposite, in tension and compression, carried over by the beam(s) and acting on the column web at the level of the joint.

The load-introduction deformability of a column web panel is defined as the component of the connection deformability relative to the local deformation of the column web in the tension and compression zones of the joint (respectively a lengthening and a shortening).

Figure 1 illustrates schematically these definitions in the particular case of a joint between one beam and one column. The deformation of the ABCD column web panel (figure 1.a.) has to be divided into two parts :
- the transversal effect of the beam flange forces F_b (statically equivalent to the beam moment M_b) results in a relative rotation ϕ between the beam and the column axes (figure 1.b.); this rotation provides a first deformability curve M_b-ϕ;
- the shear effect due to the shear force V_n results in a relative rotation γ between the beam and the column axes (figure 1.c.); this rotation makes it possible to establish a second deformability curve V_n-γ.

2. Numerical investigations

An important parametric study has been realized recently at the Polytechnic Federal School of Lausanne and at the University of Liège. All the results and all the conclusions of this study may be found in [1].

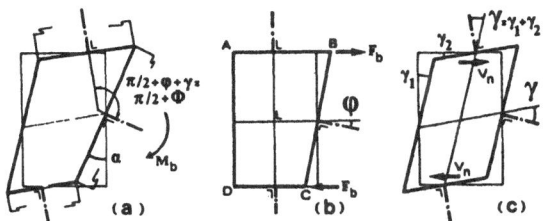

Figure 1 - Global deformation of the column web (a) decomposed into the
load-introduction effect (b) and the shear effect (c).

This study is based on numerical simulations with the non-linear FE-program
FINELG [2] of the loading up to failure of welded beam-to-column joints.
Material and geometrical non-linear effects are taken into account. The
specimens of the chosen joints are analysed in three dimensions by using
"shell" finite elements to model the webs and flanges of the profiles and
"beam" finite elements to model stiffeners. The adopted finite element
meshes are shown on figure 2, respectively for a "T" joint (one column, one
beam) and a "cross" joint (one column, two beams). Complete data may be
found in [1].

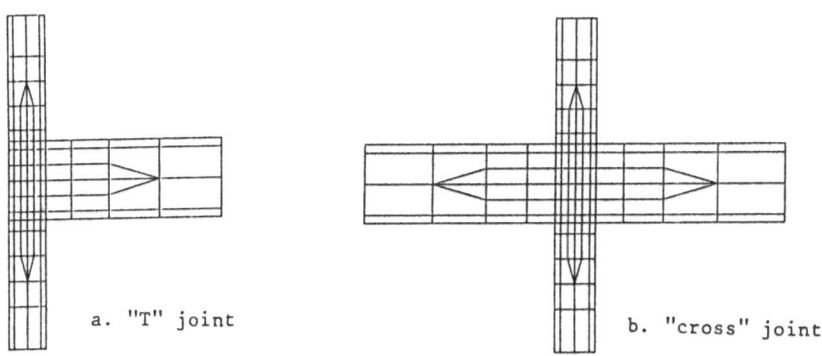

a. "T" joint   b. "cross" joint

Figure 2 - Types of beam-to-column joints studied numerically.

The numerical simulations allow to study the propagation of the plasticity
in the profiles and to observe the exact failure modes.

The good agreement between the numerical simulations and results of
experimental tests on joints is shown in [3].

The moment-rotation curves characterizing the shear deformability and the
load-introduction deformability of the column web panel have been reported
for every simulation.

The following parameters have been taken into account in the parametric
study of the joints :
a) the type of the beam(s);
b) the type of the column;
c) the loading of the joint;
d) the initial out-of-flatness of the column web;
e) the presence or not of transverse stiffeners on the column web.

Only the conclusions relative to the load-introduction behaviour of the
unstiffened column web panels are presented here.

a) The M_b-ϕ curve for a given joint depends on the actual loading of the joint.

Let us subject three similar unstiffened welded joints ("T" arrangement) to different types of loading and let us report the characteristic M_b-ϕ curves in a common diagram (figure 3). A similarity exists only in the elastic range of the web panel behaviour.

The differences between the M_b-ϕ curves in the nonlinear range of behaviour cannot be neglected.

In reality an unstiffened column web panel experiences three types of stresses in its most stressed zone (figure 4) :
- the shear stresses τ;
- the normal stresses σ_n resulting from the compression force and the bending moment in the column;
- the normal stresses σ_i resulting from the introduction of beam loads into the column web.

The load-introduction behaviour of a web panel shall obviously be affected, except in the elastic range, by the relative importance of each of these stresses according to the type of joint loading.

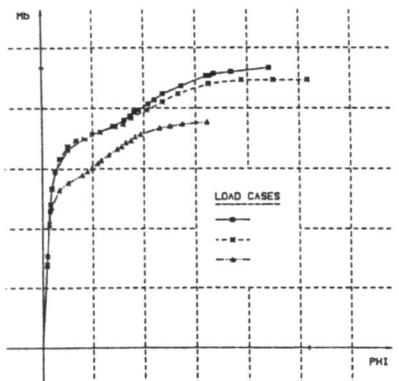
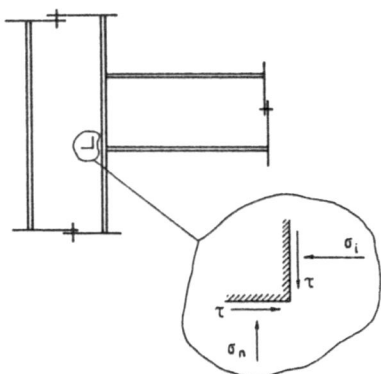

Figure 3 - Characteristic   Figure 4 - Different types of stresses
M_b - ϕ curves   in a web panel

b) It is not allowed to define the plastic capacity of a web subject to transverse loads as this may be done for sheared column web panels : the propagation of the plasticity in the column web transversally loaded does not end indeed in the apparition of an horizontal yield plateau - when the strain-hardening is omitted in the numerical simulation -, but rather in the development, till the attainment of the ultimate buckling load, of a zone characterized by a progressive increase of the resistance and the deformability of the web (figure 5.a.).

It will be consequently referred to the so-called pseudo-plastic moment, M_{bppl}, that is associated to a limit state of the column web panel due to the effect of transverse loads (figure 5.a.). This characteristic load level may obviously be similarly defined for the M_b-ϕ curves relative to "T" joints (see figure 5.b.).

c) The propagation of the plasticity in a web subject to transverse loads is not affected by the presence of σ_n stresses in the web insofar as their maximum value does not exceed a relatively high limit which should have to be explicitly determined.

206

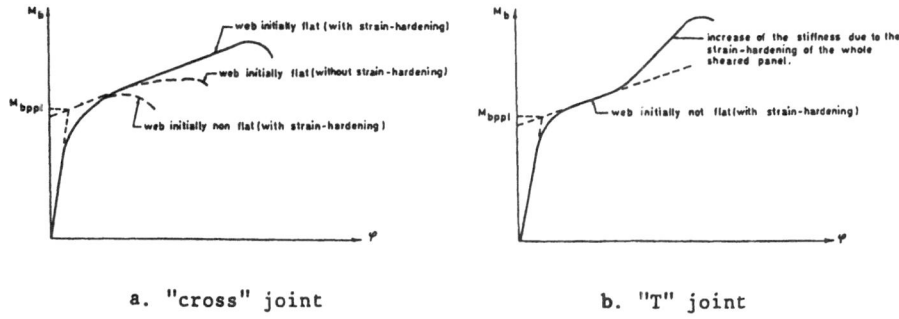

a. "cross" joint   b. "T" joint

Figure 5 - Definition of the pseudo-plastic capacity of a web subject to transverse loads.

This conclusion seems to confirm the result of an experimental study carried out in the Netherlands [4] and which tends to show that the influence, on the "plastic capacity" of the web, of σ_n stresses not exceeding 50 % of the column web yield stress is not significant. ZOETEMEIJER [4] proposes, for larger values of $\sigma_n (\sigma_n > 0.5 \ f_y)$, to reduce the "plastic capacity" by means of a factor e given by :

$$e = 1.25 - 0.5 \ \frac{|\sigma_n|}{f_y} \tag{1}$$

KATO considers for his own [5] that the attaintment of stresses σ_n greater than 0.5 f_y is not of practical interest because of the relative low values of the loads transmitted to the column in order to prevent instability.

d) The amplitude of the column web out-of-flatness influences only the shape of the M_b-ϕ curves of joints whose collapse is linked up to the buckling of the web; this initial out-of-flatness affects the value of the web ultimate buckling load in a significant way but modifies very slightly the deformability of the web as far as the collapse load is not reached (figure 5.a.)

e) The comparison (figure 6) of M_b-ϕ curves relative to a "T" joint (figure 2.a.) and to the corresponding (same column and same type of beam) "cross" joint (figure 2.b.) shows clearly the similarity of both web behaviours in the elastic range (it is not possible to compare the curves in the non elastic range on account of the different stresses interacting in the column webs).

This leads to the conclusion that the introduction of transverse loads in a column web constitutes, as far as the stability of the web is not concerned with, a local phenomena limited to the vicinity of the column flanges. The influence of the joint loading on the shape of the M-ϕ curves - as discussed in (a) - is seen to be very significant in this case.

3. Prediction of the web panel deformability curves

Mathematical models for the prediction of the characteristics V_n-γ and M_b-ϕ curves of stiffened and unstiffened column column web panels are proposed in [1]. They have been validated in [6] for joints with welded, extended end plate and flange cleated connections by means of comparisons with numerical and experimental results.

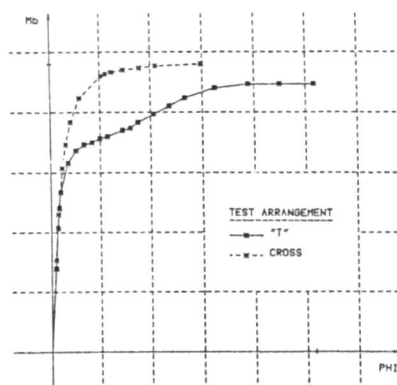

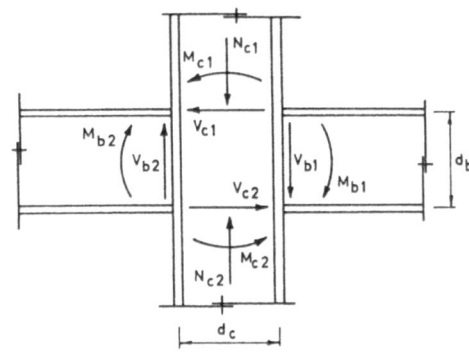

Figure 6 - Comparison between "T"
and "cross" joint
behaviour (M_b-ϕ curves)

Figure 7 - Loading of an interior
joint

4. Assessment of the ultimate strength for unstiffened web panels

The ultimate strength of a column web may be associated to one of the three following types of collapse :

- the shear collapse of the column web panel, V_{nbu}, to which corresponds a moment M_{nbu} in the beam by the following formula (figure 7) :

$$V_n = \frac{M_{b1} + M_{b2}}{d_b} - \frac{1}{2} (Q_{c1} + Q_{c2}) \tag{2}$$

- the excessive yielding of the web under transverse loads, M_{buy};
- the instability of the web under compression transverse loads, M_{bub}.

The ultimate moment M_{bu} in the beam corresponding to the collapse of the column web panel is equal to :

$$M_{bu} = \min (M_{nbu}; M_{buy}; M_{bub}) \tag{3}$$

Shear of the web panel

The ultimate carrying capacity of a sheared web panel is given by :

$$V_{nbu} = \tau^c_u \cdot A_{sh} \tag{4}$$

where A_{sh} represents the column sheared area. τ^c_u is the ultimate shear stress evaluated by means of the von MISES criterion allowing for the interaction between σ_n and τ stresses - it has been shown in [1] that the load-introduction (stresses σ_i) constitutes only a local phenomena which has no direct influence on the global shear behaviour of the web panel - but based on the attainment of the ultimate stress f_u in the column web. τ stresses are simply obtained by dividing the shear force V_n (formula 2) by the column sheared area A_{sh}.

Excessive yielding of the web

The ultimate resistance associated to the excessive yielding of a web subject to transverse loading is given by :

208

$$M_{buy} = \sigma_{iu}^c \cdot s_c \cdot l_p \cdot d_b \qquad (5)$$

σ_{iu}^c, the maximum permissible compression stress in the web, results from the consideration of the local interaction between σ_i and τ stresses by means of the von MISES criterion based, as for the shear resistance, on the attainment of the ultimate stress f_u in the web. Expression of σ_i stresses versus the beam moment are proposed in [6] for the column web of joints with welded, end plate and flange cleated connections. d_b is the distance between the compression and tensile forces acting on the web. l_p is the length of diffusion, given in [6], of the compression and tensile forces through the connection elements, the flange and the radii of fillet of the column. s_c is the column web thickness.
This ultimate resistance is associated to the collapse of the tensile zone or the compression zone according to which is the most determining.

Web instability

The instability of the web under compression affects either the whole depth of the column (web buckling) or the region located just under the beam flange (web crippling).
The associated instability load is given by :

$$M_{bub} = M_{bb} \nless M_{bppl} \qquad (6.a.)$$

with :

$$M_{bb} = \sqrt{M_{by} \, M_{bcr}} \qquad (6.b.)$$

M_{bppl} is the pseudo-plastic moment of the web evaluated by formula (5) in which σ_{iu}^c is simply replaced by σ_{iy}^c which is based on the attainment of the yield stress f_y in the web.
M_{by} is the elastic resistance of the transversally loaded web which corresponds to the onset of yielding in the compression or tensile zones of the web. It results from the study of a "beam" (column flange) on an "elastic foundation "(column web) [1,6]; only the interaction between σ_i and τ stresses must be considered.

The elastic linear instability load of the web, M_{bcr}, is expressed as :

$$M_{bcr} = (h_c - 2.t_c).s_c.d_b.k. \frac{\pi^2 \cdot E}{12.(1-\nu^2)} \cdot \left(\frac{s_c}{h_c - 2.t_c}\right)^2 \qquad (7)$$

h_c is the total depth of the column while t_c and s_c are respectively the flange and web thicknesses of the column.

The values to give to the k coefficient as well as the physical explanation of formula (7) will be found in the following sub-sections.

It is important to mention that the buckling strength M_{bb}, contrary to the pseudo-plastic moment M_{bppl}, is strongly dependent on the initial out-of-flatness of the web (see for instance figure 5.a.). The values of this imperfection have been chosen on base of rolling tolerances and those of the k coefficient, which will be defined in the next sub-sections, have been calibrated accordingly. Measurements of the actual values of the initial out-of-flatness are generally found lower than the tolerances; that results in a too safe theoretical approximation of the actual buckling load. Numerical simulations have shown that the variation of the buckling load may reach 25 - 30 % according to the value of the out-of-flatness.
This has not to be forgotten when comparisons between theory and experiments are performed.

4.1. Interior HE columns

For what concerns the interior columns (figure 2.b.) with HE sections, it has been shown [1] that the pseudo-plastic moment, M_{bppl}, constitutes a lower bound value for the web buckling load.
This may be easily explained by referring to figure 8.

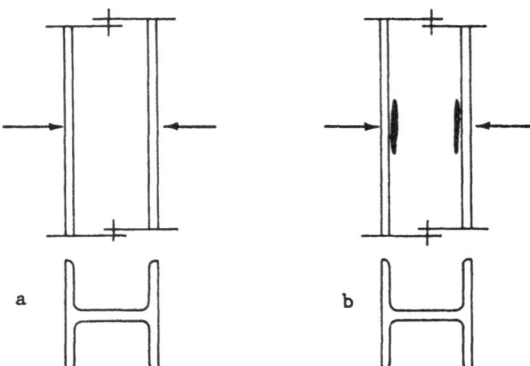

Figure 8 - Support conditions of the column web according to the load level

In the elastic range of the web behaviour (figure 8.a.), the column web may be considered as rigidly connected to the column flanges : its instability load is consequently high. The increase of the loading and the resultant yielding of the web just under the column flanges (figure 8.b.) lead to a modification of the support conditions for the web and consequently to a considerable decrease of the buckling load.
A collapse by instability in the elastic range behaviour has never been encountered, even for relatively slender column webs (HEA) and the numerical simulations have shown that the actual buckling load is always greater or equal to the pseudo-plastic moment M_{bppl} which then constitutes a lower bound value for the instability load.

The buckling strength M_{bb} given by formula (6.b.) is consequently based on the assumption that the web is pinned on the column flanges; it has to be compared to M_{bppl} in order to determine the actual buckling load of the web, M_{bub} :

$$M_{bub} = M_{bppl} \text{ if } M_{bppl} \geq M_{bb} \qquad (8.a.)$$

$$M_{bb} \text{ if } M_{bppl} < M_{bb} \qquad (8.b.)$$

The coefficient k which accounts in formula (7) for the type of loading - cross nodes symmetrically loaded - and for the web support conditions - ideal hinges - may be taken equal to 1.0.

4.2. Exterior HE columns

The fact that transverse loads are only applied to one side of the column web increases significantly the buckling strength M_{bb} of the web, whereas the pseudo-plastic moment M_{bppl} is independent of the node arrangement (cross or "T"). The k coefficient in formula (7) will consequently be chosen equal to 2.0 for "T" node.

The resulting formula (6) for the assessment of the web instability load, which has been discussed in the previous sub-section, may be applied in the case of exterior columns with HE sections.

4.3. Extension to IPE columns

Profiles with IPE sections being usually used as beams and not as columns, specific numerical simulations of joints with IPE columns have not been performed. Some test results (joints with end plate connections) are however available and the application of formula (6) has allowed to demonstrate its validity.

It must however be noted that the pseudo-plastic moment M_{bppl} corresponds rather to a web crippling than to a web buckling.

5. Comparison with results of numerical simulations and experimental tests

The here above proposed formulae have been validated in [6] through comparisons with numerous results of numerical simulations on welded joints and of experimental tests on joints with bolted connections.

6. Concluding remark

More information relative to the use of the formulae and also to the application of the mathematical model for the prediction of the characteristic deformability curves of joints may be found in the reports [1] and [6] which may be afforded to any interested people. An extensive comparison between the available numerical and experimental results and previously existing formulae for the assessment of the column web resistance and stability is included in [1].

7. References

1. ATAMAZ SIBAI, W. and JASPART, J.P., 'Etude du comportement jusqu'à la ruine des noeuds complètement soudés' Internal Report IREM, Polytechnic Federal School of Lausanne, N° 89/7, and MSM, University of Liège, N° 194.
2. De VILLE de GOYET, V., RICHARD, C., BERTARINI, I., TAQUET, F., BOERAEVE, P. and LOGNARD, B., 'FINELG, non linear finite element analysis program - user's manual", MSM, University of Liege, Fourth-up-to-date edition, 1990.
3. ATAMAZ SIBAI, W. and FREY, F., 'Numerical simulation of the behaviour up to collapse of two welded unstiffened one-side flange connections', Connections and the Behaviour, Strength and Design of Steel Structures, Elsevier Appl. Sc. Publ., 1988, pp. 85-92.
4. ZOETEMEIJER, P., Report 6-80-5, Stevin Laboratory, Delft, February 1980.
5. KATO, B. 'Beam-to-column connection research in Japan', Journal of Structural Division, ASCE, vol. 108, n° ST2, February 1982, pp. 343-360.
6. JASPART, J.P., 'Study of the shear and of the load-introduction deformability and resistance of column web panels in strong axis beam-to-column joints', Internal Report MSM, n° 202, University of Liège, April 90.

211

BEHAVIOUR OF THIN-WALLED COLD-FORMED WIDE PROFILES UNDER CONCENTRATED LOADING

Jiří STUDNIČKA, PhD

Czech University of Technology
Department of Steel Structures
Thákurova 7, Praha 6
Czechoslovakia

1. INTRODUCTION

Cold-formed steel multiple web deck sections are used frequently in building construction. Where these sections are supported by end or interior bearing plates, or are subjected to a concentrated load at some point in the span, failure of the deck can occur by web crippling.

The ultimate web crippling load capacity is a function of a number of parameters, namely, the web slenderness ratio H, the inside bend radius ratio R, the bearing length ratio N, the angle of web inclination θ (Fig. 1) and the yield strength

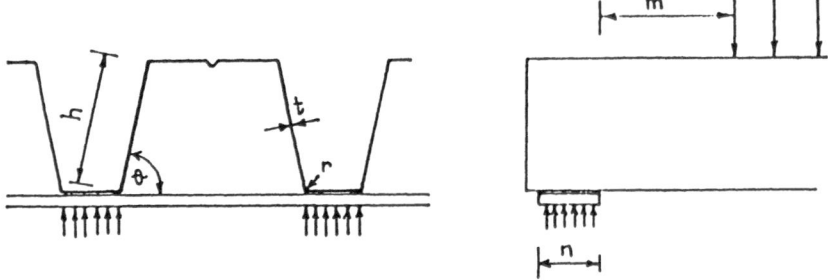

Fig. 1 Cross section of multiple web deck. $H = \frac{h}{t}$, $R = \frac{r}{t}$, $N = \frac{n}{t}$

of steel F_y. Load capacity also depends on the distance between the bearing edges of adjacent opposite concentrated loads or reactions. When this distance m is greater than 1.5 h_w one flange loading is considered to occur. Two flange loading occurs when the clear distance between bearing edges is equal to or less than 1.5 h_w. When uniform loading acts on beam, the case is always classified as one flange loading.

The substantial difference is between end and interior reaction loadings, of course.

The objective of this study was to determine, by experimental way, the load resistance of multi-web deck sections subjected to end and interior reaction loading P, as shown in Fig. 2. An experimental test program was to provide experimental data so that existing methods of computation could be compared and evaluated.

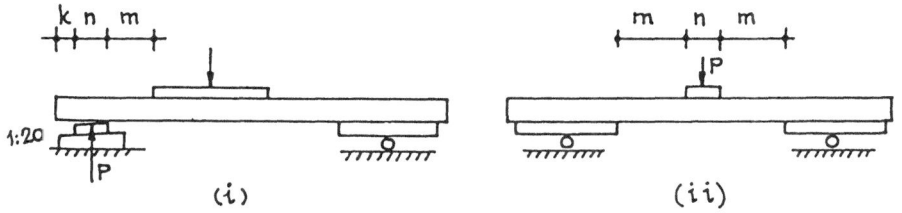

Fig. 2 Test setup for (i) end reaction, (ii) interior reaction

2. TEST PROGRAM

The test program was designed to encompass the most important parameter variations that influence the web crippling resistance of multi-web deck sections subjected to end and interior reaction loading. Test specimens were obtained from a Czechoslovak manufacturer, VSŽ Košice. The two types of specimens are shown in Fig. 3. Spreading of the webs during loading was, for some tests, prevented by transverse tie rods which were bolted to the bottom flange of profiles, see Fig. 4

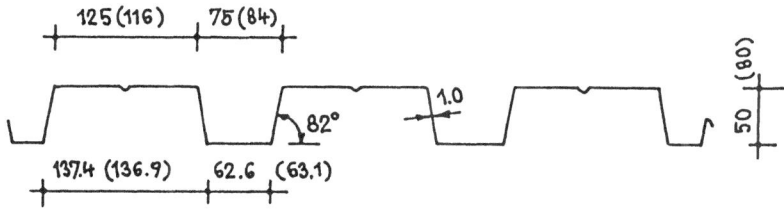

Fig. 3 Tested wide deck sections. VSŽ type 12 002 (type 12 102)

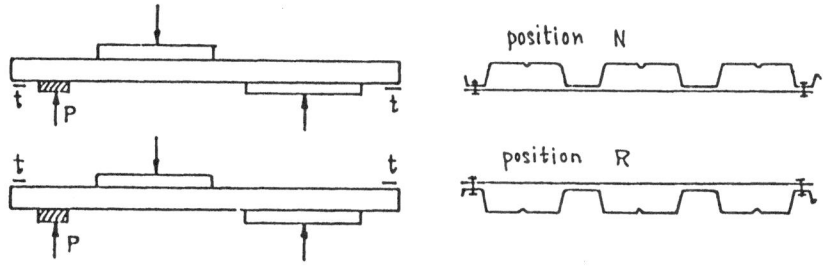

Fig. 4 End support test. Positions of profile. Tie rods t
        bolted to the flange

The specimens were simple supported at the ends and the load was applied at the centre. Relatively large end bearing plates were used for these tests to ensure that failure would occur at the interior load position and vice versa for end load position. The distance m from the edge of the interior bearing plate to the interior edge of the exterior bearing plate was changed to obtain the conditions for one flange and two flange loadings. However, the main changeable dimension was the bearing width n, see Fig. 1. For tests at end support an inclined steel bearing plate was used following the ECCS Recommendations. The deck specimens were tested in both positions N and R, see Fig. 4.

The test load P_t was taken either as the largest load the specimen was able to sustain (after which a sudden decrease in load was experienced), or as the load which a residual deformation of 1.0 mm developed, whichever was lesser.

3. TEST RESULTS

3.1 Interior support

Altogether 40 specimens were tested. The following comments may be made regarding the results:
  (i)   test loads are not substantially different for the both positions of deck
  (ii) test results are almost linearly influenced by bearing width
  (iii) test loads for specimens with ties are greater that without ties.

3.2. End support

Altogether 76 specimens were tested. The same three comments as for the interior support results can be made and a further two can be added:
  (iv) influence of distance between edges of adjacent opposite concentrated loads on test results is very small
  (v) when distance between end of deck and end of bearing plate is increased, the test load also increases. But the influence is not very strong.

4. COMPARISON OF TEST LOADS AND COMPUTED LOADS

The AISI Specification and the Canadian code were used to compute the ultimate web crippling load P_c. Since the method of permissible stress is used in AISI Specification, multiplying all formulae by safety factor 1.85 was necessary. As the Canadian standard is in limit state terms, there no corrections were necessary except for the missing resistance factor ϕ_s. Terms in the AISI Specification were converted to SI units.

Measured yield strength F_y = 260 MPa was used for computation of P_c, of course.

4.1 Interior support

Only one flange loading was examined for interior support tests. Comparison of test results P_t with ultimate computed web crippling loads P_C (using AISI - 1986 Specification and Canadian - 1984 Standard) is shown in Fig. 5 and Fig. 6. The solid line in the figures represents perfect correlation ($P_t = P_c$); the dashed lines are + 20 % limits which are acceptable scatter limits for tests of this type, based on previous research. One can see that better conformity was reached in Fig. 6 in which almost all results are within + 20 % limits over full range of the bearing lengths. However, the mean value of P_t/P_c of 0.856 with coefficient of variation 0.106 is somewhat disturbing. A small number of tests makes it impossible to draw more firm conclusions.

4.2 End support

Both one and two flange loadings were used in these tests. Comparison of test results P_t with computed load P_c shown that the test results are in reasonably harmony with predicted values of web crippling end support load, according to both American and Canadian code. However, increasing of distance k over boundary value 1.5 h_w did not increase the test load in such way to be comparative with the computed load for interior support.

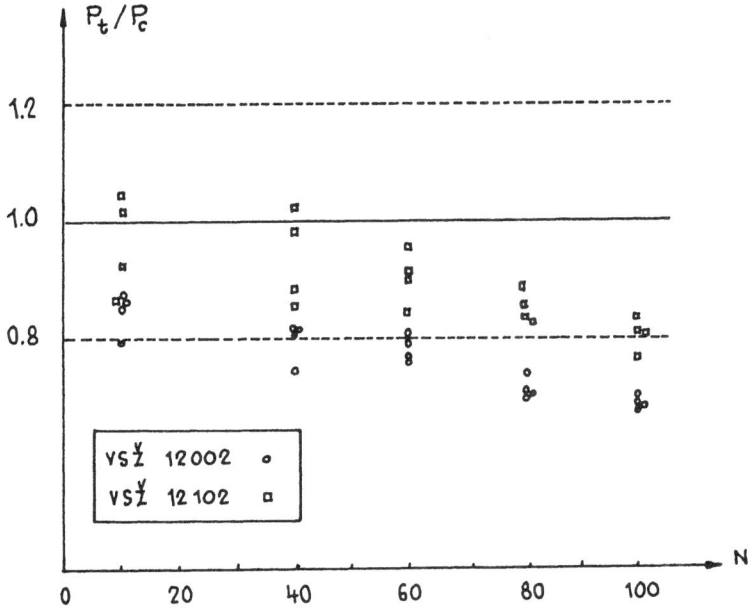

Fig. 5 Test load P_t vs. computed load P_c. Interior support
        (P_c according to AISI Specification)

215

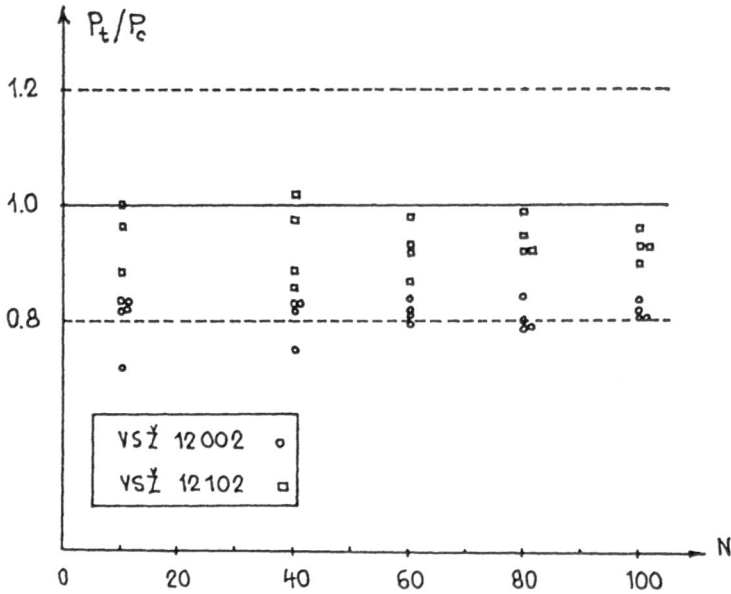

Fig. 6   Test load P_t vs. computed load P_c . Interior support
(P$_c$ according to CAN code)

Using the test results, a new slightly modified expression
was developed, following the Canadian Standard

$$P = 10\ t^2 F_y (\sin\ \Theta)(1-0.1\ \frac{F_y}{230})(1-0.1\sqrt{R})(1-H/500)(1+K/15H)$$

$$(1+0.005N) \tag{1}$$

Eq. 1 predicts the web crippling capacity for end reaction
for both one and two flange loading, if $K \leqslant 3H$. The other
limitations from CAN3 - S136 are still fully valid.

Comparison of eq. 1 with test data in Fig. 7 shows very
good prediction of those expression. This is indicated by the
values of the mean (1.018) and coefficient of variation (0.136)
of P_t/P_c.

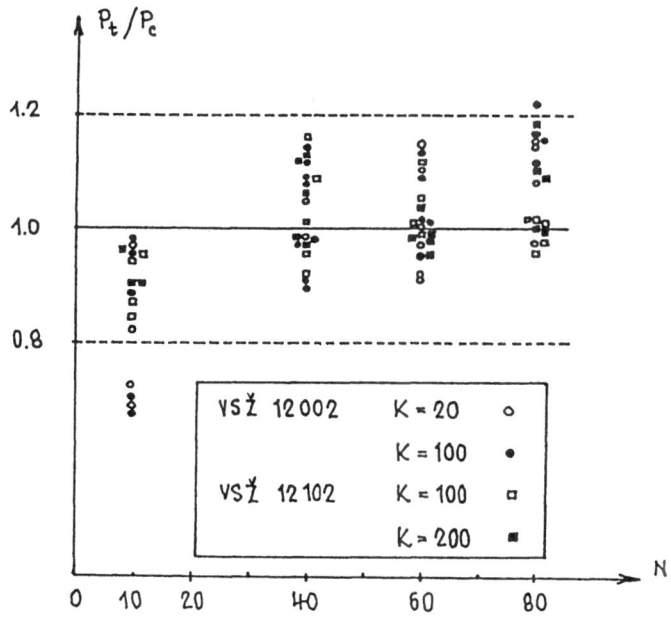

Fig. 7 Test load P_t vs. computed load P_c . End support .

5. CONCLUSIONS

Based on the comparison of the results of 40 interior support and 76 end support tests with different methods of computation, the following conclusions were made:

(i) The Canadian Cold-Formed Steel Specification predicted the web crippling capacities reasonably well for interior support conditions.

(ii) The use of Eq. 1 resulted in a better prediction of the web crippling capacity for end support than any of the existing methods. Eq. 1 is equally valid for one flange loading and two flange loading.

(iii) Tests with end support did not confirm the Canadian and AISI provision that the increasing of the distance from edge of the bearing plate to the end of multi-web deck can bring substantial increasing of web crippling capacity.

A FINITE ELEMENT FORMULATION FOR THE ANALYSIS OF LOCAL EFFECTS

A VENKATESH and J. JIROUSEK

LSC, Department of Civil Engineering,
Swiss Federal Institute of Technology,
CH-1015 Lausanne, Switzerland

Abstract

The paper reviews the progress achieved in the analysis of geometry and load dependent local effects, based on an alternative FE formulation known as the hybrid-Trefftz (HT) FE model. A brief description with examples presents the developments which have lead to a simple p-adaptive analysis of structures with stress-raisers (e.g., holes, cracks and corners) in presence of multiple load cases, including local concentrated loads.

1. Introduction

Stress-concentration due to geometry or load dependent local effects in structures is usually realized through cumbersome mesh-refinement in the conventional FE analysis, which often raises questions about element distortion. In the case of load governed stress-concentration, the problem is further complicated when one deals with different load cases, each with differently located concentrated loads. Such a situation which is common in practical structural analysis, calls either for a mesh-refinement nearly everywhere in the structure or an expensive remeshing for each load case separately. Such problems have successfully been tackled by the alternative FE formulation based on the Trefftz's approach [1] in which, unlike in the conventional FE model, the assumed displacement field $u_e = \overset{\circ}{u}_e + \sum_{j=1}^{m} \phi_j\, c_j = \overset{\circ}{u}_e + \Phi_e c_e$ (Trefftz field) exactly satisfies the non-homogeneous governing differential equations $Lu + \overline{b} = 0;\ L\phi_j = 0\,,\ L\overset{\circ}{u}_e = \overline{b}$ on Ω_e. The interelement continuity and the boundary conditions are satisfied in a weak sense through the use of an auxiliary conforming frame field which can (thanks to the exact nature of the Trefftz field) directly be defined along the element boundaries [2,3,5-11,13-15].

The first Trefftz type element has been presented in 1977 (Jirousek and Leon [2]) who also observed that, the Trefftz's functions represent an optimal expansion basis for hybrid elements in which, the interelement continuity need not a priori be satisfied. Different alternative variational formulations for HT elasticity elements have been presented in 1978 (Jirousek [3]). Such elements were found to be substantially more accurate than the conventional ones presenting the same nodal DOF. Delicate mathematical aspects of some non-structural Trefftz type element formulations have been studied in 1978 by Babuska, Oden and Lee [4]. Various Trefftz type formulations were also later (1985) investigated by Zielinsky and Zienkiewicz [5] based on the classical problem of Poisson's equation. Though the principle of developing p-version finite elements based on the Trefftz's approach has been described in 1982 (Jirousek and Teodorescu [6]), the actual demonstration of its efficiency has only been presented in 1988 (Jirousek [7]). The variable accuracy frame field is defined in terms of a basic polynomial field (associated with element corner nodes) augmented by additional hierarchic modes (fig.1 plane elasticity element) whose amplitudes are associated with the element mid-side nodes as optional DOF (numbering M). The minimum number m of the Trefftz's functions is linked to the total number N of unknowns of the hierarchic frame field (DOF at corner nodes + DOF at mid-side nodes) through the customary stability condition of the hybrid elements. The optimal number m of Trefftz's functions from the performance point of view needs to be determined by numerical experimentation. Very recently (1988), a one-step adaptive approach (Jirousek and Venkatesh [8]) which applies indiscriminately to single as well as multiple load cases, based on a simple pointwise stress error estimator (Jirousek and Venkatesh [9] - 1989), has been proposed for the HT finite element analysis.

By definition, the Trefftz's basis accounts for the applied loads through the particular solution to the governing differential equations and this has been found to lead to, an accurate analysis of problems with stress-concentration due to localized loads [2,3,6,10,11]. It has also been recognized (Rao, Raju and Murthy [12] - 1971, Lin and Tong [13] - 1980, Jirousek and Teodorescu [6] - 1982, Piltner [14] - 1982, Gerhardt [15] - 1984) that various geometry dependent local effects such as due to holes, cracks and corners can be accounted for efficiently by making use of special purpose Trefftz's functions which a priori satisfy the boundary conditions at the stress-raisers e.g., stress-free condition along the contour of a hole. It must be noted that Gerhardt [15] deals with the anisotropic materials while Refs. 6, 12, 13, 14 are restricted to isotropy. In Refs. 12-15, special purpose Treffftz's elements are developed in order to isolate the stress-raisers while, the rest of the structure is discretized by conventional finite elements. As opposed to this commonly used approach, Jirousek and Teodorescu [6] use a single Trefftz type element formulation for the whole domain where the finite elements simply choose, special or ordinary functions from the library of optional Trefftz's functions depending upon whether the element is near a stress-raiser or not. A unified concept for hybrid-Trefftz finite element analysis now seems to exist in which - (1) crude quasi-uniform meshes of p-converging finite elements are used, (2) the precision of the finite elements can be increased by modifying a single variable in the input, (3) strong geometry dependent local effects are treated by choosing suitable Trefftz's bases from the library of optional functions, (4) stress-concentration due to concentrated loads can be analyzed by using meshes which are independent of the load location, (5) numerous load cases which include a large number of arbitrarily located concentrated loads can be analyzed without remeshing.

The formulation of the HT finite elements has been presented in a number of papers [6,7,10,11] to which the reader should refer for clarity.

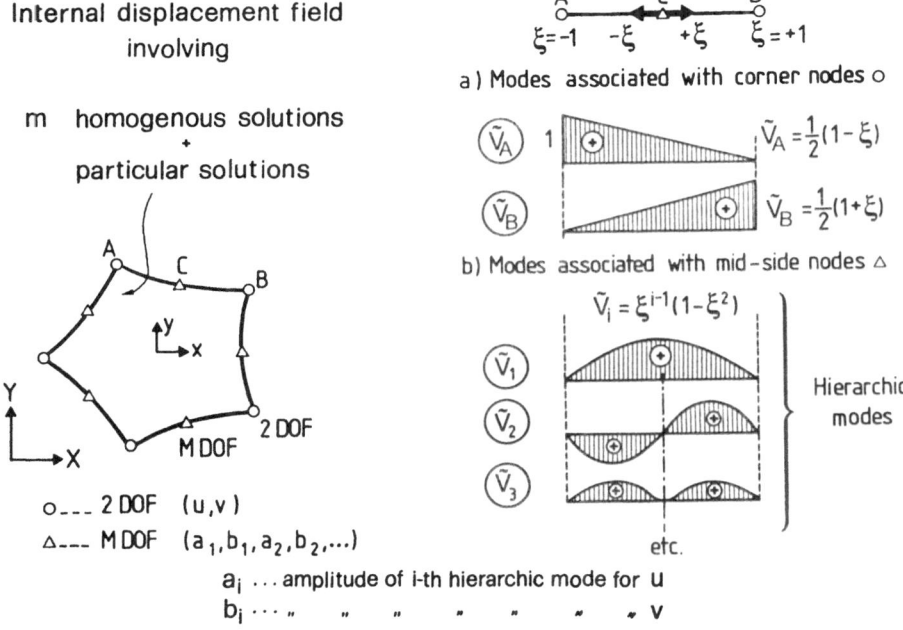

Fig. 1. p-version plane elasticity element.

219

2. Superior Accuracy of HT Elements

The HT elements externally resemble conventional finite elements, in performance though, they are much superior. This is clearly demonstrated through the example of an internally pressurized thick cylinder, analysed by using quadrilateral elements (fig.2) based on the conventional displacement model as well as the HT model. The exact solution for the problem is known and consequently the convergence of the energy norm

$$e = \sqrt{1 - U_{FE}/U_{Exact}} \tag{1}$$

could be studied comparatively against the effective number Na of the DOF in the problem. The fact that the real advantage of the HT model lies in its p-extension is very much evident from this comparison.

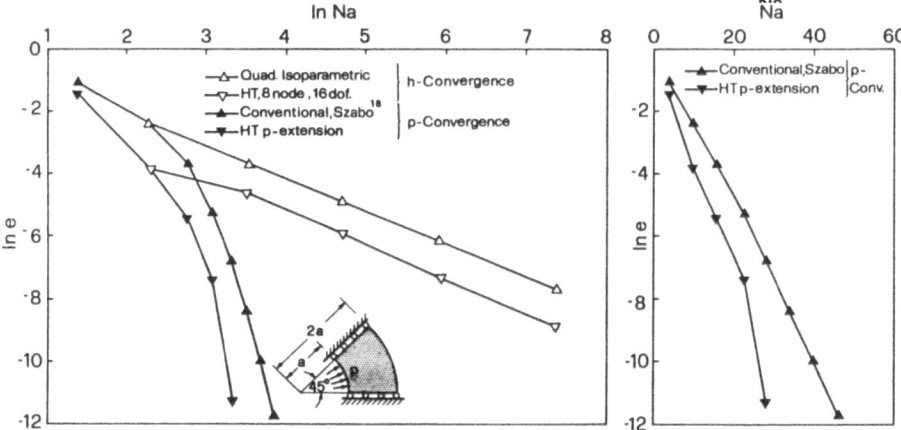

Fig. 2. Comparison of HT and conventional FE (thick cylinder under internal pressure $v = .3$).

3. Accurate Prediction of Local Effects

The key in the HT model to precise prediction of stress-concentration under local loads is the use of suitable load dependent particular part of the Trefftz field. For point loads we have e.g.

1. Thin plate bending

$$\overset{\circ}{w}_e = P\, r_p^2\, Ln\, r_p^2 / 16\, \pi D, \quad r_p^2 = (x-x_p)^2 + (y-y_p)^2 \qquad \text{Point load in the interior} \tag{2a}$$

2. Plane elasticity

$$\overset{\circ}{u}_e = \frac{P_x}{\pi Et}\left\{\frac{-(1+v)x^2}{(x^2+y^2)} + Ln\left(\frac{x^2+y^2}{a^2}\right)\right\} + \frac{P_y}{\pi Et}\left\{\frac{-(1+v)xy}{(x^2+y^2)} - (1-v)\, Tan^{-1}\, (y/x)\right\}$$

Point load at the boundary (3a)

$$\overset{\circ}{v}_e = \frac{P_x}{\pi Et}\left\{\frac{-(1+v)xy}{(x^2+y^2)} - (1-v)\, Tan^{-1}\, (y/x)\right\} + \frac{P_y}{\pi Et}\left\{\frac{-(1+v)y^2}{(x^2+y^2)} + Ln\left(\frac{x^2+y^2}{a^2}\right)\right\}$$

$$\overset{\circ}{u}_e = \frac{P_x(1+v)}{4\pi Et}\left\{\frac{(3-v)}{2}\, Ln\left(\frac{x^2+y^2}{a^2}\right) - \frac{(1+v)x^2}{(x^2+y^2)}\right\} - \frac{P_y(1+v)}{4\pi Et}\left\{\frac{(1+v)xy}{(x^2+y^2)}\right\}$$

Point load in the interior (3b)

$$\overset{\circ}{v}_e = \frac{P_x(1+v)}{4\pi Et}\left\{\frac{-(1+v)xy}{(x^2+y^2)}\right\} + \frac{P_y(1+v)}{4\pi Et}\left\{\frac{(3-v)}{2}\, Ln\left(\frac{x^2+y^2}{a^2}\right) - \frac{(1+v)y^2}{(x^2+y^2)}\right\}$$

220

The particular solutions for loads of practical importance such as line loads and patch loads can be derived by integrating the particular solutions for point loads. In the case of thin plate bending, the following particular solutions have been derived [16].

1. Line Load (fig.3b)

The vector of generalized displacements $\overset{\circ}{v}_e$ and conjugate tractions $\overset{\circ}{T}_e$ are :

Case i

$$\overset{\circ}{v}_e = \left\{ \begin{array}{c} \overset{\circ}{w}_e \\ -\dfrac{\partial \overset{\circ}{w}_e}{\partial n} \\ -\dfrac{\partial \overset{\circ}{w}_e}{\partial s} \end{array} \right\} = \dfrac{pd^2}{16\pi D} \left\{ \begin{array}{c} d\left| (\chi + \dfrac{\chi^3}{3}) \left\{ \ln \dfrac{d^2(1+\chi^2)}{a^2} - \dfrac{2}{3} \right\} - \dfrac{2\chi}{3} + \dfrac{4}{3} \, \text{artan} \, \chi \right|^{\theta_B}_{\theta_A} \\[2mm] 2\left| n_x\chi + \left\{ n_x\chi + n_y \dfrac{(1+\chi^2)}{2} \right\} \ln \dfrac{d^2(1+\chi^2)}{a^2} + 2n_x \, \text{artan} \, \chi \right|^{\theta_B}_{\theta_A} \\[2mm] 2\left| n_y\chi + \left\{ n_y\chi - n_x \dfrac{(1+\chi^2)}{2} \right\} \ln \dfrac{d^2(1+\chi^2)}{a^2} + 2n_y \, \text{artan} \, \chi \right|^{\theta_B}_{\theta_A} \end{array} \right\} \quad (4a)$$

$$\overset{\circ}{T}_e = \left\{ \begin{array}{c} \overset{\circ}{Q}_n \\ \overset{\circ}{M}_n \\ \overset{\circ}{M}_{ns} \end{array} \right\} = \dfrac{p}{8\pi} \left\{ \begin{array}{c} 4\left| n_x \, \text{artan} \, \chi + \dfrac{n_y}{2} \ln \dfrac{d^2(1+\chi^2)}{a^2} \right|^{\theta_B}_{\theta_A} \\[2mm] -d\left| \left\{ (1+\nu)\chi + 2(1-\nu) \, n_x n_y \right\} \ln \dfrac{d^2(1+\chi^2)}{a^2} + \right. \\[1mm] \left. + 2\left\{ (1+\nu) - (1-\nu) \, n_y^2 \right\} \text{artan} \, \chi + (1-\nu) \, \chi \right|^{\theta_B}_{\theta_A} \\[2mm] d(1-\nu) \left| (n_x^2 - n_y^2) \ln \dfrac{d^2(1+\chi^2)}{a^2} - 2n_x n_y \, (2 \, \text{artan} \, \chi - \chi) \right|^{\theta_B}_{\theta_A} \end{array} \right\} \quad (4b)$$

Case ii

$$\overset{\circ}{v}_e = \dfrac{p}{16\pi D} \left\{ \begin{array}{c} \dfrac{a^3\rho^3}{9} \, (2 - 3 \ln \rho^2) \left|^{\rho_B}_{\rho_A} \right. \\[2mm] a^2\rho^2 \cos\gamma \left|^{\rho_B}_{\rho_A} \right. \\[2mm] a^2\rho^2 \sin\gamma \left|^{\rho_B}_{\rho_A} \right. \end{array} \right\} \quad (4c)$$

$$\overset{\circ}{T}_e = \dfrac{p}{8\pi} \left\{ \begin{array}{c} 2 \cos\gamma \ln \rho^2 \left|^{\rho_B}_{\rho_A} \right. \\[2mm] -\left| \left\{ (1+3\nu) - 2(1+\nu) + 2(1-\nu) \cos^2\gamma \right\} a\rho + (1+\nu) \, a\rho \ln \rho^2 \right|^{\rho_B}_{\rho_A} \\[2mm] 2(1-\nu) \, a\rho \sin\gamma \cos\gamma \left|^{\rho_B}_{\rho_A} \right. \end{array} \right\} \quad (4d)$$

221

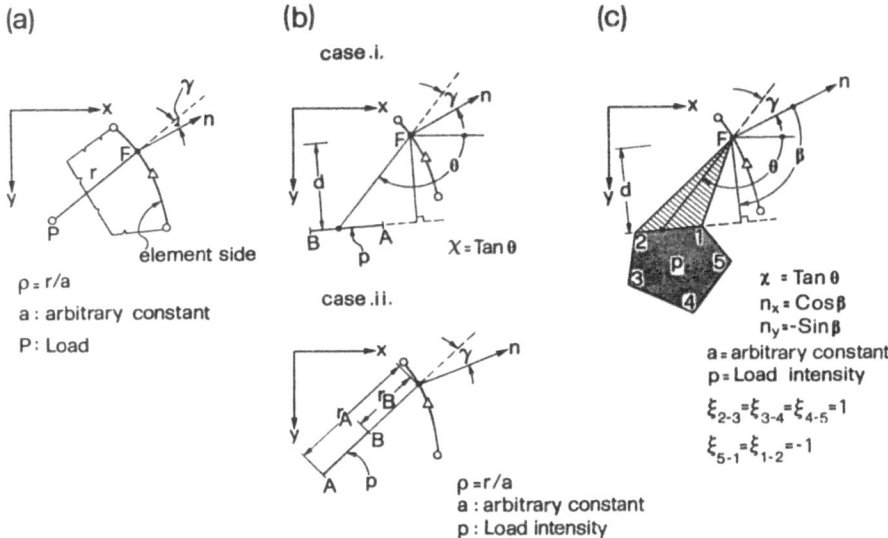

Fig. 3. Details of (a) point load, (b) line load, (c) patch load.

2. Patch load (fig.3c)

In the case of a uniformly distributed polygonal patch load, the particular solution is obtained by summing the contributions of the triangles formed by each side of the load polygon with the integration point at the finite element boundary. It is clear that each of these triangles is associated with a sign which depends on the sense of numbering of the load polygon corners w.r.t. the point of integration. For any one of these triangles, the expressions of $\overset{\circ}{v}_e$ and $\overset{\circ}{T}_e$ are the following:

$$\overset{\circ}{v}_e = \frac{pd^3}{24\pi D}\xi\left\{\begin{array}{l} \dfrac{3d}{8}\left|\left\{\ln\dfrac{d^2(1+\chi^2)}{a^2}-\dfrac{7}{6}\right\}\left(\chi+\dfrac{\chi^3}{3}\right)-\dfrac{2\chi}{3}+\dfrac{4}{3}\text{ artan }\chi\right|_{\theta_A}^{\theta_B} \\[1em] \left|\left\{\dfrac{(1+\chi^2)}{2}n_y+\chi n_x\right\}\ln\dfrac{d^2(1+\chi^2)}{a^2}-\dfrac{5\chi}{3}n_x+2n_x\text{ artan }\chi-\dfrac{\chi^2}{3}n_y\right|_{\theta_A}^{\theta_B} \\[1em] \left|\left\{\dfrac{(1+\chi^2)}{2}n_x-\chi n_y\right\}\ln\dfrac{d^2(1+\chi^2)}{a^2}+\dfrac{5\chi}{3}n_y-2n_y\text{ artan }\chi-\dfrac{\chi^2}{3}n_x\right|_{\theta_A}^{\theta_B} \end{array}\right\} \quad (5a)$$

$$\overset{\circ}{T}_e = \frac{pd}{8\pi}\xi\left\{\begin{array}{l} 4\left|n_x\text{ artan }\chi+\dfrac{n_y}{2}\ln\dfrac{d^2(1+\chi^2)}{a^2}\right|_{\theta_A}^{\theta_B} \\[1em] -d\left|\left\{1-(1-\nu)n_y^2\right\}(2\text{ artan }\chi-\chi)+\left\{\dfrac{1}{2}(1+\nu)\chi+(1-\nu)n_xn_y\right\}\ln\dfrac{d^2(1+\chi^2)}{a^2}\right|_{\theta_A}^{\theta_B} \\[1em] d(1-\nu)\left|\dfrac{(n_x^2-n_y^2)}{2}\ln\dfrac{d^2(1+\chi^2)}{a^2}-n_xn_y(2\text{ artan }\chi-\chi)\right|_{\theta_A}^{\theta_B} \end{array}\right\} \quad (5b)$$

222

By defining the particular solutions for local loads in the global coordinates, one can use finite element meshes which are entirely independent of the load location. For computational economy, the definition can be semi-global i.e., the contribution of a local load is assigned only to those elements which are in its immediate (automatically determined by the element subroutine) neighbourhood. Moreover, the strong stress-concentration is realized as a part of full structure analysis and no secondary analysis of the neighbourhood is required. In figure 4, the efficient handling of line loads in plane elasticity and square patch loads in plate bending is shown. The change in load location does not call for any change in the finite element mesh. In the rhombic plate problem, the obtuse corners have been isolated by HT plate elements which use special purpose Trefftz's functions for simply supported corners (a case clubbing the load and the geometry governed local effects). The smoothness of the contours of transverse shear resultants shows the level of the precision in the analysis.

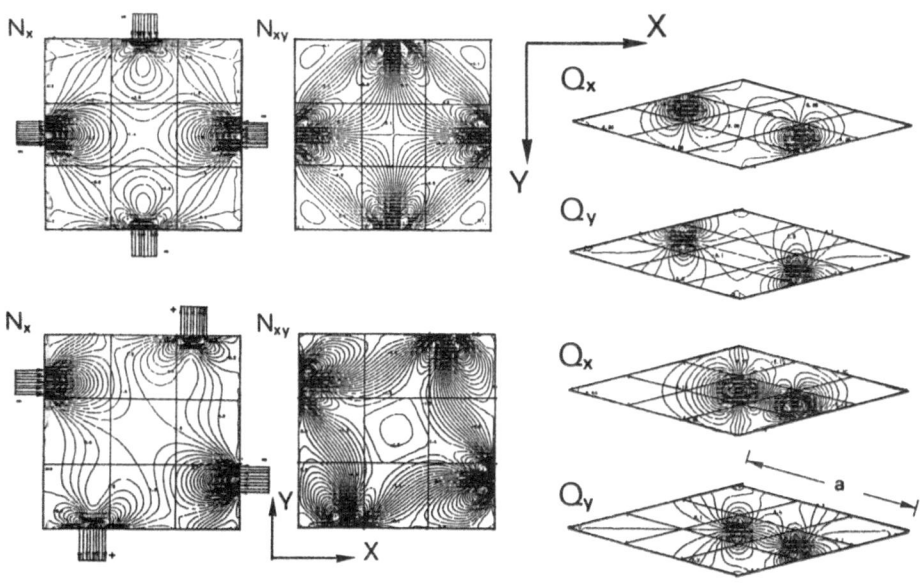

Fig. 4. a. Plate under in-plane line loads. b. Simply supported rhombic plate under transverse patch loads, $v = .3$.

Though the direct use of suitable particular solutions for local concentrated loads accurately predicts the neighbouring stress-concentration, the excessive smoothness of the polynomial frame field in comparison, adversely affects the precision when strongly concentrated loads are placed close to the element boundaries. This problem can be bypassed through an alternative strategy similar to the one suggested in the past for conventional elements [17]. The solution is taken as the sum of a known singular part (defined over an infinite or a semi-infinite domain) and a regular FE solution, the aim of the latter being to restore the actual boundary conditions. The two steps are combined into a single one and it reflects in the load vector calculation where, the singular displacement term is replaced by an 'equivalent' regular term (consistent with the frame field) which at nodes yields the same displacements as would the singular one. The effect of such a replacement is that in global FE solution the singular part of the solution is well represented while the trouble of matching the singular particular solution and the polynomial frame field is eliminated. For more details Refs. 11, 17 should be consulted.

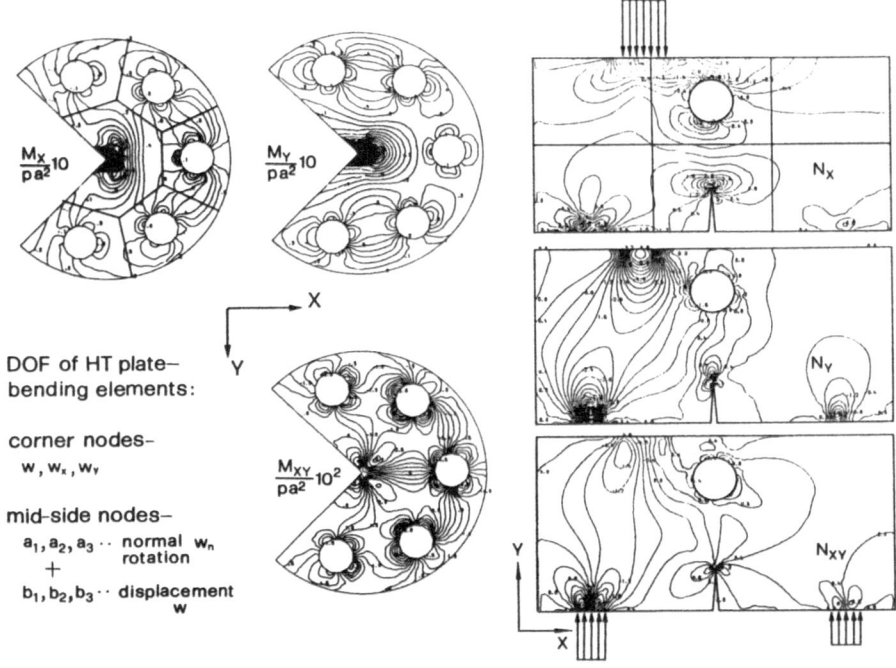

Fig. 5. a Simply supported 3/4 circular plate with holes under uniform transverse load.
b Wall-beam with a hole and a crack under inplane line loads.

As mentioned in the introduction, the geometry governed local stress-concentration can be represented by making use of optional special purpose Trefftz's functions which apart from exactly satisfying the governing differential equations, also a priori satisfy those boundary conditions which are actually responsible for steep gradients e.g., the stress-free sides of cracks or
contours of cutouts [6,10-15]. In Fig. 5 examples involving such geometry governed stress-
concentration are shown. In the zones of acute stress concentrations i.e. near corners, cracks
and cutouts, special purpose Trefftz's functions have been used while the rest is discretized by
HT elements whose Trefftz's functions only satisfy the governing differential equations.

4. Error Estimation and Adaptive Analysis

The estimation of error and the development of a simple adaptive analysis is relatively
uncomplicated with the HT finite elements since, the error in each element is concentrated in a
narrow band along the element sides [8,9] and a large superconvergent zone remains inside the
element. The worst error occurs always at the element corners and an effective error measure
can be established by comparing the results calculated by the FE program at the element corner
nodes and the "best guess" there of the exact results. The "best guess" is calculated based on
the method known as "Krigeing" which uses the results calculated at a regular grid of sampling
points in the superconvergent zones of the HT elements. The value at any point (e.g., corner
node) is written in terms of the results at a certain number of sampling points in the
neighborhood as:

$$M_c = \sum_i M_i \lambda_i \tag{6}$$

where the weights λ_i can be obtained by solving a small number of equations (for details see Ref. 8 where other possibilities have also been discussed for establishing the best guess of the exact solution for error estimation). The error measure which has proved to be very useful is

$$\eta = \frac{\Delta M_{eqv}}{M_{eqv}\,(100\%)} \times 100 \tag{7}$$

defined in terms of the generalized stress M_{eqv} (for plate bending)

$$M_{eqv} = \sqrt{2DU_0} = \sqrt{[M_x^2 + \nu M_x M_y + M_y^2 + 2(1+\nu)M_{xy}^2]/(1-\nu^2)} \tag{7a}$$

where U_0 is the strain energy density. Here, ΔM_{eqv} is the generalized stress corresponding to the error in the computed stresses

$$\Delta M_{eqv} = \sqrt{[\Delta M_x^2 + \nu \Delta M_x \Delta M_y + \Delta M_y^2 + 2(1+\nu)\Delta M_{xy}^2]/(1-\nu^2)} \tag{7b}$$

where

$$\Delta M_x = M_x - M_x^{FE}, \quad \Delta M_y = M_y - M_y^{FE}, \quad \Delta M_{xy} = M_{xy} - M_{xy}^{FE}$$

are the differences at a node between the results calculated by the FE program and the results which serve as the best guess of the exact solution. $M_{eqv}(100\%)$ is the generalized stress which is used as a suitably chosen reference e.g, a stress value which is used to calculate the thickness of the plate.

In order to achieve in a very simple way, the solution for which the error measure does not exceed a certain specified value, the following average error measure is introduced:

$$A\nu\eta = \sqrt{\frac{1}{n}\sum_{IE=1}^{NE}\sum_{IC=1}^{NC(IE)}\eta^2_{IE,IC}} \quad \text{with} \quad n = \sum_{IE=1}^{NE} NC(IE) \tag{8}$$

where $\eta_{IE,IC}$ are the relative percentage errors (13) of the raw unaveraged FE results at the element corners. NE is total number of elements in the assembly and NC(IE) are the number of corner nodes in each of the elements. The adaptive solution necessitates the knowledge of the convergence rate in $A\nu\eta$. For plate bending problems for example, the variation of $A\nu\eta$ w.r.t. the total number of DOF (restrained included) on a Log-Log scale yields a near straight line whose slope oscillates about the ratio 8:3. This rate appears to be insensible to the gradients in the exact solution as long as the HT elements which include the geometric singularities use suitable optional special purpose Trefftz's bases and the localized loads are appropriately represented through their respective particular solutions. If a nominal rate of 7:3 rather than 8:3 is assumed in order to account for the fluctuation of the actual convergence rate, one can estimate the p-refinement level required for a specific value of the average error measure by using the following relationship:

$$N_M = N_1 \left[\frac{A\nu\eta_1}{(A\nu\eta)_{Allow.}}\right]^{3/7} \tag{9}$$

where N_1 is the total number of DOF in an initial analysis with the lowest p-refinement level and $A\nu\eta_1$ is the corresponding average error measure. $(A\nu\eta)_{Allow.}$ is the allowable value of the average error measure and N_M the total number of DOF estimated to attain the allowable error measure. Numerical experiments have shown that, a value of 3-5% for the average error measure $A\nu\eta$ is largely sufficient. This is due to the fact hat the error measures $A\nu\eta$ and η are both defined in terms of the results calculated at the element corners where worst predictions can be expected. The corresponding results at points inside the superconvergent zones (the values which are finally used for graphical postprocessing) show errors which are as a rule, one or more orders of magnitude lesser than the η at nodes. The example of a plate with a re-entrant corner and a circular hole subjected to uniformly distributed load (Fig. 6) demonstrates the simple adaptivity described above. The contour lines presented in the figure show the intensity of the stress-concentration in the problem due to geometry-governed local effects.

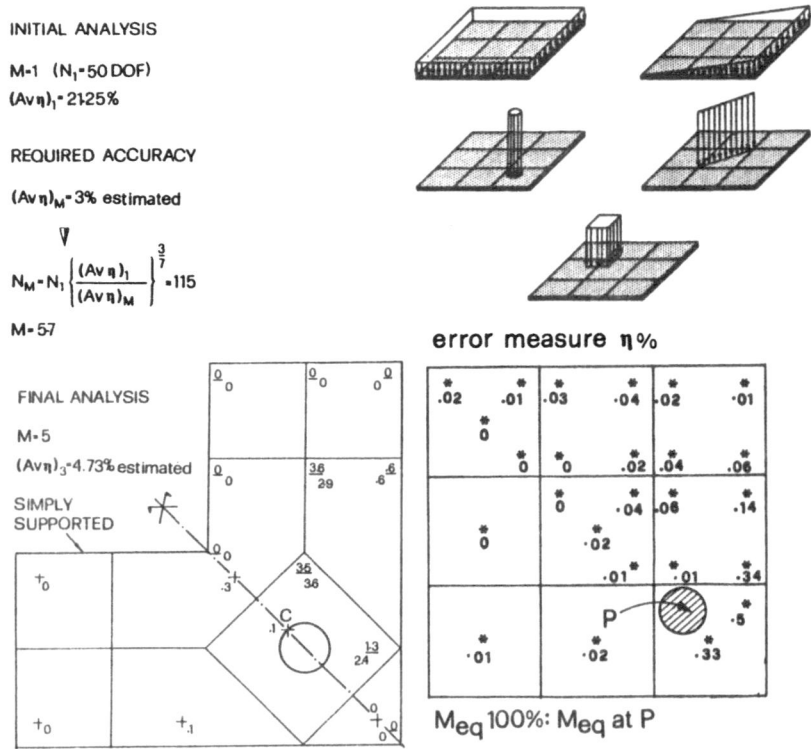

INITIAL ANALYSIS

M-1 $(N_1\text{-}50\,DOF)$

$(Av\eta)_1\text{-}21.25\%$

REQUIRED ACCURACY

$(Av\eta)_M\text{-}3\%$ estimated

$$N_M\text{-}N_1\left\{\frac{(Av\eta)_1}{(Av\eta)_M}\right\}^{\frac{3}{7}}\text{-}115$$

M-57

FINAL ANALYSIS

M-5

$(Av\eta)_3\text{-}4.73\%$ estimated

SIMPLY
SUPPORTED

error measure $\eta\%$

M_{eq} 100%: M_{eq} at P

Fig. 6. Simply supported L-shaped plate with a hole under uniform load.

Fig. 7. Multiple load cases - clamped square plate under transverse loading.

The low sensitivity of the HT FE model to load dependent local effects enables a mesh definition which is independent of the loading specified in several independent load cases. A unique uniform p-refinement level M which ensures for all load cases that the allowable stress error is not exceeded is again obtained in a single step, by simply adjusting the number of hierarchic DOF based on the largest $Av\eta_1$ in the initial solution among all the load cases. The clamped square plate shown in Fig. 7 has been analyzed for five separately taken load cases and, the following initial $Av\eta$ has been observed ($M_{eqv}(100\%)$ = maximum M_{eqv})

Uniform load	12.9%	Hydrostatic load	13.11%
Circular patch load	8.97%	Uniform line load	8.19%
Square patch load	9.75%		

Using (15), the total DOF needed to reduce the largest $Av\eta_1$ of 13.11% to the allowable 3% would be N_M = 136. On selecting M = 5 (total 168 DOF, while M = 3 only leads to 120 DOF), the required accuracy is attained in all cases except in the case of the circular patch load which has an $Av\eta_5$ = 5.86%. The very low errors in the element interiors however show that even this analysis is sufficiently accurate.

226

5. References

1. E. Trefftz. Ein Gegenstück zum Ritzschen Verfahren. Proc. 2nd Intl. Cong. Appl. Mech., Zürich, pp. 131-137 (1926).
2. J. Jirousek and N. Leon. A powerful finite element for plate bending. Comp. Meths. Appl. Mech. Engrg. 12, 1, pp. 77-96 (1977).
3. J. Jirousek. Basis for development of large finite elements locally satisfying all field equations. Comp. Meth. Appl. Mech. Engrg. 14, 1, pp. 65-92 (1978).
4. I. Babuska, J.T. Oden and J.K. Lee. Mixed hybrid finite element approximations of second-order elliptic boundary value problems, Part 2 - Weak-hybrid methods. Comp. Meth. Appl. Mech. Engrg. 14, 1, pp. 1-22 (1978).
5. A.P. Zielinski and O.C. Zienkiewicz. Generalized finite element analysis with T-complete boundary solution functions. Int. J. Num. Meth. Engrg. 21, pp. 509-528 (1985).
6. J. Jirousek and P. Teodorescu. Large finite element method for the solution of problems in the theory of elasticity. Comp. Struc. 15, pp. 575-587 (1982).
7. J. Jirousek and Lan Guex. Hybrid-Trefftz finite element model and its application to plate bending. Int. J. Num. Meth. Engrg. 23, pp. 651-693 (1986).
8. J. Jirousek and A. Venkatesh. Adaptivity in hybrid-Trefftz finite element formulation. Int. J. Num. Meth. Engrg. 29, pp. 391-405 (1990).
9. J. Jirousek and A. Venkatesh. A simple stress error estimator for hybrid-Trefftz p-version elements. Int. J. Num. Meth. Engrg. 28, pp. 211-236 (1989).
10. J. Jirousek. Hybrid-Trefftz plate bending elements with p-method capabilities. Int. J. Num. Meth. Engrg. 24, pp. 1367-1393 (1987).
11. J. Jirousek and A. Venkatesh. Hybrid-Trefftz plane elasticity elements with p-method capabilities. Int. J. Num. Meth. Engrg. To appear.
12. A.K. Rao, I.S. Raju and A.V.K. Murthy. A powerful hybrid method in finite element analysis. Int. J. Num. Meth. Engrg. 3, 3, pp. 389-403 (1971).
13. K.Y. Lin and P. Tong. Singular finite elements for the fracture analysis of v-notched plate. Int. J. Num. Meth. Engrg. 15, pp. 1343-1354 (1980).
14. R. Piltner. Spezielle finite Elemente mit Löchern, Ecken und Rissen unter Verwendung von analytischen Teillösungen. Fortschritt-Berichte der VDI Zeitschriften 1, 96, VDI-Verlag GmbH, Düsseldorf (1982).
15. T. D. Gerhardt. A hybrid/finite element approach for stress analysis of notched anisotropic materials. J. Appl. Meth. Trans. ASME 51, pp. 804-810 (1984).
16. A. Venkatesh and J. Jirousek. Accurate FE analysis of thin plates under concentrated loading. IREM Internal Report 89/6, Dept. of Civil Engineering, Swiss Federal Institute of Technology, Lausanne, June 1989.
17. J. Jirousek. Implementation of local effects into conventional and non-conventional finite element formulations. Local Effects in the Analysis of Structures. Ed. P. Ladevèze. Elsevier, pp. 279-298 (1985).
18. B.A.Szabo. Estimation and control of error based on p-convergence. Accuracy Estimates and Adaptive Refinements in Finite Element Computations. Ed. I. Babuska, O.C. Zienkiewicz, J. Gado and E.R. de A. Oliveira. Wiley, pp. 61-78 (1986).

ACKNOWLEDGEMENT

The research reported in this paper was supported in part by the Swiss National Science Foundation under the Grant No. 20-25269.88.

DISCUSSIONS

EARLY PRAGUE TESTS ON WELDED PLATE GIRDER WEBS UNDER PARTIAL EDGE LOADINGS

M. DRDÁCKÝ

Central Laboratory for Experimental Mechanics,
Institute of Theoretical and Applied Mechanics,
Vyšehradská 49
128 49 Prague 2
Czechoslovakia

1. Introduction

In 1964-65 a serie of 74 test girder panels was investigated in the Institute of Theoretical and Applied Mechanics of the Czechoslovak Academy of Sciences. The experiments were carried out by Vladimír Březina - the man who very substantially developed the Czechoslovak stability research in sixties but because of he left the country, his name was erased from books by former totality regime and his works were not allowed to be referred to. This short communication not only informs about his research into the behaviour of slender webs under partial edge loadings but also reminds the need of his scientific rehabilitation.

2. Test Girders

Three groups of plate girder panels were tested. In each of them the flange and vertical stiffeners dimensions were constant, whereas the web thickness and width varied. So the bending rigidity of flanges and bending and torsional rigidity of both flanges and girders varied in a range given in Table 1. The details of the test girders are presented in Fig. 1.

All girders were of welded construction. The web thicknesses and the yield point stresses f_{yw} of the web material, which were determined by means of tensile tests on coupons cut off from each girder, are summarized in Table 2.

Table 1.

Test Girder	Flange flexural rigidity $[10^3 mm^4]$	Torsional rigidity of a test girder $[GI_t . 10^{-5} \ N \ mm^2]$
A	109.227	1,11
B	109.227	8,91
C	10,617.446	30,88

TEST GIRDERS

Fig. 1

DIMENSIONS OF WEB PANELS

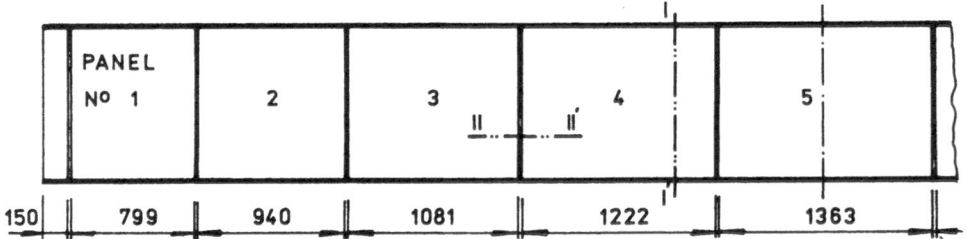

| PANEL Nº 1 | 2 | 3 | 4 | 5 |

150 799 940 1081 1222 1363

TEST GIRDER "A"

SECTION I-I'

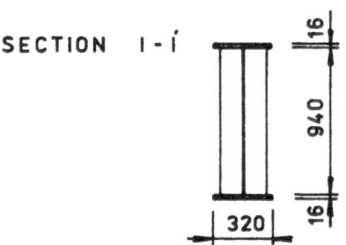

SECTION II-II'

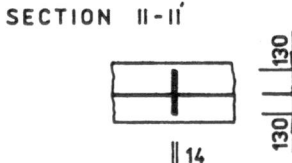

TEST GIRDER "B"

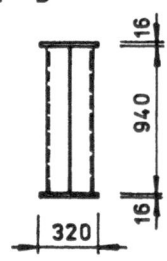

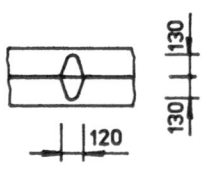

TEST GIRDER "C"

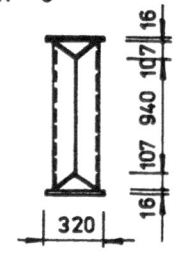

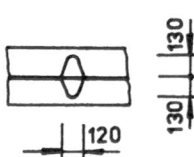

Table 2.

Test girder	Web thickness t_w [mm]	Yield stress f_{yw} of web material [MPa]
A1, A2	5,9-6,1	291
A3, A4	4,2	348
A5	3,1	366
B1, B2	6	280,5
B3, B4	4,2	363,5
B5	3,1	378
C1, C2	6	267
C3, C4	4,2	365,5
C5	3,1	390

3. Test Arrangement

All experimental girders were subjected to a compressive partial edge loading acting on the top flange at the mid-distance of the vertical stiffeners. In most tests the load was transmitted to a flange through a 500 mm long rail of rectangular cross-section of 80/80 mm, in some cases of 60/60 mm or through a plate of 100 mm wide.

Membrane strains directly under the loaded flange were measured by means of strain-gauges T1 - T3. Deflections of a web in the central vertical section were measured using a set of dial gauges H1 - H7.

Figs.2a, b give general details of the test set-up.

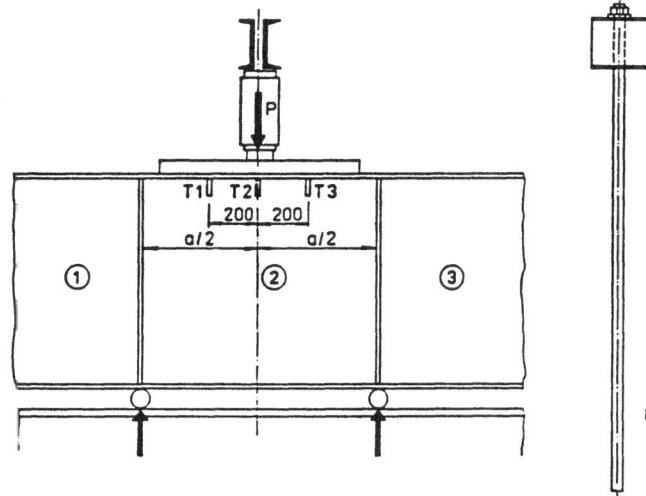

Fig. 2a

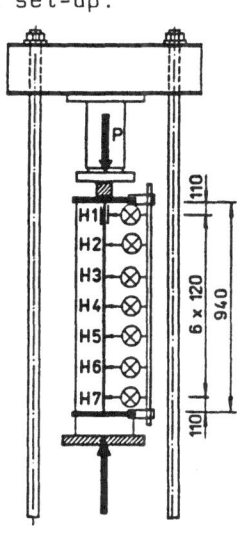

Fig. 2b

4. Test results

The main results are shown in Fig. 3. In his original report [1] the author presented the following conclusion.
  i) There were no remarkable changes in the web behaviour under critical loads, either in stresses nor in deflection
  ii) The thinner the web, the higher postcritical reserve. The load carrying capacity of thin webs may be 20 times higher

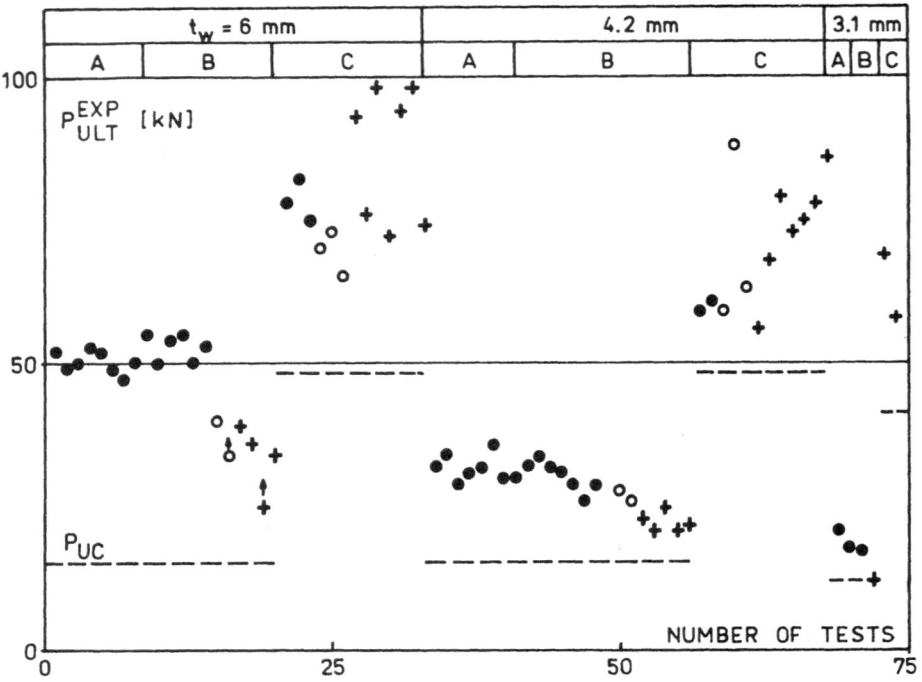

Fig. 3

than their critical loads.

iii) For loads transmitted into a web through a sufficiently
stiff and long distribution bar (rail) the allowable
value can be estimated by means of the well known
Girkmann´s formula

$$P_{UC} = 3,27 \; t_w^2 \cdot f_{yw} \sqrt[3]{J_{fb}/t_w^4}.$$

5. References

[1] Březina, V.: Buckling tests of plate girder webs subjected
    to concentrated loads between vertical stiffeners, Report
    of UTAM CSAV, (in Czech), Prague 1965.

234

COMMENTS ON CONSTRUCTIONS OF EMPIRICAL FORMULAS

M. DRDÁCKÝ

Central Laboratory for Experimental Mechanics,
Institute of Theoretical and Applied Mechanics,
Vyšehradská 49
128 49 Prague 2
Czechoslovakia

1. Introduction

The complexity of theoretical solutions of many mechanical problems leads to a construction of empirical or semiempirical formulas being utilized in practical structural design. This is also the case at estimations of load-carrying capacity of slender plate girder webs subjected to a partial edge loadings. In this field a quite rich activity of researchers or engineers can be noted. Let us remind empirical formulas proposed by Granholm (1960), Ostapenko-Yen-Beedle (1968), Rockey-Elgaaly (1971), Bergfelt (1971), (1973), (1974), (1979), Herzog (1974), Drdácký (1974), (1978), Roberts-Rockey (1978), Roberts (1981), Weimar, Ramm (1987), Spinassas-Raoul-Virlogeux (1990). The formulas are derived using different approaches which are not correct in some cases.

2. Correct construction of empirical formulas

2. 1. Dimensional homogenity

Empirical or semiempirical formula describing a physical entity or phenomenon must be created as a dimensionally homogeneous equation in order to conserve full model similarity. Deriving formula (7) or (9) in [1] the author has taken advantage of the Buckingham theorem and has expressed a failure load as a relation between a complete set of linearly independent dimensionless numbers composed of structural, loading and possibly environmental parameters. Such an approach ensures the dimensional homogenity of empirical formula. Fig. 1 presents a comparison of achieved results with dimensionally nonhomogeneous formula from the very valuable paper [2] in light of experimental results.

2. 2. Parametric study of influences

The influence of different structural and loading parameters on the behaviour of plate girder webs is not clearly apparent because of many interactions. Figs. 2 a, b, c, d show the scatter of experimental data for selected structural characteristics. Their influences should be studied at keeping all other parameters constant. Such an approach was adopted to obtain e. g. relations presented in Figs. 13, 14 [1] , which

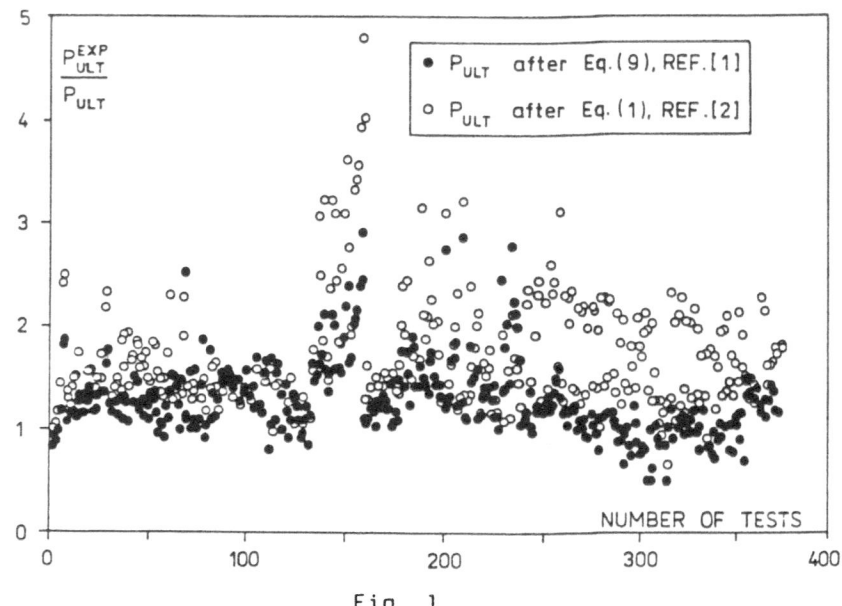

Fig. 1

assumed to have special test specimens [3] . Even though the
similarity laws are valid, it need not be sufficient to
conserve constant dimensionless numbers at experimental
investigations. In other words, in the problem under discussion,
studying the influence of slenderness ratio it is better to
keep constant the web thickness than the web depth, [4] .

2. 3. Construction of dimensionless numbers

There is an infinite number of possibilities in
construction of dimensionless numbers from structural and
loading parameters which are to be taken into account. The
better way seems to be an adoption of a reasonable theoretical
background, as e. g. in [1] , than a simple statistical
handling with experimental data. In the latter case the result
can be unreasonable, as it is in the case of Herzog s formula
[5] , where an increase of web aspect ratio leads to an
increase of load-carrying capacity of the web.

2. 4. Nonlinear regresion

Relations between dimensionless numbers are mostly
nonlinear. According to the authors experience it is generally
better to search constants or exponents by nonlinear regresion
methods which avoid a logaritmic linearization.

2. 5. Validity of empirical formulas

Any empirical or semiempirical formula should be
supplemented by a definition of the range of its validity.
This range must be based on bounds of experimental data which
were utilized for construction of the formula.

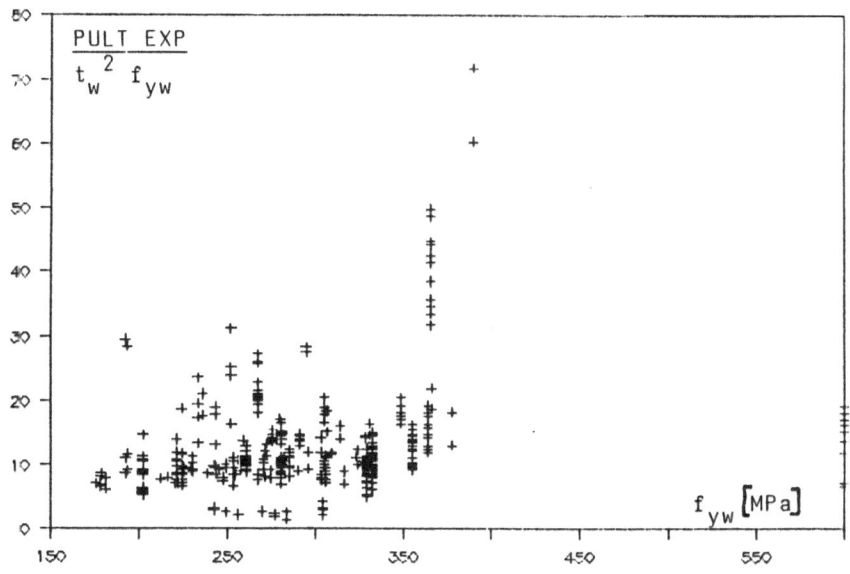

Fig. 2 a

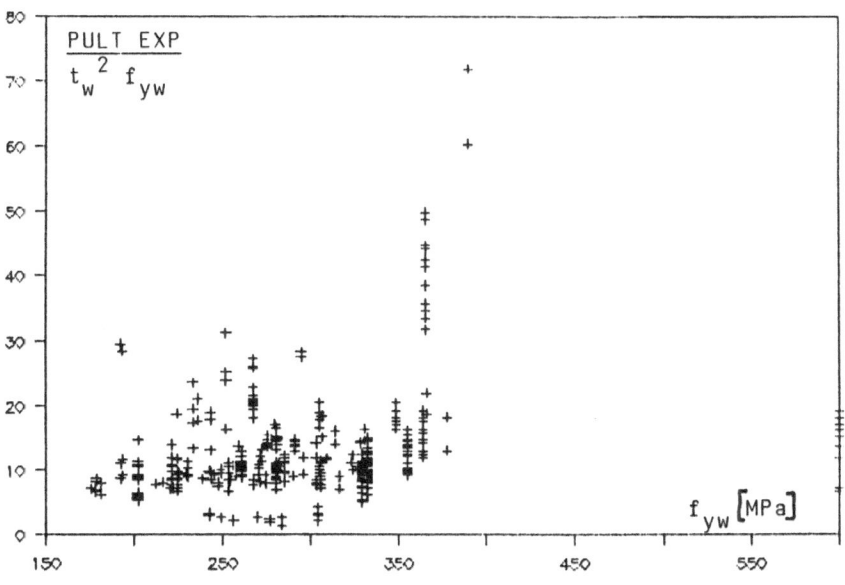

Fig. 2 b

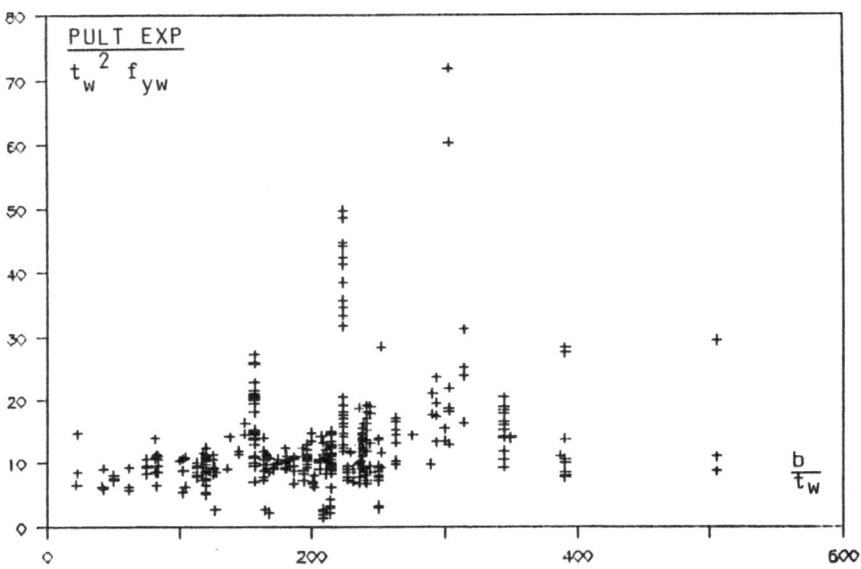

Fig. 2 c

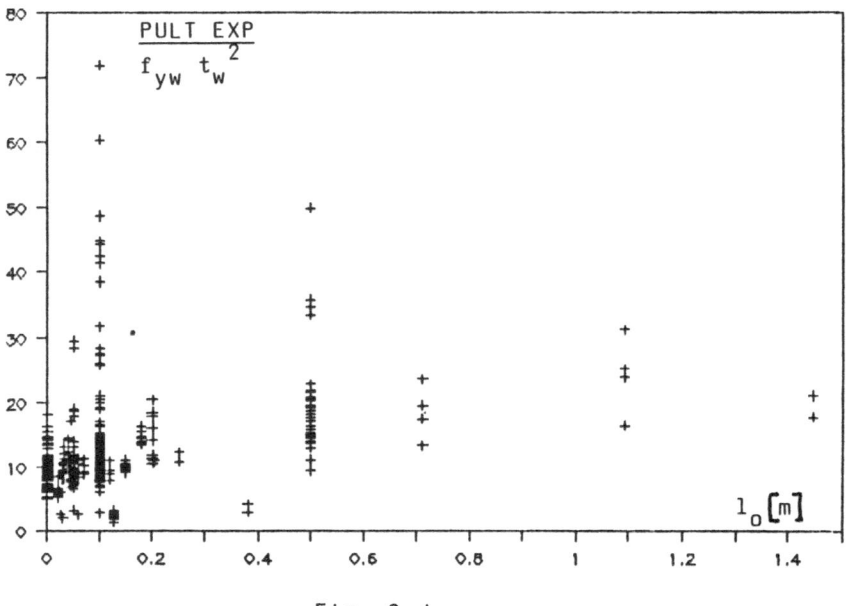

Fig. 2 d

3. References

[1] Drdácký, M.: Non-stiffened steel Webs with Flanges under Patch Loading, this book, pp.

[2] Spinassas, I.; Raoul, J.; Virlogeux,M.: Parametric study on Plate Girders Subjected to Patch Loading, this book, pp.

[3] Drdácký, M.: The Thin Web of a Plate Girder with Non--Rectangular Flanges Subjected to Concentrated Loads, Proc. of the Int. Sci. Conference "Metal Constructions", Vol.II, pp. 277-287, Katovice 1979.

[4] Drdácký, M.: Test of the Influence of Different Structural and Loading Parameters on the Behaviour of Thin-Walled Plate Girders, proc. of the Int. Sci. Conference "Metal Constructions", Vol. 4, pp. 53-60, Gdaňsk.1989.

[5] Herzog, M.: Die Krüppellast sehr dünner Vollwandträgerstege nach Versuchen, Der Stahlbau, 43, s. 26-28, 1/1974.

MODELS OF THIN-WALLED BEAM CONNECTIONS

S. Krenk
Institute of Building Technology and Structural Engineering
University of Aalborg, DK—9000 Aalborg, Denmark

L. Damkilde
Department of Structural Engineering
Technical University of Denmark, DK—2800 Lyngby, Denmark

INTRODUCTION

In thin—walled beam theory the kinematics of the beam is described by 6 degrees of freedom for a rigid body displacement of each cross—section plus an additional degree of freedom for cross—section warping. Torsion and warping are coupled, and this coupling requires special attention at joints. In general the transfer of warping through a joint will create deformation of the cross—section. Although distortion in the form of cross—section deformation is not accounted for in classical thin—walled beam theory this effect is often local, and for cross—sections containing two main flanges — such as I, U and C—profiles — simple joint models compatible with classical thin—walled beam theory can be developed.

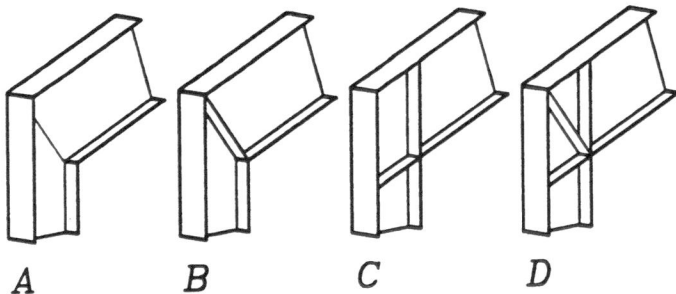

Fig. 1. Four different types of I—beam joints.

In the present paper models are presented for the four I—beam joints shown in Fig. 1. These joints have been considered by Vacharajittiphan & Trahair (1974), who represented the effect of the joint as a numerically calibrated warping restraint stiffness. In the present approach the key issue is to identify suitable kinematical continuity conditions, and then — if necessary — to include any additional stiffness associated with distortion of the beams and torsion of the plates added to the joints of type B, C and D.

The height of the beams is h_1 and h_2, respectively. At the joints the warping intensity of the beams is denoted by θ_1 and θ_2. The warping of a cross–section of beam 1 according to classical thin–walled beam theory consists of a relative inclination of the two flanges of magnitude $h_1\theta_1$ in the plane of the flanges as shown in Fig. 2. At the joint the continuity of the inside and outside flanges leads to mutual inclination of the flanges of magnitude ψ_1 in the plane of the cross–section as shown in Fig. 3. At any of the four types of joint the four kinematical parameters θ_1, ψ_1, θ_2 and ψ_2 are related by two continuity conditions. Thus the deformation of the beam cross–sections at the joint can be expressed in terms of the warping intensities θ_1 and θ_2. In joint types B and C an additional equation is obtained from the very large in–plane bending stiffness of the added plates, while in type D two additional equations are obtained from the plates in the joint. Thus type A has two free warping parameters, types B and C have one, while warping as well as cross–section deformation is fully restrained in type D.

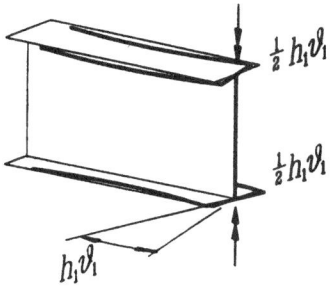

Fig. 2. Warping of beam 1.

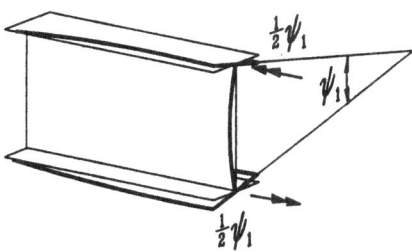

Fig. 3. Distortion of beam 1.

FLANGE CONTINUITY

The displacement of a joint may be considered as consisting of a rigid body motion and a deformation of the joint permitting warping and distortion of the associated beams. The present analysis is only concerned with the latter part. The rotation of the intersection line of the inner flanges must be the same whether described in terms of the parameters $h_1\theta_1$, ψ_1 or $h_2\theta_2$, ψ_2. Projection of the rotation vectors shown in Fig. 4 then gives the continuity conditions

$$\begin{bmatrix} \psi_1 \\ \psi_2 \end{bmatrix} = \frac{1}{\sin\alpha} \begin{bmatrix} \cos\alpha\, h_1 & -h_2 \\ h_1 & -\cos\alpha\, h_2 \end{bmatrix} \begin{bmatrix} \theta_1 \\ \theta_2 \end{bmatrix} \tag{1}$$

This relation enables elimination of the distortion parameters ψ_1 and ψ_2 locally at the joint, provided the beam distortion does not couple to the neighbouring joints.

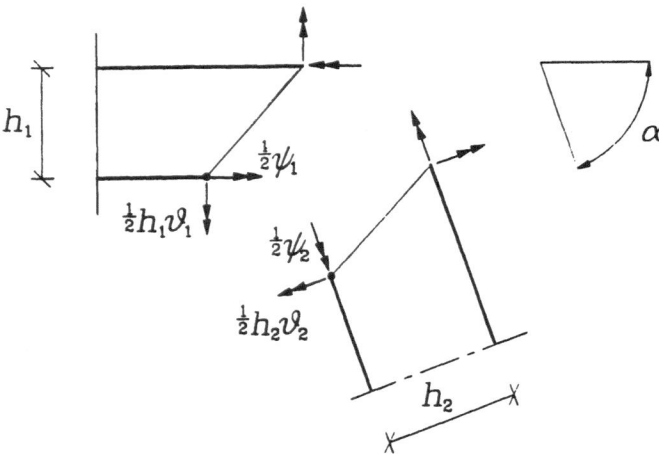

Fig. 4. Flange continuity at joint.

ENERGY AND LENGTH SCALES

The elastic energy per unit length of beam associated with warping is

$$W_\theta = \tfrac{1}{2} E C \, \theta'^2 + \tfrac{1}{2} G K \, \theta^2 \tag{2}$$

where E is the modulus of elasticity and G is the shear modulus. K is the St. Venant torsion constant of the full beam cross–section, and $C = h^2 I_f/2$ is the warping stiffness when I_f is the in–plane bending stiffness of one flange.

The elastic energy associated with distortion can be approximated by a similar expression.

$$W_\psi = \tfrac{1}{2} G \, 2K_f(\tfrac{1}{2}\psi')^2 + \tfrac{1}{2}\tfrac{1}{h} D_w \, \psi^2 \tag{3}$$

K_f is the St. Venant torsion stiffness of one flange, and $D_w = E t_w^3/12(1-\nu^2)$ is the bending stiffness of the web.

The Euler equations corresponding to (2) and (3) are

$$E C \, \theta'' - G K \, \theta = 0 \tag{4}$$

$$G \tfrac{1}{2} K_f \, \psi'' - (D_w/h) \, \psi = 0 \tag{5}$$

These equations have exponential solutions with parameters given by

$$k_\theta^2 = \frac{GK}{EC} \,, \quad k_\psi^2 = \frac{2D_w}{GK_f h} \tag{6}$$

The parameters k_θ and k_ψ determine the attenuation of warping and distortion with distance from the joint, respectively. Usually k_ψ is greater than k_θ implying faster attenuation of the distortion mode.

In the following the distortion modes are assumed to be local. The elastic distortion energy in each of the beams can then be represented by the value corresponding to a semi–infinite beam, e.g.

$$E_{\psi_1} = \tfrac{1}{2} \sqrt{\frac{D_{w1}}{2h_1} \, GK_{f1}} \, \psi_1^2 \tag{7}$$

243

This expression is accurate to within 10 pct for $k_{\psi_1}l \geq 1.5$.

UNSTIFFENED JOINTS, TYPE A

In the unstiffened joint of type A the warping intensities θ_1 and θ_2 are used as independent parameters, while the distortion parameters ψ_1 and ψ_2 are expressed by the relation (1) as

$$\begin{bmatrix} \psi_1 \\ \psi_2 \end{bmatrix} = [A] \begin{bmatrix} \theta_1 \\ \theta_2 \end{bmatrix} \tag{8}$$

The distortion energy associated with the joint then follows from (7) in the form

$$E_\psi = \tfrac{1}{2}(\psi_1, \psi_2) \begin{bmatrix} \sqrt{(D_{w1}/2h_1)GK_{f1}} & 0 \\ 0 & \sqrt{(D_{w2}/2h_2)GK_{f2}} \end{bmatrix} \begin{bmatrix} \psi_1 \\ \psi_2 \end{bmatrix} \tag{9}$$

When the diagonal stiffness matrix is denoted [D], the distortion energy is expressed in terms of the warping parameters as

$$E_\psi = \tfrac{1}{2}(\theta_1, \theta_2) [A]^T [D] [A] \begin{bmatrix} \theta_1 \\ \theta_2 \end{bmatrix} \tag{10}$$

Thus the unstiffened joint of type A acts in warping as a hinge with two independent warping parameters (θ_1, θ_2) and an additional 2 by 2 elastic spring stiffness matrix given by the matrix product in (10).

STIFFENED JOINTS, TYPE B

In stiffened joints of type B the in−plane bending stiffenss of the cross plate in the joint is usually sufficiently stiff to effectively prevent in−plane deformation. This leaves only one parameter to describe the combined warping and distortion of this type of joint.

The geometry of the joint is shown in Fig. 5, where the parameters are determined by

$$a_1 = (h_2 - h_1 \cos\alpha)/\sin\alpha \tag{11}$$

$$a_2 = (h_1 - h_2 \cos\alpha)/\sin\alpha \tag{12}$$

and

$$\tan\alpha_1 = a_1/h_1 \tag{13}$$

$$\tan\alpha_2 = a_2/h_2 \tag{14}$$

The length of the cross plate is

$$h_c = h_1/\cos\alpha_1 = h_2/\cos\alpha_2 \tag{15}$$

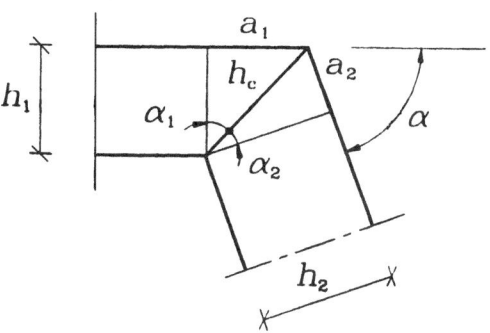

Fig. 5. Geometry of type B joint.

It is convenient to introduce the parameters θ_c and ψ_c representing warping and in–plane bending of the cross plate, see Fig. 6. Projection of the rotation vectors gives the following expression for the parameters of beam 1.

$$\begin{bmatrix} h_1\theta_1 \\ \psi_1 \end{bmatrix} = \begin{bmatrix} \cos\alpha_1 & \sin\alpha_1 \\ -\sin\alpha_1 & \cos\alpha_1 \end{bmatrix} \begin{bmatrix} h_c\theta_c \\ \psi_c \end{bmatrix} \tag{16}$$

The similar formula for $(h_2\theta_2, \psi_2)$ follows by exchange of α_1 with $-\alpha_2$.

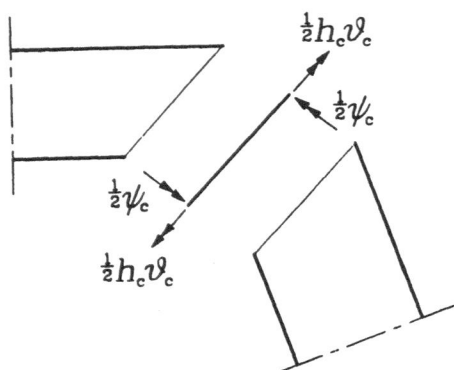

Fig. 6. Joint warping θ_c and in—plane bending ψ_c.

In practice the in—plane bending stiffness of the cross plate is much larger than the warping stiffness, and thus $\psi_c = 0$ is a good approximation. By (16) and the similar formula for beam 2 this implies

$$h_1\theta_1/\cos\alpha_1 \; = \; h_2\theta_2/\cos\alpha_2 \; = \; h_c\theta_c \qquad (17)$$

Elimination of h_1 and h_2 by use of (15) then gives the continuity condition

$$\theta_1 \; = \; \theta_2 \; = \; \theta_c \qquad (18)$$

Thus constraint of the in—plane bending of the cross plate $-$ $\psi_c = 0$ $-$ implies that the two beams have identical warping parameters at the joint $-$ $\theta_1 = \theta_2$.

There are two stiffness contributions associated with the warping parameter θ_c $-$ warping of the cross plate and distortion of the beam cross—sections. The elastic energy required for warping of the cross plate with dimensions $h_c \times b_c \times t_c$ is

$$E_\theta \; = \; \frac{1}{2}\left\{\frac{1}{3}G \; h_c b_c t_c^3\right\} \theta_c^2 \qquad (19)$$

The distortion parameters ψ_1 and ψ_2 follow from (1) with $\theta_1 = \theta_2 = \theta_c$. When the geometric parameters a_1 and a_2 shown in Fig. 5 are introduced from (11) and (12)

$$\begin{bmatrix} \psi_1 \\ \psi_2 \end{bmatrix} = \begin{bmatrix} a_1 \\ a_2 \end{bmatrix} \theta_c \tag{20}$$

This relation is substituted into the energy expression (9) to give the energy of distortion

$$E_\psi = \tfrac{1}{2} \left\{ a_1^2 \sqrt{(D_{w1}/2h_1)GK_{f1}} + a_2^2 \sqrt{(D_{w2}/2h_2)GK_{f2}} \right\} \theta_c^2 \tag{21}$$

The only dependence on the joint angle α is through the lengths a_1 and a_2.

STIFFENED JOINTS, TYPE C

In stiffened joints of type C the warping is conveniently expressed in terms of the out–of–plane displacement $\pm\Delta$ of the four points of flange intersection. The analysis has been carried out by Krenk et al. (1990). The result is that there is a single warping parameter

$$\theta_c = 4 \frac{\Delta}{A} \tag{22}$$

where $A = h_1 h_2/\sin\alpha$ is the web area enclosed by the flanges. The warping at the beam ends is determined by

$$\theta_1 = -\theta_2 = \theta_c \tag{23}$$

Thus warping is of the same magnitude, but of different sign in the two beams.

There are two stiffness contributions — warping of the additional flange lengths, and distortion of the beam cross–sections. The additional energy from warping of flanges with dimensions $b_1 \times t_1$ and $b_2 \times t_2$ is

$$E_\theta = \tfrac{1}{2} \left\{ \tfrac{1}{3} G \frac{h_2 b_1}{\sin\alpha} t_1^3 + \tfrac{1}{3} G \frac{h_1 b_2}{\sin\alpha} t_2^3 \right\} \theta_c^2 \tag{24}$$

The distortion energy is computed by introducing the common warping parameter θ_c from (23) into (1) and (9). Experience indicates that the main point is the enforcement of the continuity condition (23), while the additional stiffness contributions are of secondary importance.

STIFFENED JOINTS, TYPE D

The stiffened joints of type D must satisfy the continuity conditions of joints of type B as well as joints of type C. This implies that

$$\theta_1 = \theta_2 = 0 \tag{25}$$

for joints of type D. The distortion determined by (1) then also vanish, and the type D joint can be considered as a full warping and distortion restraint.

EXAMPLES AND FINITE ELEMENT RESULTS

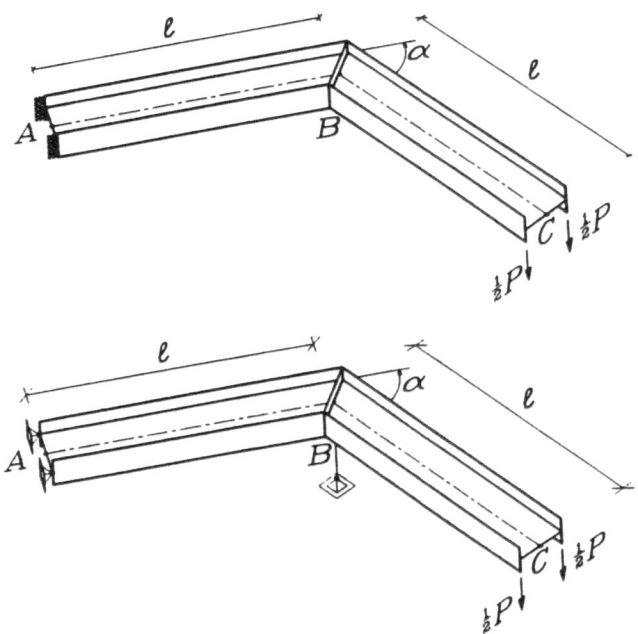

Fig. 7. Cantilever and simply supported angle beams.

The continuity and stiffness conditions for warping of thin–walled I–beam joints of types A and B have been investigated using the angle beam configurations shown in Fig. 7a and 7b for joint type B. Two identical I–beams of length ℓ are joined at an angle α, and a transverse load P is applied at the point C. Two support conditions are considered: in case a) the beam AB is rigidly fixed at A, and in case b) the flanges of the beam AB are simply supported at A, while a single simple support is introduced at B.

In the classical theory for thin–walled beams of open cross–section warping is determined by vanishing shear strain in the mid–surface. The warping parameter θ then equals minus the rate of twist φ'.

$$\theta = -\varphi' \tag{26}$$

Torsion and warping of the beams AB and BC in Fig. 7 are then governed by the differential equation

$$(EC\varphi'')'' - (GK\varphi)'' = 0 \tag{27}$$

and the appropriate boundary and continuity conditions.

In the following the accuracy of the continuity conditions and stiffness contributions for joints of type A and type B is evaluated by comparing the rotation φ_B of beam AB at B predicted by the present theory with three–dimensional finite element calculations. The results are normalized with respect to the rotation that would result if the beam AB were free to warp at both ends and loaded by the torsional moment

$$M_0 = P \ell \sin\alpha \tag{28}$$

Thus the results are expressed in terms of the non–dimensional parameter $\varphi_B GK/\ell M_0$.

The cross–section dimensions in all cases are h = 200 mm, b = 100 mm and t_f = 8.5 mm, t_w = 5.6 mm. With $\nu = 0.3$ the k–parameters are

$$k_\theta = 1.36 \; 10^{-3} \; \text{mm}^{-1} \qquad k_\psi = 4.88 \; 10^{-3} \; \text{mm}^{-1}$$

Thus the approximation of distortion as a local effect can be used for $k_\vartheta l \geq 0.4$ for this cross-section.

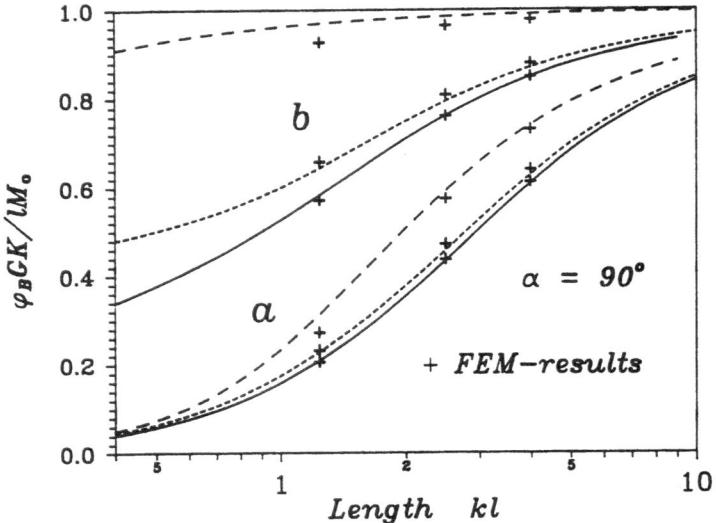

Fig. 8. Normalized angle φ_B from beam theory and FEM, $\alpha = 90^0$. $(- -)$ type A, $(\cdot\cdot\cdot)$ type B without joint stiffness, $(-\!\!\!-)$ type B with full stiffness.

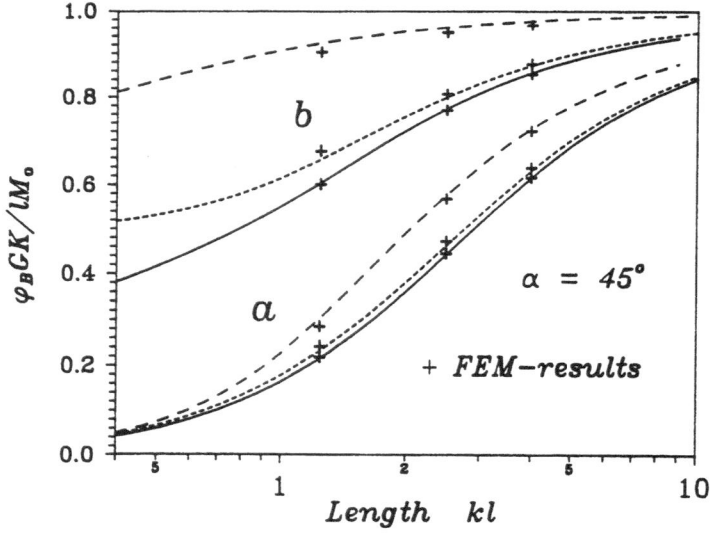

Fig. 9. Normalized angle φ_B from beam theory and FEM, $\alpha = 45^0$. $(- -)$ type A, $(\cdot\cdot\cdot)$ type B without joint stiffness, $(-\!\!\!-)$ type B with full stiffness.

The results according to the present theory have been evaluated in closed form for joints of type B by Krenk et al. (1990), and all four joint models have been incorporated in a finite element program for thin–walled frames by Damkilde et al. (1990). These results are given by curves in Figs. 8 and 9 for $\alpha = 90^0$ and 45^0, respectively. Note the limited influence of the angle α.

The results of the beam theory have been compared with three–dimensional finite element analyses using he PAFEC program. Three different lengths were considered with $k_\theta \ell = 1.25$, 2.5, 4.0. The element model used four approximately square 8–node shell elements for the web and four similar elements across each flange. A typical mesh is shown in Fig. 10.

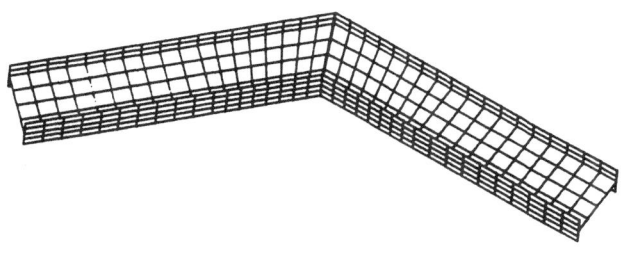

Fig. 10. Finite element mesh.

CONCLUSIONS

Continuity conditions and stiffness properties have been derived for warping and distortion of the four types of thin–walled I–beam joints shown in Fig. 1. The unstiffened joint, type A. has two independent warping parameters, and distortion of the joint appears as a 2 by 2 spring stiffness matrix. The stiffeners in joints of type B and C enforce warping of equal magnitude in both beams. In addition there are stiffness contributions from the added plates and disortion. Joints of type D act as full warping constraints.

The theory is expressed in tems of simple, explicit formulas suitable for use in connection with classical thin–walled beam theory. Three dimensional finite element calculations indicate high accuracy of the beam–type models of the joints.

ACKNOWLEDGMENT

This paper is part of a project supported by the Danish Technical Research Council.

REFERENCES

Krenk, S., Petersen, P. & Damkilde, L. (1990): Warping of Joints in I–Beam Assemblages, University of Aalborg, Aalborg, Denmark.

Damkilde, L., Petersen, P. & Krenk, S. (1990): Combined Bending–Torsion Instability of Thin–Walled Frames with Deformable Joints, Technical University of Denmark, Lyngby, Denmark.

Vacharajittiphan, P & Trahair, N. S. (1974): Warping and Distortion of I–Section Joints, Journal of the Structural Division, ASCE, Vol. 100, ST3, pp. 547–564.

PART III.

POSTERS

ON STABILITY OF CRUSHED COLUMNS UNDER NON-CONSERVATIVE CONTACT LOADING

BOGACZ R., IMIEŁOWSKI Sz.

Institute of Fundamental Technological Research
Polish Academy of Sciences
ul. Świętokrzyska 21, 00-049 Warsaw, Poland

Summary - The paper deals with the problem of stability of axially crushed thin-walled columns. To determine the behaviour of speciment in first stage of crushing process, the model of Reut column with single elastic transverse-slidable or hinge-joint as idealization of first local buckling wave is used. There were distinguish two sectors of columns in which the critical force, determined from kinetical criterion of stability, is lower than for uniform column. One of this lies near the base of column and the second one above its centre. This sectors are pointed as a place of localization of first deformation lobes of speciment. The presented results of numerous experimental tests confirm this result. Analitical solution to the problem is found by the transfer matrix technique.

1. Introduction

The considerably activity on the behaviour of thin-walled tubes follow mainly from the viewpoint of energy absorption during post-buckling behaviour with large strains and deflections. Thin-walled steel tubes are often used as a energy absorption members of automobile, trains or buses. A rapid progress of investigations caused a development of various empirical, analitical and numerical method of calculation. A mean value of crushing load, effective crushing distance or strain rate as well as typical deformation modes and mechanisms leading to global buckling of tube are selected and described, [3]÷[6].

It seems that the localization of the first local buckling wave is important from the viewpoint of design of energy absorption members . As far as the authors are aware, the localization of first local buckling wave for the column without any imperfections and with weld of uniform thickness is not defined. In order to find this localization a model of a structure, with already created first lobes of deformation, idealized as transverse-slidable or hinge-joint, was assumed . The dependence of critical load versus joint location for various values of its stiffness was shown. In cross-sections localized near by both ends of column the critical force appeared to be smaller than for uniform column and this regions were pointed as a place of localization of first deformed lobes. Numerous experimental results confirm that in this localizations the first buckling wave is formed.

2. Assumptions

The observations of postbuckling behaviour of crushed tube were taken into account to create the model of speciment. There were considered tests of dynamically as well as statically loaded tubes because in both kinds of loading the same modes of deformations were observed [5]. The same basic deformation modes appears in speciment of diffrent cross-sections, in that number square, triangular, hexagonal and rhombic and of various lehghts [6]. The symmetric or asymmetric collapse modes were selected. Symmetric deformation is suitable to determination of the behaviour associated with the axial crushing of circular tubes [4] and is also used to describe some kinds of deformations of polygon thin-walled column, the asymmetric mode of deformation appears in polygon tubes, [5], [6]. All details at the deformation modes for the example of square tubes are presented in [5].

First created lobes of deformation can be considered as the layer of local loss of speciment stiffness. Both, the local loss of bending stiffness as well as loss of shear rigidity are observed. In the case of the asymmetry of deformation modes a shift of the upper edge of lobe relative to the lower one in the direction of horizontal axies as well as the inclination of the undeformed part of the column relative to the vertical axies is well seen in the high-speed film in Fig.1, [5]. A transition from progressive axial crushing to overall bending therefore could occur in a column of sufficient asymmetric lobes developed to produce instability in the sense of Euler as it is shown in Fig.2a, [5], for square tubes. It seems, that in the deformed layer, as the first the loss of shear rigidity appears. It induce mutual displacement of upper and lower undeformed parts of speciment. Eccentric of loading is then developed and as consequence a transition to bending mechanism is observed. A transition to overall bending of a square tube may also develop following symmetric crushing as illustrated in Fig. 2b.

All this observation induce the authors to assumed that in the first stage of process, the axially crushed tube may be modeled as the column consisting of two segments connected by a tranverse-slidable or hinge-joint as idealization of already formed first local buckling wave. The local buckling wave is understand as all individual lobes of one layer. In this first stage of process, when one tries to predict the initialization of buckling wave only, the assumption of elastic model of joint as well as elastic model of segments seems to be proper.

In experimental tests, of axially loaded column the crushing force act always in the direction of the undeformed axis of speciment,Fig. 3. This is a case of nonconservative loading. The uniform elastic column under such a loading is known as Reut column [7]. As far as in postbuckling behaviour of tube the usually assumed Shanley model as idealization of crushed speciment is proper one, it seems that in the first stage of process, when the column behave as elastic structure, the model of Reut column will be better. Also the boundary conditions given good approximation in this case.

Finally, in order to find the localization of first local buckling wave, the Reut model of column with single elastic transverse-slidable or hinge joint is assumed. For such a idealization of allready formed first layer of lobes the cross-section, in which the smallest value of critical force will be loocked for.

3. Formulation of the problem

The structures are shown schematically in Fig.3 . Each of its consists of two segments connected by elastic transverse-slidable or hinge joint, placed at position x_1 and characterized by the flexibility parameters γ_{Rj}, γ_{Sj} respectively (index R - rotary and S - shear). Sometimes more convenient in calculations are stiffness parameters $\kappa_{Rj} = 1/\gamma_{Rj}$, $\kappa_{Sj} = 1/\gamma_{Sj}$. Analitical solution of the problem was found by the transfer matrix technique.

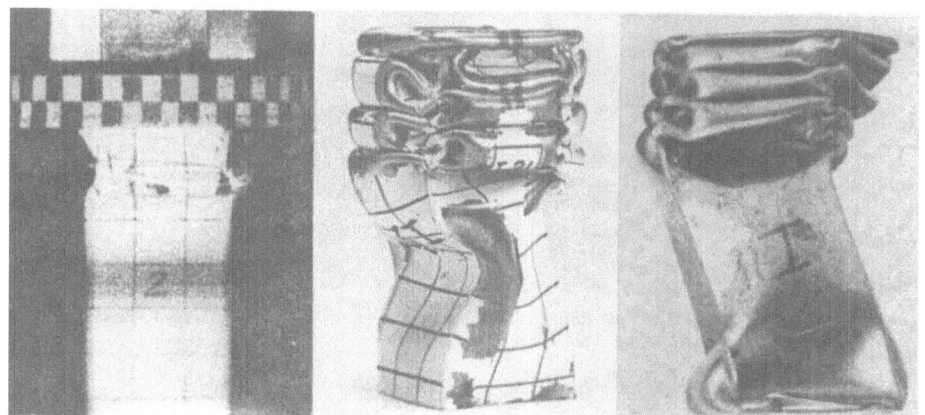

Fig.1. Frame from a high-speed film
of dynamically crushed square tube

Fig.2. Initiation of overall bending
after development of a) asymmetric
b) symmetric lobes of deformation

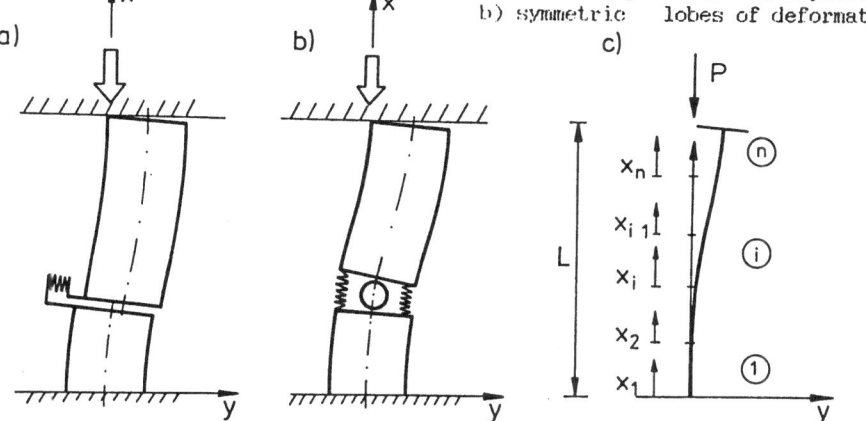

Fig.3. Reut column with localized elastic
a) transverse-slidable joint b) hinge joint

c) Segmentation of column

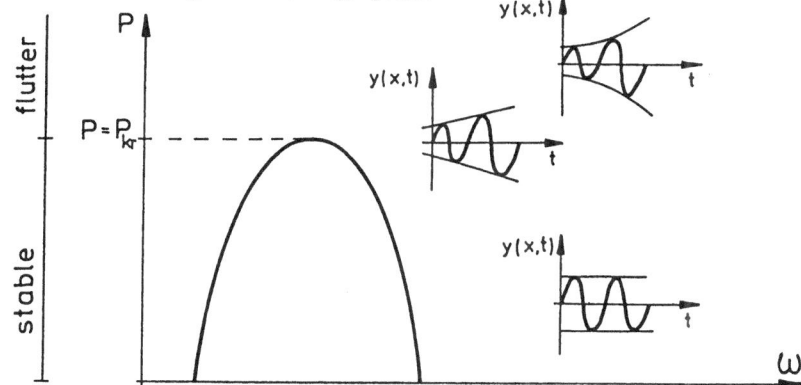

Fig.4. The configuration of characteristic curves for Reut column
and stability discussion

The equation of motion for a segment of column is assumed as:

$$\frac{\partial^2}{\partial x^2} (EJ \frac{\partial^2 y}{\partial x^2}) + P \frac{\partial^2 y}{\partial x^2} + \rho A \frac{\partial^2 y}{\partial t^2} = 0 \quad . \tag{1}$$

The boundary conditions for column (Reut problem)

$$x = 0 \qquad\qquad y = 0 \quad , \qquad\qquad \frac{\partial y}{\partial x} = 0 \quad , \tag{2}$$

$$x = L \qquad EJ \frac{\partial^2 y}{\partial x^2} + Py = 0 \quad , \quad \frac{\partial}{\partial x} (EJ \frac{\partial^2 y}{\partial x^2} + Py) = 0 \quad . \tag{3}$$

The problem is adjoint with Beck problem [7]. The model of Beck column with single elastic hinge-joint was considered in Ref. [9]÷[11].

The exact solution of Eq.(1) i.e. for the segment of constant mass and stiffness distribution has the form

$$y(x,t) = e^{i\omega t}(A_1 sh\lambda_1 x + A_2 ch\lambda_1 x + A_3 sin\lambda_2 x + A_4 cos\lambda_2 x) \quad , \tag{4}$$

where:

$$\lambda_{1/2} = \left[\pm \frac{P}{2EJ} + \sqrt{(\frac{P}{2EJ})^2 + \frac{\rho A \omega^2}{EJ}} \right]^{1/2} \quad . \tag{5}$$

All dependent variables y , ϕ , M , Q , have a similar constitutive form and they are closed in state vector $\underline{G}$ defined as

$$\underline{G} = [\ y, \phi, M, Q\ ]^T = [\ y, y', -EJy'', -EJy'''\ ]^T \quad . \tag{6}$$

Two states vectors on both ends of i-th segment, Fig.3c, are connected by the partial trasfer matrix $\underline{T}_i$

$$\underline{G}^o_{i+1} = \underline{T}_i\ \underline{G}^o_i \quad , \tag{9}$$

where: $\underline{G}^o_i = \underline{G}\ (x_i = 0)$ and $\underline{G}^o_{i+1} = \underline{G}\ (x_{i+1} = 0)$. For the complete structure we have

$$\underline{G}^o_{n+1} = \underline{T}_n \cdots \underline{T}_2\ \underline{T}_1\ \underline{G}^o_1 = \underline{T}\ \underline{G}^o_1 \quad . \tag{10}$$

The partial transfer matrix for the segment can be found using the solution (4) of Eq. (1) and is given in [8] .

For the the transverse-slidable and hinge joint we have respectively

$$\Delta y\ (x_j)\ = Q(x_j)/\kappa_{sj}\ = Q(x_j)\ \gamma_{sj} \quad , \tag{11}$$

$$\Delta\phi\ (x_j)\ = M(x_j)/\kappa_{Rj}\ = M(x_j)\ \gamma_{Rj} \quad ,$$

where κ_{Rj} or κ_{sj} is the stiffness and γ_{Rj} or γ_{sj} the flexibility of the joint.

According to (9) nonzero elements of the transfer matrix for the transverse-slidable joint

$$t_{11} = 1 \quad , \qquad t_{14} = \kappa_{sj} \tag{12}$$

andforthehingejoint

$$t_{11} = 1 \quad , \qquad t_{23} = \kappa_{Rj} \quad . \tag{13}$$

Satysfying the boundary coditions we get a characteristic equation as the relation betweem force and frequency, the same as in the case of Beck column

$$\begin{vmatrix} t_{33} & t_{34} \\ t_{43} & t_{44} \end{vmatrix} = G\ (P, \omega)\ = 0 \tag{14}$$

4. Results of numerical calculations

A typical configuration of curves on the frequency-load plane for Reut column is given in Fig.4 . The column is stable when, during the process, the amplitude of vibrations don't increase, otherwise the structure loss its stability by flutter. This is explained in Fig.4 too . Appearence of displacement discontinuity causes qualitative and quantitative changes of the shape of characteristic curves and is connected with changes of critical load.

The nondimensional critical force P_{cr}^{*} , $P_{cr}^{*} = P_{cr} l^2/EJ$ versus joint location x_1 , $x_1 = x/l$ for various hinge-joint flexibilities γ_R^{*} , $\gamma_R^{*} = \gamma_R EJ/l$ is shown on graph in Fig.5 and for various γ_S^{*} , $\gamma_S^{*} = \gamma_S EJ/l^3$ in Fig.6 . It can be seen in both cases, that in two sectors, near the base of column and above its centre, the critical force for finite value of flexibility is lower than for uniform column. This two sectors are the shorter the point flexibility is higher. The minimum value of P_{cr} in both this sectors is similar for column with hinge-joint. When transverse-slidable joint is place in the base of column, P_{cr} diminishes considerably. On the upper end P_{cr} take always the value as for uniform column.

When the joint is placed in the centre of column, for a small values of its stiffness, near fourfold increase in critical force for the case of hinge-joint and near twofold increase for the case of transverse-slidable joint is observed. For each localization of joint above the centre of column a discontinuous changes of critical force are shown. This phenomenons are describe in details for the case of column with elastic or viscoelastic joints in Ref. [9] ÷ [12] . As it is explained in Chap.5 it don't be considered now.

5. Discussion

In really speciment it is difficult to define the joint flexibility in sense of Eq.(11). In the first stage of process the flexibility of this layer of speciment from which in postbuckling the first buckling wave will be intialized is near equal the flexibility of whole the column. For the model in which this layer is considered as a joint its flexibility will be very small, near equal zero. The graphs for $\gamma_R^{*} = 0.2$, 1.0, 2.0 and for $\gamma_S^{*} = 0.05$ are drawed in Fig.5, Fig.6 . For this values of γ_R^{*} the reduction of critical force occurs in two sectors of column, for $x_1 < 0.14$ and $x_1 > 0.58$, Fig.5 and $x_1 < 0.46$ and $x_1 > 0.78$, Fig.6 . In one of this sectors should appears the first deformed layer. When it is created as asymmetric deformation, idealized by transverse-slidable joint, Fig.3a , it should appears near the base of speciment. When first local buckling wave is symmetric one, Fig.4b , the first lobes of deformation occur near the base or near the upper end of the speciment. The exact localization depends probably of material or geometrical imperfections. Additionally the effect of elastic waves which appears in dynamical tests caused that some deviation from this principle may occur. The experimental tests on speciments of variuos cross-sections confirmed this assumption.

The influence of geometrical imperfections on localization of first lobes is well seen on example of cylindrical grooved tubes [14], Fig.7 . The first local buckling wave is placed near the base or near the upper end of column, what depends on distance betweem grooves. In Fig.8a, from Ref. [6], the initialization of first deformations of hexagonal tube is is placed near the base and above the centre of column[1]. In the next example of rhombic tube, Fig.8b, [6], the first lobe are created near upper end of speciment[2]. The creation of first deformations of square tube loaded dynamically is shown in the high-speed film in Fig.1 , [5].

The mentioned in Chap.4 jump phenomenon don't influence on process of creating of first buckling wave because the flexibility of hinge-joint is near equal zero. This phenomenon appears for higher values of joint flexibility and for higher than for uniform column values of P . The consideration closed above are related to the first stage of process only. The strong assumptions don't let to say anything about further process. See the footnotes [1] [2]

259

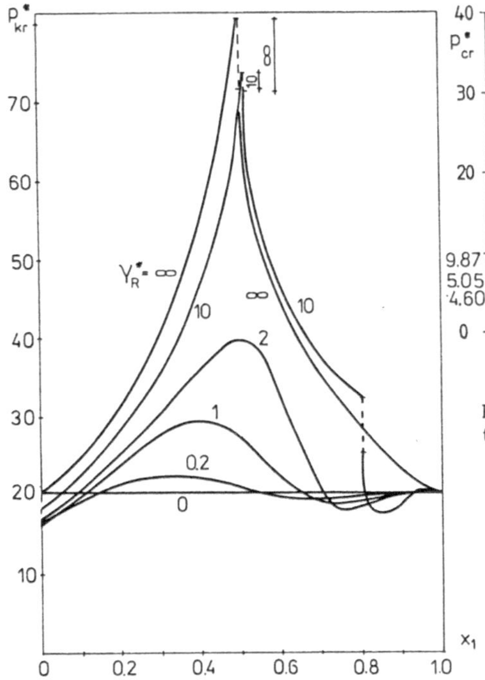

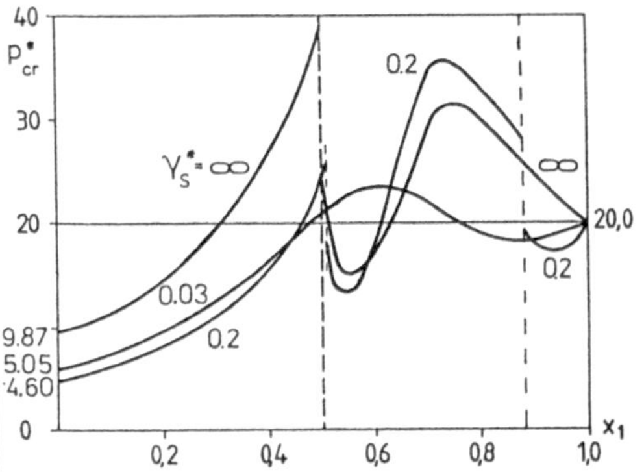

Fig.6. Critical force versus elastic transverse-slidable joint location

Fig.5. Critical force versus elastic hinge-joint location

Fig.7. Influence of geometrical imperfect, Fig.8. Process of creation of first on localization of first lobes of deform. deformations for examples of in the case of cylindrical grooved tubes a) hexagonal tube b) rhombic tube

[1] In this speciment, after creating the first asymmetric layer of lobes near the base of column, the symmetric ones were created in progressive axial crushing process.

[2] In this case the transition to an Euler type of instability after initiation of first deformed lobe was observed.

6. Conclusion

The results obtained in this paper shows that for prediction of localization of first deformed layer of thin walled tubes, the model of Reut cclumn with single elatic joint, as idealization of local buckling wave, is sufficient. The local loss of shear or bending stiffness which is connected with creation first deformations of speciment is modeled as transverse-slidable or hinge joint respectively. In many experimental tests of axially loaded thin-walled speciments of various cross-sections and various lenght, the deformations accurs near the ends of tube as it follows from ccnsiderations supported on such assumption.

The question of behaviour of axially crushed thin-walled tubes in the first stage of process, particulary from the viewpoint of prediction of overall bending will be investigate.

References

1. JONES N. and WIERZBICKI T., Eds, Structural Crashworthiness, Butterworths, London, 1983

2. DAVIES G.A.O., Ed. Structural Impact and Crashworthiness, Elsevier, Applied Science Publishers, London, 1984

3. WIERZBICKI T., ABRAMOWICZ W., On the crushing mechanics of thin-walled structures, J. Appl. Mech., 50, 727-734, 1983

4. JOHNSON W., SODEN P.D., Al-HASANI S.T.S., Inextensional collapse of thin-walled tubes under axial compression, J. Strain Anal., 12, 317-330, 1977

5. ABRAMOWICZ W., JONES N., Dynamic axial crushing of square tubes, Int. J. Impact Ingng., vol.2, no.2, 179-208, 1984

6. ABRAMOWICZ W., IMIEŁOWSKI Sz., WĄSOWSKI A.O., Quasi-static crushing of polygon metall column, IFTR Reports, 1-22, 5/1985 (in Polish)

7. NEMAT-NASSER S., HERRMAN G., Adjoint systems in nonconservative problems of elastic stability, AIAAJ, 4, 2221-2222, 1966

8. PESTEL E.C., LECKIE F.A., Matrix Methods in Elastomechanics Mc. Graw-Hill Book Company, New York, 1963

9. BOGACZ R., MAHRENHOLTZ O., On stability of column under circulatory load, Archives of Mechanics, vol.3, 281-287, 1986

10. BOGACZ R., NIESPODZIANA A., On stability of continous Beck column with localized loss of rigidity, IFTR Reports, 27/1987 (in Polish)

11. BOGACZ R., IMIEŁOWSKI Sz., Stability of columns with hinge or transverse-slidable joint subjected to circulatory loading, Proc. of 28th Symp. "Modelling in mechanics", Gliwice-Wisła, 65-70, 1989 (in Polish)

12. BOGACZ R., IMIEŁOWSKI Sz., On the discontinuous changes of critical force for columns with transverse-slidable joint under follower load, Prel. Rep. Int. Coll. "Stability of steel structures", Budapest, vol.1, 245-252, 1990

13. BOGACZ R., IMIEŁOWSKI Sz., MAHRENHOLTZ O., On the generalization of Beck-Reut problem for the case of discontinuities of displacements and their derivatives, Proc. of 28th Polish Solid Mechanics Conference, Kozubnik, 1990

14. MAMALIS A.G., VIEGELAHN G.L., MANOLOAKOS D.E., JOHNSON W., Experimental investigations into the axial plastic collapse of steel thin-walled grooved tubes, Int. J. Impact Ingng., vol.4, no.2, 117-126, 1986

INTERACTION OF STRUCTURES

PETR BROŽ

Building Research Institute
Technical University Prague
Šolínova 7
166 08 Prague 6
Czechoslovakia

ABSTRACT

For the contact problem of two bodies, we mention a comparison
of model solutions accordingly to methods of the finite element
and the boundary one. A number of applications of this problem
may be found, e. g. a simulation of the tunnel lining action
appears to be the most characteristic, and on a tunnel wall
carried on a bedrock or a beam elastically supported, the contact between plates and the use of discontinuous elements.
We use the condensation of parameters outside the contact to
transform the problem into the compatibility balance on the
contact in order to make possible an analysis according to
Coulomb´s law.

Further, a problem of the half-plane to which pressure by
means of a punch of the infinite rigidity is applied, is being
under consideration. At the same time, the shape of the punch
is defined by the symmetrical function, determined except for
a constant, of the argument running along the half-plane
boundary. There are zero friction forces on the contact. Use
is made of Green´s function of the displacement conponents
which is appurtenant to the half-plane problem.

Besides, by means of the Boundary Element Method it is possible to solve up-to-date and requested problems of the plates
and shells to advantage, viz. their resistence and reliability,
e. g. for circular rings with a variable stiffness, shells of
revolution, interaction of the foundations and both the cylindrical and supported shells, ones filled with the elastic material, loaded with concentrated loads all.

INTRODUCTION

Contact problems include the solution of a contact between two
bodies, i. e. of the common boundary of two domains Ω_1 and Ω_2.

This joint domain exists either before the deformation occurs,
or it becomes real only after deformation.

An interesting survey of the mechanics of contact between solid
bodies was given by J. J. Kalker in yr. 1977 account for the
classical formulation of the contact problem as well as for the
variational one which has been especially for numerical
calculations, for instance with the Finite Element Method.
In most contact problems the contact area is a function of the

external forces. When friction has to be taken into account, the whole load history has to be followed, it is necessary to solve the equations for increments of the quantities (/1/). For **every** load increment **interactions** have to be performed to find the slip and adhesion area respectively. A lot of iterations have to be performed and it is important that the system matrix in the iterations is small. This makes it fruitful to use the Boundary Element Method for solving contact problems, expecially, a powerful tool may be to apply a superelement technique.

2. PROBLEMS

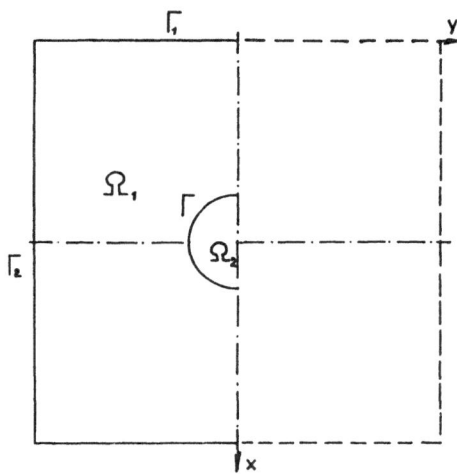

Fig. 1

A. Let us assume the domain Ω_1 to be an elastic two-dimensional wall containing a circular hole - the domain Ω_2 (see Fig. 1). Symmetry about the axis x is supposed so that normal displacement along x are zero in both domains. The following assumptions in regard to the physical situation are made:

(1) A vertical load acts along Γ_1.

(2) There are no normal displacements and no tangential forces along the bottom boundary of Ω_1 and Γ_2 , respectively.

(3) Both domains are loaded by their own weight.

Let us now denote by u^i, t^i , X^i (for i = 1,2) the vector of displacements, boundary and volume forces over the domains Ω_1 and Ω_2, respectively. The components of displacements satisfy Lame´s equation

$$(\lambda+\mu)v_i + \mu\nabla^2 u_i = X_i , \quad v = u_{k,k} , \qquad (2.1)$$

where λ , μ are constants. The indices following the comma denote the derivation and $\nabla^2 = \Delta$ is Laplace´s operator. At the same time u^1 satisfied Eg. (2.1) over Ω_1 and components u^2 the same equation over Ω_2 .

Let $n = (n_x, n_y)$ be the outward normal to Γ with respect to Ω_2. Then the functions $u_n^l = n_x u_x^l + n_y u_y^l$, l = 1,2 are displacements in the n-direction. With respect to the condition of small displacements, it is natural to satisfy on Γ the following condition:

$$u_n^1 - u_n^2 = [u]_n \leqq 0.$$

Let us now define the discretization of the boundary of both domains including the contact Γ . We replace these curves by a broken line; the elements lie on this respective line and vertices are the nodes.

Suppose that the components u^i, t^i, i = 1,2 are piecewice contact with respect to the elements, viz. on the boundary of the domains of definition. After discretization we obtain

$$U^1 t^1 = T^1 u^1 + X^1 , \qquad (2.2)$$
$$U^2 t^2 = T^2 u^2 + X^2 . \qquad (2.3)$$

263

The size of the matrices U^i and T^i is $n_i * n_i$; t^i, u^i, X^i are vectors having n_i components, where n_i is number of elements dividing the boundary of the domain Ω_i. Eg.(2.2) describes the situation along the boundary of Ω_1, Eg.(2.3) on the boundary of Ω_2. The vectors X_1 , X_2 are established by integration of the product of the kernel u^* and the volume forces over Ω_1 and Ω_2 (in our case the dead weight is considered as the volume force).

Eg.(2.2) is solvable uniquely. If the vertical components of either the weight or the loading on the boundary Ω_2 are nonzero, the solution of Eg.(2.3) is not unique. However, it will do to choose fixed value of the displacement in the direction X at one point, then there exists a unique solution to Eg.(2.3).

It is possible to renumerate the boundary elements so as to terminate the numbering on the contact line Γ . Because of the regularity of the matrices U^i, $i = 1,2$; their inversion is possible. We multiply both Egs.(2.2) and (2.3) by these inverted matrices and get

$$t^1 = H^1 u^1 + \overline{X}^1 ,$$

$$\tag{2.4}$$

$$t = H^2 u^2 + \overline{X}^2 ,$$

$$\tag{2.5}$$

where

$$H^i = (U^i)^{-1} T^i \quad , \quad \overline{X}^i = (U^i)^{-1} X^i$$

and the superscript - 1 denotes the inversion. After applying "suitable" forces, we require the interaction of both domains along the contact in the following sense:
- There are no shear forces along Γ (i.e. in the direction of the tangent line).
- The condition $[u]_n \geq 0$ is satisfied, no intersection of both bodies Ω_1 and Ω_2 occurs.
- If it holds that $[u] > 0$, the radial forces are zero along Γ .
For both domains Ω_1 and Ω_2 the problem is linear, the situation is more complicated on the contact line.

B. Let a body $\Omega = \{(x,y) \in R_2 ; x \in (-\infty;\infty); y > 0\}$ with the boundary $\Gamma_1 = \{(x,y) \in R_2 ; x \in (-\infty, \infty), y = 0\}$ be given. We make identical Ω_1 with an elastic half-plane. There is a rigid punch acting on the boundary Γ_1 as a given body of infinite stiffness, the shape being described by a function of the argument x on the contact with Γ_1 . Let us further assume zero both the friction forces and the own weight, i. e. $X_i = 0$, $i = 1,2$.

The starting equations of the problem integral formulation will have the form:

$$u_i - \int_{-\infty}^{\infty} V t_{2i}^* \, dx = 0 \; , \; i = 1,2,$$

$$\tag{2.6}$$

for $\Gamma = \Gamma_1$ and $\tau = 0$. After expanding (2.6) and inserting for forces due to the unitary loading, we obtain:

$$u(\xi,\eta) - \int_{-\infty}^{\infty} 2V(x)(\alpha \frac{\overline{X}}{r^2} + \beta \overline{x} \frac{\eta^2}{r^4}) \, dx = 0 ,$$

$$v(\xi,\eta) - \int_{-\infty}^{\infty} 2V(x)(\alpha \frac{\eta}{r^2} - \beta \frac{\eta^3}{r^4}) \, dx = 0, \; (\xi,\eta) \in \Omega_1 , \tag{2.7}$$

where

$$r^2 = \overline{X}^2 + \eta^2.$$

Now, the problem may be detailed. The rigid punch is indented to the half-plane by the force P parallel to the axis y

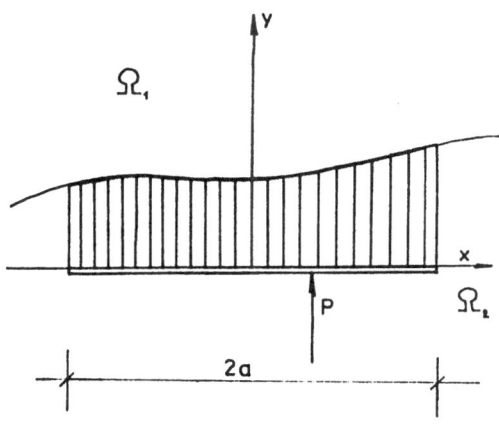

Ω_1

P

Ω_2

$2a$

Fig. 2

(see Fig. 2), i. e. v is the function symmetric to x /it holds $v(-x)=v(x)$/. The contact exists on Γ_1 at the interval $x \in (-a,a)$, where the real number a is prescribed. Consequently, the function v is determined at the interval $(-a,a)$ with exception an even contact b, i. e. $V(x) = v(x) + b$
We determine the constant from the equilibrium condition between the applied force P and a reaction resulting from the half-plane resistance against identation of the punch. Hence, the function F for , is defined un ambiguously

which it holds $F'(x) = V(x)$; $x \in \Gamma_1$
in the interval $(-a,a)$. Outside the interval $|x|<a$ we get:

$$\bar{\sigma}_y(x,0) = 0 \qquad (2.8)$$

and $\tau(x,0) = 0$ in interval $-\infty < x < \infty$. Indeed Green´s function expressed in accordance with /2/, by the relations

$$u_{11}^* = A\left[M \log(rR) - (\tfrac{\bar{x}^2}{r^2} + \tfrac{\bar{x}^2}{R^2}) \right], \qquad u_{22}^* = A\left(M \log \tfrac{r}{R} + \tfrac{R\bar{y}}{R^2} - \tfrac{\bar{y}^2}{r^2} \right),$$
$$u_{12}^* = A\bar{x}\left(\tfrac{R\bar{y}}{R^2} - \tfrac{\bar{y}}{r^2} \right),$$
$$u_{21}^* = -A\bar{x}\left(\tfrac{\bar{y}}{r^2} + \tfrac{R\bar{y}}{R^2} \right), \qquad A = \tfrac{-L_1}{4\pi\mu L_2}, \quad M = \tfrac{L_3}{L_1} \qquad (2.9)$$

satisfies the latter condition.
It is suitable to modify the formulae (2.7), by means of the part integration, to the form

$$u(\xi,\eta) = -\int_{-\infty}^{\infty} F(x)\left[\alpha \log((x-\xi)^2 + \eta^2) - \beta \tfrac{\eta^2}{(x-\xi)^2+\eta^2} \right] dx \qquad (2.10)$$
$$v(\xi,\eta) = -\int_{-\infty}^{\infty} F(x)\left[\tfrac{1}{\pi} \operatorname{arctg} \tfrac{\bar{x}}{\eta} + \beta \tfrac{(x-\xi)\eta}{(x-\xi)^2+\eta^2} \right] dx$$

Where, till now, the course of the function F has not been determined outside the interval $|x|<a$. The following stress components correspond with Egs (2.10):

$$\bar{\sigma}_x(\xi,\eta) = \tfrac{-2\mu L_1}{\pi L_2} \int_{-\infty}^{\infty} F(x) \tfrac{\bar{x}(\eta^2-\bar{x}^2)}{r^4} dx,$$
$$\bar{\sigma}_y(\xi,\eta) = \tfrac{2\mu L_1}{\pi L_2} \int_{-\infty}^{\infty} F(x) \tfrac{\bar{x}(\bar{x}^2+3\eta^2)}{r^4} dx, \qquad (2.11)$$
$$\tau(\xi,\eta) = \tfrac{-2\mu L_1}{\pi L_2} \int_{-\infty}^{\infty} F(x) \tfrac{\eta(\bar{x}^2-\eta^2)}{r^4} dx$$

It is possible to define the function F on Γ_1 by means of the condition (2.8), the second Eg. (2.11) and from the equilibrium contact condition. According to the condition (2.8), after putting $\bar{\sigma}_y(\xi,0) = 0$ for $|\xi|>0$ in (2.11_2), we will obtain:

265

Hence

$$\int_{-\infty}^{\infty} \frac{F(x)}{x-\zeta}\,dx = 0 \quad , \quad |\zeta| > q .$$

$$\int_{|x|>q} \frac{F(x)}{x-\zeta}\,dx = f(\zeta), \quad |\zeta| > q ,$$

(2.12)

where

$$f(x) = -\int_{-q}^{q} \frac{F(x)}{x-\zeta}\,dx$$

(2.12a)

is a well-known quantity.
A solution of this Cauchy's integral equation has the form:

$$F(\zeta) = \frac{1}{\sqrt{(\zeta^2-q^2)}}\left[A + \frac{1}{\pi^2}\int_{|t|>q} \frac{\sqrt{(t^2-q^2)}}{t-\zeta} f(t)\,dt \right] , \quad |\zeta| > q ,$$

where A = const. Substituting for f from (2.12a) and by
integration with respect to t we will come to the expression:

$$F(\zeta) = \frac{1}{\sqrt{(\zeta^2-q^2)}}\left[A + \frac{1}{\pi}\int_{-q}^{q} \frac{\sqrt{(q^2-x^2)}}{x-\zeta} F(x)\,dx \right] , \quad |\zeta| > q. \quad (2.13)$$

To specify A , we apply the aforementioned reactive condition:

$$P = \int_{-q}^{q} \tau_Y(\zeta,0)\,d\zeta .$$

(2.14)

Determine the course of the function τ_Y in interval $|\zeta| < q$ on Γ_1 .
It follows that

$$A = \frac{-PL_2}{2\pi T L_1} .$$

(2.15)

After inserting A to (2.13), a course of F for the arguments
$|\zeta| < q$ will be solved and, consequently, all over Γ_1 . At any
point of the half-plane Ω_1 the displacement components are
defined by the expressions (2.10) and the state of stress in
compliance with Egs.(2.1).

C. Recently, algorithms for the solution of shells subjected to
the concentrical loading have been developed. At the same time,
difficult problems of the contact interaction of shells with
supports are the matter, e.g. cylindrical shells stiffened by
both the longitudinal and transverse ribs or shells of revolution isotropic in one direction.

In the field of shallow shells, the Boundary Element Method is
worked out properly and in the case of more curved structures,
for the time being, in approximative sense.

REFERENCES

/1/ Anderson, T., Fredriksson, B., Persson, A. B. G.: The
     Boundary-Element Method Applied to Two-Dimensional Problems.
     In: Boundary Elements VIII, Springer Verlag and Computati-
     onal Mechanics pp. 247-263

/2/ Brož, P., Procházka, P.: Boundary Element Method in
     Engineering, (In Czech), State Publishing House for
     Technical Literature, Prague, 1967

THE INFLUENCE OF EDGE EFFECTS ON THE STRESS CONCENTRATION AROUND A HOLE IN A THIN SPHERICAL SHELL

V.Z.GRISTCHAK, A.N.PISANKO

Applied Mathematics Department
Dnepropetrovsk State University,
Gagarin av.,72,
320625, Dnepropetrovsk,
USSR

The paper is devoted to the stress concentration around a circular hole in a thin elastic spherical shell analysis in which the derivation and the solution of the basic boundary value problem have been dealt with extensively. The stresses obtained are consist of the following items: the membrane and bending stresses of the solution of the shell problem under classical and modified boundary conditions, the warping stress associated with non-uniforn torsion and stresses due to corresponding plane strain problem. Some questions pertaining to the accuracy and the limitations of the analysis developed have been treated while also a few relatively simple numerical examples have been given for the isotropic and transverse-isotropic shells.

The most interesting case from the authors point of view is connected with the last case of a transvers-isotropic spherical shell. Relatively complex governing equation has been forced to use an approximate (instead of a close form solution for the isotropic shell) asymptotic WKB-method technique. This approach has been used to calculate a stress concentration factors around a hole in a spherical shell, loaded by a self-equilibrium bending and twisting edge moments. Futhermore it did **enable** one to determine consistent corrections to the stresses distribution.

In recent years a number of authors have published theoretical, numerical and experimental results of the analysis of stresses around holes in thin walled shells.

They left no doubt about the fact that there is a considerable influence of the curvature of the shell and boundary effects on the stress distribution as it is affected by the presence of the hole. So the application of known stress concentration factors valid for a flat plates with a hole or obtained on the foundation of a linear (especially shallow) shell theory with the classical Kirchoff type boundary conditions can yield results that cannot be trusted and may even be completely wrong.

It is well known that the main problem in the shell theory consists of a reduction of three-dimensional equations of the theory of elasticity to the two-dimensional equations. The estimates of the errors involved are only valid if the edge tractions in the dynamic boundary conditions for the three-dimensional problem are specified in accordance with the statically admissible stress distribution derived from the shell theory. However, the corresponding stress distributions (linear - for the stress parallel to the middle surface and parabolic - for the shear stress perpendicular to the middle surface) are known for the case of a free edge only, where the edge tractions are zero. Moreover, the concrete numerical results, which can confirm of refute a general error estimates [1] are obtained for a special cases of a plates and cylindrical shells [2] .

We consider here a thin walled spherical shell (radius R , wall thickness h) that is loaded mechanically only its boundaries. The material of the shell is assumed to be linearly elastic(Young's modulus E , Poisson's ratio v). Under mentioned assumptions Koiter's simplest possible equation for a sphere is valid [3]

$$\left(\Delta + \frac{2}{R^2}\right)\left(\Delta + \frac{1}{R^2} + \frac{2ic}{hR}\right)\psi = 0 \qquad (1)$$

,

where
$$\psi = w + \frac{2ic}{Eh^2}F \qquad (2)$$

Δ is Laplacian, ψ and F are curvature and stress resultantant functions, $C = \sqrt{3(1-v^2)}$

Our aim is to calculate the bending and membrane stress concentration factors around a nominally free edge. A few important essentials of the analysis will be mentioned here. Forces, moments and stresses are expressed in the form of the accociated Legendre functions of the first and second kinds of integral order m and of complex degree with respect to the spherical meridional coordinate θ (it has be convenient to introduce spherical coordinates θ , φ such that $\theta = 0$, π specify the poles). The most important case is connected with the shell that has one loaded $\theta = \theta_1$ and one free $\theta = \theta_0$ edge. We distinguish between two variants in this case, namely classical and modified dynamic boundary conditions at the free edge. The term "nominally" means that at the free edge the following factors are presented

$$N_\nu + \frac{1}{R} M_\nu \qquad ,$$

$$N_t - \frac{1}{R} M_\nu \qquad ,$$

$$\frac{1}{R \sin \theta} \frac{dM_\nu}{d\varphi} \mp Q \qquad , \qquad (3)$$

$$M_t \qquad ,$$

where N_ν , N_t and Q are respectively the normal, tangential and transverse loads per unit length, whereas M_ν and M_t represent the torsional and bending moments per unit length.

In Fig.1 the results for the spherical shell using the modified boundary conditions in the case of edge twisting are compared to the classical case. As the curves show, the normal stresses in the corner points of the free edge for the thinner sphere behave more like the classical sphere than the thicker sphere.

In conclusion, the results obtained for a transverse-isotropic spherical shells using asymptotic WKB-method [4] show more strong influence of the boundary conditions modification on the stress and strain components near the hole. The differences between modified and classical results are of the order 20-30 percent for the problem of edge twisting and 40-50 percent for the problem of edge bending for some determined values of elastic constants of transverse-isotropic material.

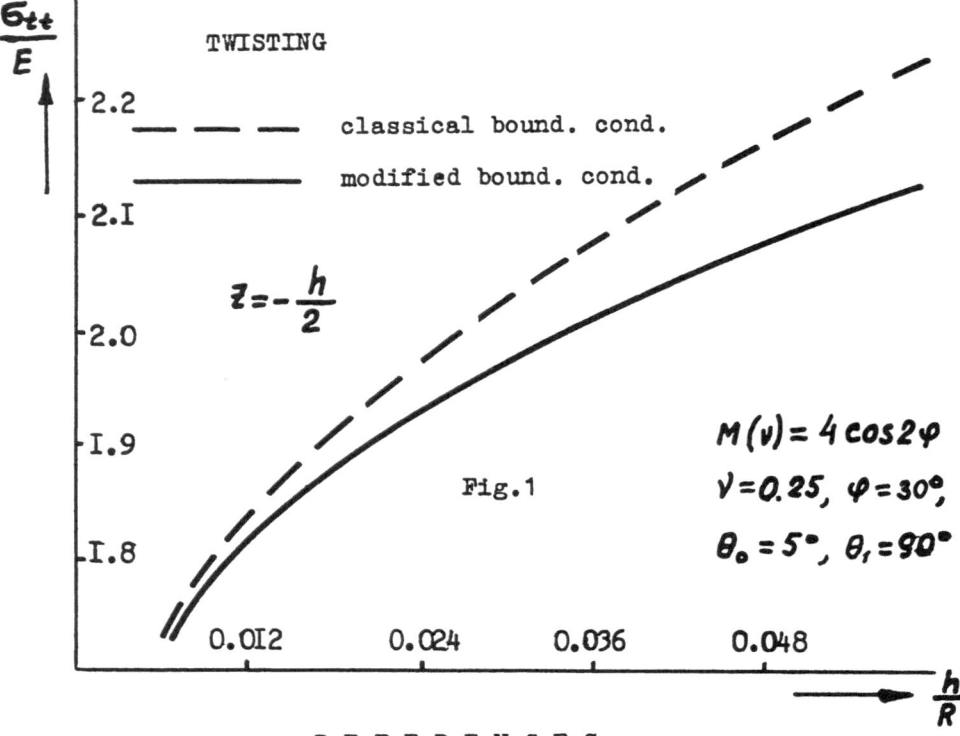

Fig.1

REFERENCES

1. W.T. Koiter and J.G. Simmonds. Foundations of shell theory.
   Rep. 473 Lab. Eng. Mech., Delft, 1972, 41pp.
2. A.M.A. Heijden van der. On modified boundary conditions for
   the free edge of a shell.-Delft Univ. Press, 1976, 131pp.
3. W.T. Koiter. A unified reduction of the intrinsic equations
   for spherical and circular cylindrical shells.-Rep.744, Lab.
   Eng. Mech.,April 1983.
4. V.Z. Gristchak and A.N. Pisanko. Statement and the solution
   method for the boundary value problem of shells and plates
   of the variable structure.-Allunion konf. on theory of shells
   and plates, Tbilisi (in Russion), 1987, pp.421-426.

LOCAL EFFECTS IN THIN-WALLED BEAMS SUBJECT TO BENDING AND SHEAR

M. HÝČA

Research Institute of Transport Equipment (VÚTZ)
Kartouzská 4
150 99 Praha 5

With the laterally loaded beam-like structural elements re-inforced by diaphragms the secondary axial membrane stresses are generated due to the shear-lag [e.g. 1]. These stresses, representing self-balanced internal force system, are superimposed on the primary axial membrane stresses, corresponding to the fictious displacement field arising at free warping of cross-sections. This phenomenon caused by the restraint of warping of cross-sections is most marked in built-in cross-sections and in cross-sections in the region of abrupt changes of the loading or internal shear force, particularly with beams at low values of the slenderness ratio or if the load is acting at a relatively small distance from a supported or built-in cross-section compared with the length of a beam.

The problem of bending stresses caused by longitudinal bending moment in beam walls was investigated by Vlasov [2] who proved that the effect of this stress is negligible. However, this conclusion may only be accepted allowing for slender beams not loaded in the vicinity of supported or built-in cross-sections in which the restraint of warping may be ignored. In other cases, especially with relatively short beams or even with slender beams loaded in relatively small distance from supported or built-in cross-sections compared with the length of the beam, neither any available theory of shear-lag nor the quoted Vlasov theory make possible to approximate theoretically the longitudinal bending stresses in beam walls. In these cases the effect of the restraint of warping may cause significant changes of the curvature of beam deflection curve at the same positions of cross-sections where the shear-lag is most marked [5] although the deflection of the beam is affected only very slightly. Thus the corresponding bending stress in the beam walls may be significant as well [6]. It superimposes on the axial membrane stress in walls corresponding to restrained warping of reinforced cross-sections. This phenomenon is most marked in outer longitudinal fibres in the region of cross-sections suffering the maximum restraint of warping [6]. The resulting high stress magnitudes may lead to a fatigue problem and in addition local buckling may become more critical [3].

Using conventional assumptions for thin-walled members with walls fully effective in supporting both axial direct stress and shear flow, except the assumption of negligible mid-surface shear strain, a consistent one-dimensional small displacement theory of straight thin-walled beam with a symmetrical open or closed cross-section as in Fig. 1, reinforced by rigid diaphragms, has been developed by the author neglecting the bending stiffness of diaphragms and ignoring local buckling [4,6].

The theory involving the effect of cross-sectional warping inclusive of its restraint makes it possible to determine the displacement and axial membrane stress fields of a beam and the longitudinal bending stress in beam walls allowing for thin-walled beams with arbitrary end conditions and subjected to a transverse loading with arbitrary discontinuities acting in the beam's longitudinal plane of symmetry.

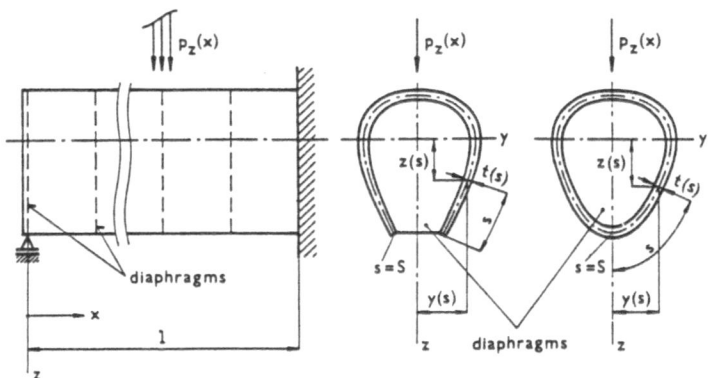

Fig. 1

Considering the n-th approximation of the beam mid surface shear strain as proposed in [6] the non-classical mathematical model in question is represented by the linear non-homogeneous boundary-value problem for n+1 simultaneous ordinary differential equations of the (2n+4)-th order with n standing for the numbers of terms in the approximation of the mid-surface shear strain

$$w_z^{IV}(x) + \sum_{i=1}^{n} {}^{i1}\beta_y \cdot {}^{i}\varrho_y^{III}(x) = \frac{p_z(x)}{EJ_y} \tag{1}$$

$$^{1j}\beta_y w_z^{III}(x) + \sum_{i=1}^{n} {}^{ij}\chi_y \cdot {}^{i}\varrho_y^{II}(x) - \frac{GF}{EJ_y} \sum_{i=1}^{n} {}^{ij}\beta_y \cdot {}^{i}\varrho_y(x) = 0 \tag{2}$$
$$(j = \overline{1,n})$$

$$\left[\delta w_z \cdot T_z\right]_{x=0,1} = 0, \qquad \left[\delta w_z' \cdot M_y\right]_{x=0,1} = 0, \tag{3}$$

$$\left[\delta {}^{i}\varrho_y \cdot {}^{i}D_y\right]_{x=0,1} = 0 \quad (i = \overline{1,n}),$$

where

$$M_y(x) = -EJ_y\left[w_z''(x) + \sum_{i=1}^{n} {}^{i1}\beta_y \cdot {}^{i}\varrho_y'(x)\right], \quad T_z(x) = M_y'(x) \tag{4}$$

$$^{i}D_y(x) = EJ_y\left[{}^{i1}\beta_y \cdot w_z''(x) + \sum_{j=1}^{n} {}^{ij}\chi_y \cdot {}^{j}\varrho_y'(x)\right] \quad (i = \overline{1,n}) \tag{5}$$

are the generalized internal forces (the bending moment M_y, the shearing force $T_z(x)$ and the warping moments ${}^{i}D_y(x)$);

272

$$^{ij}\beta_y = \frac{1}{F}\int_F \Pi^{2(i+j-1)}(s).dF \qquad (i,j = \overline{1,n}) \qquad (6)$$

$$^{ij}\chi_y = \frac{1}{J_y}\int_F {}^i\Lambda_y(s).{}^j\Lambda_y(s).dF \qquad (i,j = \overline{1,n}) \qquad (7)$$

are the dimensionless cross-sectional properties and

$$^i\Lambda_y(s) = \int_0^s \Pi_y^{2i-1}(\eta)d\eta + \frac{1}{F}\int_F F(\eta).\Pi_y^{2i-1}(\eta)d\eta \qquad (i = \overline{1,n}) \quad (8)$$

represent the unit warping functions of cross-sections where

$$\Pi_y(s) = \frac{F.S_y(s)}{J_y.t(s)} \quad , \qquad S_y(s) = \int_0^s z\ t\ d\eta \ , \qquad J_y = \int_F z^2.dF \ ,$$

$$F(s) = \int_0^s t.d\eta \ , \qquad F = F(s = S) \ . \tag{9}$$

The symbol w_z stands for the beam deflection and the symbols $^i\varrho_y(x)$ denote the unknown distribution functions of restrained cross-sectional warping.

The membrane stress components 6_x^m and τ_{xs}^m are expressed as follows

$$6_x^m(x,s) = M_y(x)\frac{z(s)}{J_y} + E\sum_{i=1}^n {}^i\varrho_y'(x)\left[{}^{ii}\beta_y.z(s) + {}^i\Lambda_y(s)\right], \quad (10)$$

$$\tau_{xs}^m(x,s) = G\sum_{i=1}^n {}^i\varrho_y(x).\Pi_y^{2i-1}(s) \ . \tag{11}$$

A glance at equations (1) to (3) shows that the presented mathematical model involves, only as a special case, the classical Bernoulli-Navier theory of bending.

Referring e.g. to rectangular box- or I-section beams, the bending stress on the upper and/or lower surfaces of flanges (for any $s \in \langle -B/2 , B/2 \rangle$) may be expressed as

$$^p6_x(x) = \pm \frac{E\ t}{2(1 - \mu^2)}\ w_z''(x) \quad \text{on the} \quad \begin{matrix}\text{upper}\\\text{lower}\end{matrix} \quad \text{surfaces of flanges} \tag{12}$$

where t is the flange thickness and μ = Poisson's ratio. Superimposing this stress on the membrane stress component (10) and referring to (4) we arrive at a final general expression for the resulting axial stress on the outer surfaces of flanges as follows

$$^p6_x(x,s) = \frac{M_y(x)}{J_y}\left[z(s) \mp \frac{t}{2(1 - \mu^2)}\right] + E.\sum_{i=1}^n {}^i\varrho_y'(x) \times$$

$$\times \left\{ {}^{11}\beta_y \left[z(s) \mp \frac{t}{2(1-\mu^2)} \right] + {}^{1}\Lambda_y(s) \right\} \quad \text{for the } \begin{array}{c}\text{upper}\\\text{lower}\end{array} \text{ flange. (13)}$$

The significance of the restraint of cross-sectional warping and the longitudinal bending stress arising on outer surfaces of box beam flanges may be quantified in percentage of the axial membrane stress corresponding to the classical theory of bending (see the first term in the expression (10)) as

$$\Delta \sigma_x^m(x,s) = \frac{\sigma_x^m(x,s) - M_y(x)\dfrac{z(s)}{J_y}}{M_y(x)\dfrac{z(s)}{J_y}} \cdot 100 \quad [\%], \quad (14)$$

$$\Delta \sigma_x^o(x,s) = \frac{{}^P\sigma_x^o(x)}{M_y(x)\dfrac{z(s)}{J_y}} \cdot 100 \quad [\%]. \quad (15)$$

Referring to the first approximation theory and considering simply supported box beams, the agreement between the predicted maximum values of the resulting axial stress ${}^P\sigma_x$ according to (13) and the results reported from the tests appears to be satisfactory [6].

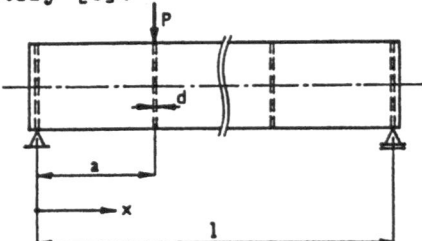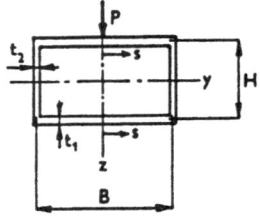

Fig. 2

Confining ourselves to the first approximation theory and referring to a simply supported beam as in Fig. 2, the following expressions may be derived for the axial membrane stress component σ_x^m and the bending stress ${}^P\sigma_x^o$ on the outer surfaces of flanges using general expressions (10) and (12):

$$\sigma_x^m(x \leqslant a ; s) = P \frac{1^2}{J_y} \left[(1 - \frac{a}{1}) \frac{x}{1} \frac{z(s)}{1} + \frac{E}{G} \frac{{}^1\theta_y}{\lambda_y^2} \frac{\text{sh}(1 - a/1){}^1\theta_y}{\text{sh}\,{}^1\theta_y} \cdot \text{sh}\,{}^1\theta_y \frac{x}{1} \times \right.$$
$$\left. \times \frac{{}^{11}\beta_y \cdot z(s) + {}^1\Lambda_y(s)}{1} \right], \quad (16)$$

$${}^P\sigma_x^o(x \leqslant a) = {}^-_+ \frac{P}{2(1-\mu^2)} \frac{t\,1}{J_y} \left[(1 - \frac{a}{1}) \frac{x}{1} + \frac{E}{G} \frac{{}^{11}\beta_y\,{}^1\theta_y}{\lambda_y^2} \frac{\text{sh}(1 - a/1){}^1\theta_y}{\text{sh}\,{}^1\theta_y} \times \right.$$

$$\times \text{sh}^1\theta_y \frac{x}{l}\Bigg] \quad \ldots \quad \text{on the } \begin{matrix}\text{upper}\\\text{lower}\end{matrix} \text{ surfaces of flanges} \tag{17}$$

where
$$\lambda_y = l/i_y = l\sqrt{F/J_y} \tag{18}$$

is a slenderness of a beam and

$$^1\theta_y = \lambda_y \cdot {}^1\vartheta_y\big/\sqrt{E/G}\;, \qquad {}^1\vartheta_y = \sqrt{{}^{11}\beta_y\big/({}^{11}\chi_y - {}^{11}\beta_y^2)} \tag{19}$$

are dimensionless quantities referring to a beam.

To illustrate the developed theory the quantities

$$\Delta 6_x^m(x = a-d/2,\ s = B/2) \quad [\%]$$
$$\text{and} \qquad \Delta 6_x^o(x = a-d/2,\ s = B/2) \quad [\%] \tag{20}$$

as defined in eqs. (14) and (15) were computed using the relations (16) and (17) considering steel beams shown in Fig. 2 in which

$$d = \frac{t_1 + t_2}{2}\;, \qquad \frac{a}{l} = 0{,}5\;, \qquad \frac{E}{G} = 2{,}62\;,$$
$$\frac{B}{H} = \frac{t_1}{t_2} = 2\;, \qquad B/t_1 \in \langle 10\,,\,50\rangle\;, \qquad l/2(B+H) \in \langle 1\,;\,4\rangle. \tag{21}$$

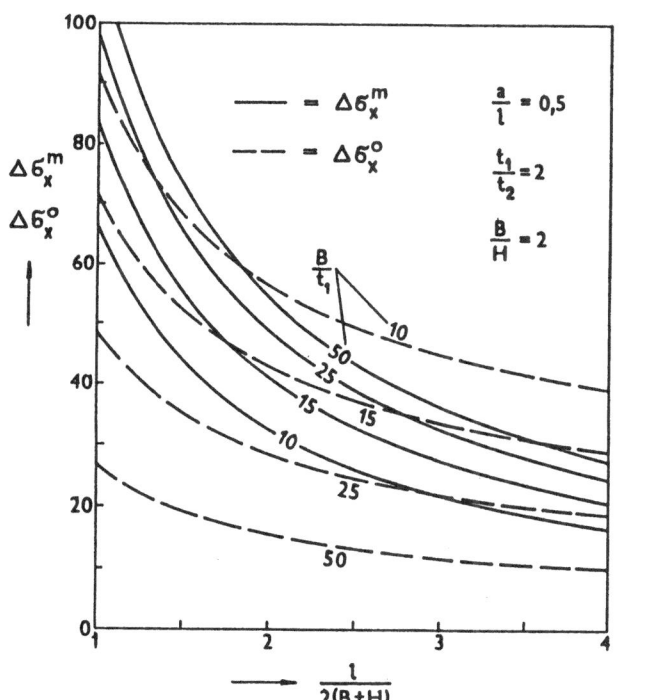

Fig. 3

275

Observing Fig. 3 reveals that with beams in question the significance of shear-lag and/or longitudinal bending stress in flanges may be the same, the effect of the first (second) phenomenon being an increasing (decreasing) function of the quantity B/t_1 and both phenomena representing decreasing functions of the beam slenderness $1/2(B+H)$.

REFERENCES

1 KLÖPPEL, K. and THIELE, F.: Analytische und experimentelle Ermittlung der Spannungsverteilung in kastenförmigen Biegequerschnitten bei örtlicher Krafteinleitung (Mittragende Breite). Der Stahlbau, 5, 1966, pp. 152 - 156

2 VLASOV, V.Z.: Thin-Walled Elastic Beams (in Russian), Gosud. Izd. Fiz.-Mat. Liter., Moskva, 1959

3 GORPICHENKO, V.M. and KULKOVA, N.N.: Fatigue Strength of Thin-Walled Box-Section Beams (in Russian), Vestnik Mashinostroeniya, No 3, 1976, pp. 30 - 33

4 HÝČA, M.: Zum Problem der Biegung bei dünnwandigen Balken mit Rücksicht auf Wölbschubverformungen. Z. Angew. Math. Mech. (ZAMM), 55, 1975, No 12, pp. 764 - 766

5 HÝČA, M.: A Functional Condition for Zero Warping of Cross-Sections of Thin-Walled Beams Under Bending. Acta Technica ČSAV, No 2, 1984, pp. 151 - 176

6 HÝČA, M.: Mathematical Model of Thin-Walled Beams Subject to Lateral Bending Allowing for Cross-Sectional Warping and Flange Bending (in Czech), Strojírenství, 36, 1986, No 6/7, pp. 339 - 345

THE NONLOCAL THEORY OF THERMOELASTIC PLATES AND SURFACE PROBLEMS

ESIN INAN
Technical University of Istanbul
Faculty of Science, Maslak 80626
ISTANBUL TURKEY,

1. INTRODUCTION

Classical continuum mechanics do not consider the influence of strains and temperature of distant points on the reference point. Nonlocal theory on the other hand, incorporates these distant effects. As a result the stress at a reference point is a functional of motion and temperatures of all points of the body. Longitudinal wave propagation in thermoelastic plates is investigated in the context of nonlocal thermoelasticity in one of the author's work |1|. Using the integral form of constitutive equations, field equations are obtained. Then the frequency equations are found for symmetric and anti-symmetric modes in generalized forms.

One of the very important problem of the nonlocal elasticity is the determination of the nonlocal moduli. This has been achived by matching the dispersion equation for plane longitudinal waves with the parallel equation known in the theory of atomic lattice dynamics. To obtain identical results to the atomic theory, dispersion relations must coincide. For homogeneous and isotropic elastic solids Eringen |2| has expressed the nonlocal moduli in the form of

$$\lambda'(|x - x'|) = \lambda\alpha(|x - x'|) \tag{1.2}$$

where

$$\alpha(x) = \frac{1}{a} (1 - \frac{|x|}{a}), \quad \frac{|x|}{a} \le 1 \quad \text{and} \quad \alpha(x) = 0, \quad \frac{|x|}{a} > 1$$

and λ characterizes any elastic modula and a is the lattice spacing. If the nonlocal moduli vary according to the relations given by(1.2),the calculations based on the non-local theory and atomic lattice dynamics will be identical.

During the derivation of the governing equations of above mentioned plate problem |1|, long-range interactions along the thickness of the plate have not been considered which disregards the surface effects. The aim of this work is to investigate the surface effects. For this purpose first we have investigated the plane longitudinal waves along one direction. As the second step we have given the results of the corresponding problem in lattice dynamics. And finally we compared the results and discussed the surface effects.

2. GOVERNING EQUATIONS

To investigate the propagation of plane longitidunal waves in a nonlocal infinite medium in one direction, say $x_1 = x$, we consider that only $u_1 = u$ component of the displacement vector $\underline{u}$ along x direction is nonzero. Then the corresponding constitutive equation is

$$t_{11} = \int_V (\lambda' + 2\mu') \frac{\partial u'}{\partial x'} dv' \tag{2.1}$$

Here λ' and μ' denote nonlocal moduli of elasticity and (') denotes the related quantity at any point x'. Second basic equation is the equation of motion

$$t_{11,1} = \rho \ddot{u} \tag{2.2}$$

Here t_{11} denotes the only non-zero stress component. Now substituting (2.2)

This work is supported by NATO Research Grant No: 86/0484 (Renewal)

277

into (2.1) together with the form given by equation(1.2)for the nonlocal modula
we arrive

$$-\frac{\partial}{\partial x}\int_V K(|x'-x|,x)\frac{\partial u'(x')}{\partial x'}\, dv'-\rho u^{..}=0 \qquad (2.3)$$

Here $K(|x'-x|,x)$ denotes the nonlocal effects and V is the volume.

To investigate the surface effects we consider an infinite plate which is defined by $0 \le x \le 2h$ and waves propagating along x direction.

The equation (2.3) may be rewritten as

$$\frac{\partial}{\partial x}\int_o^{2h}K(|x-x'|,x)\frac{\partial u'(x')}{\partial x'}\, dx'-\rho u^{..}=0 \qquad (2.4)$$

Considering the general form of nonlocal kernel $K(|x-x'|,x)$ given by Eringen |2|, it will be convenient to investigate the present problem in different three regions,I, II,III as shown in the Figure . Then the non-local kernel in region I becomes a function of x which is the location of point under consideration as well as the distance of this point to any point

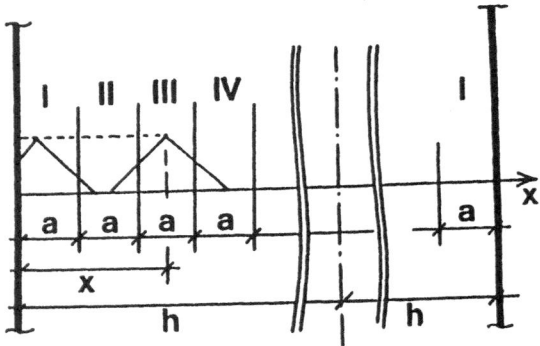

of the medium $|x-x'|$. Thus $K(|x-x'|,x)$ may be written as

$$K(|x-x'|,x)=(\lambda+2\mu)\alpha(|x-x'|)H(x)=(\lambda+2\mu)\frac{1}{a}(1-\frac{|x-x'|}{a})H(x) \qquad (2.5)$$

Here H(x) denotes Heaviside step function. Nonlocal kernels for next two regions may be written as

$$K(|x'-x|,x)=(\lambda+2\mu)\alpha(|x-x'|) \qquad x\epsilon II,III; \; x'\epsilon I,II,III \qquad (2.6)$$

which is the function of $|x-x'|$ only.
Then eq.(2.4) may be written for region I,II and III as followings.

$$u^{..}_o=\frac{c_1^2}{a}\frac{\partial}{\partial x}\{\int_o^x(1-\frac{\xi}{a})\frac{\partial u(x-\xi)}{\partial \xi}\, d\xi+\int_o^{a+x}(1-\frac{\xi}{a})\frac{\partial u(x+\xi)}{\partial \xi}\, d\xi\} \qquad x\epsilon I \qquad (2.7)$$

$$u^{..}_{1,2}=\frac{c_1^2}{a}\{\int_o^a(1-\frac{\xi}{a})[\frac{\partial^2 u(x-\xi)}{\partial \xi^2}+\frac{\partial^2 u(x+\xi)}{\partial \xi^2}]\, d\xi \qquad x\epsilon II,III \qquad$$

Here $c_1^2=\frac{\lambda+2\mu}{\rho}$. $\qquad (2.8)$

After some mathematical manipulations we find following difference-differential equations for displacement u;

$$u^{..}_o-\frac{c_1^2}{a^2}\left[u_1(x+a)-2u_o(x)\right]+\frac{c_1^2}{a^2}u_o(0)=0 \qquad ; \qquad x\epsilon I$$

$$u^{..}_1-\frac{c_1^2}{a^2}\left[u_2(x+a)-2u_1(x)+u_o(x-a)\right]=0 \qquad ; \qquad x\epsilon II \qquad (2.9)$$

$$u^{..}_2-\frac{c_1^2}{a^2}\left[u(x+a)-2u_2(x)+u_1(x-a)\right]=0 \qquad ; \qquad x\epsilon III$$

For $x\ge 3a$ we may write in general $(u_3=u)$

$$u^{..}-\frac{c_1^2}{a^2}\{u(x+a)-2u(x)+u(x-a)\}=0 \qquad x>3a \qquad (2.10)$$

which is the equation for the longitudinal wave propagation for infinite medium. Then we may at least conclude that surface effects cease beyond this distance which is 3a. Now writting $u_j=U_je^{i\omega t}$, we obtain

278

$$\left[\left(\frac{a\omega}{c_1}\right)^2 - 2\right] U_0(x) + U_1(x+a) + U_0(0) = 0 \qquad ; \qquad x \epsilon I \qquad (2.11)$$

$$\left[\left(\frac{a\omega}{c_1}\right)^2 - 2\right] U_j(x) + U_{j+1}(x+a) + U_{j-1}(x-a) = 0 \; ; \qquad x \epsilon II, III, IV, \; j = 1,2,3$$

Here considering $U(x) = \delta e^{ikx}$, we find for $x > 3a$ $\qquad\qquad\qquad (2.12)$

$$\{\left(\frac{a\omega}{c_1}\right)^2 - 2 + e^{ika} + e^{-ika}\}\delta e^{ikx} = 0 \qquad\qquad\qquad (2.13)$$

which gives us the following conditional equation

$$\left(\frac{\omega}{k}\right)^2 = c_1^2 \frac{\sin^2 ka/2}{(ka/2)^2} \qquad\qquad\qquad (2.14)$$

This is a well known dispersion equation for plane longitudinal waves found in nonlocal theory and also obtained in lattice dynamics by considering one dimensional monoatomic lattice model which is known as Born-Karman model.

3. LATTICE MODEL

The calculation of vibration frequencies of crystal lattices has generally made use of socalled "cyclic" or "periodic" boundary condition. The cyclic condition has great advantage that it produces considerable simplification into the calculations of vibration frequency. But the cyclic condition has disadvantage that its use prevents one from investigating surface effects. The vibration frequency of monoatomic one-dimensional lattice with free ends and nearest neighbor Hooke's law interactions have been discussed by Born. All modes of vibration are wave-like in character, and all frequencies lie in a single band. Then we thought that the investigation of the vibrational frequency of finite diatomic one-dimensional lattices and obtaining the results of monoatomic model as the limit case of diatomic model may give a sight to the present problem. The atoms composing a given lattice may have either of two different masses and are arranged in an alternating way. Only nearst neighbor Hook's law interactions are considered. The end atoms constitute what may be called the the surface of the lattice. Considering only longitu-dinal vibrations, following equations of motion for the atoms are obtained

$$m\ddot{u}_1 = \alpha(u_2 - u_1), \quad M\ddot{u}_{2N} = \alpha(u_{2N-1} - u_{2N})$$

$$m\ddot{u}_{2j} = \alpha(u_{2j+1} + u_{2j-1} - 2u_{2j}) \qquad 1 \le j \le N-1 \qquad (3.1)$$

$$M\ddot{u}_{2j-1} = \alpha(u_{2j} + u_{2j-2} - 2u_{2j-1}) \qquad 2 \le j \le N$$

where u_j is the displacement of an atom from its equilibrium position and the subscripts $2j-1$ and $2j$ specify the positions of light and heavy atoms, m and M respectively. If one makes the substitutions

$$u_{2j-1} = U_{2j-1} e^{i\omega t} , \quad u_{2j} = U_{2j} e^{i\omega t} \qquad\qquad (3.2)$$

the equations of motion are transformed to a system of linear homogeneous algebraic equations. Then the seculer determinant can be written as $|3|$

$$D_{2N}(u,v) = \begin{vmatrix} u+1 & 1 & 0... \\ 1 & v & 1 & 0... \\ 0 & 1 & u & 1..0.. \\ .... \\ ....0..1 & v+1 \end{vmatrix}_{2N} = 0; \quad \begin{array}{l} u = (\omega^2/q_1 - 2), \\ v = (\omega^2/q_2 - 2), \\ q_1 = \alpha/m , \; q_2 = \alpha/M \end{array} \qquad (3.3)$$

If one introduces the quantity θ defined by $(uv)^{\frac{1}{2}} = -2\cos\theta$, we find

$$D_{2N} = (uv + u + v) \; \sin 2N\theta/\sin 2\theta$$

which gives us following equations.

$$\sin 2N\theta/\sin 2\theta = 0, \quad uv + u + v = 0 \qquad\qquad (3.4, a-b)$$

The relationship between the frequency ω and the variable θ can be found as

$$\omega^2 = \sigma\{1 \pm (\cos^2\theta + \varepsilon^2\sin^2\theta)^{\frac{1}{2}}\}; \quad \sigma = q_1 + q_2; \quad \varepsilon = \frac{M-m}{M+m} \tag{3.5}$$

The solutions of equations (3.4) are then

$$\omega^2 = 0, \quad \omega^2 = \sigma \quad \{\theta = n\pi/2N, \quad 1 < n < N-1)\} \tag{3.6}$$

The zero frequency corresponds to translation of the lattice. The frequency given by $\omega^2 = \sigma$ lies in the region between the acoustical and optical branches which contains no frequency if the cyclid condition is used. Because the range of the frequency for acoustical branch is $(\theta = 0 \to \pi/2)$ $0 < \omega < \sqrt{2\alpha/m}$ while that of the optical branch is $\sqrt{2\alpha/m} < \omega < \sqrt{2\alpha(m+M)/mM}$. Then the remaining normal mode has a frequency $\omega = \sqrt{\sigma} = \sqrt{\alpha(m+M)/mM}$ which is in the gap. The value of θ corresponding to eq.(3.6.b) is complex and is given by

$$\theta = \frac{\pi}{2} + i Sh^{-1}(p/2), \quad p^2 = (M-m)^2/Mm$$

The mode given by $\omega^2 = \sigma$ can be then considered as a surface mode i.e. the maximum displacements of the atoms decrease roughly exponentially from the end having lighter atom. All other modes are wave-like in character with parameter θ playing the role of a reciprocal wavelength. When the masses are equal, the surface mode passes into a wave-like mode. Then the solution of eq.(3.4.b) gives

$$v = u = 0 \to \quad \omega^2 = 2\alpha/m \quad \text{and} \quad u = -2 \to \quad \omega^2 = 0 \tag{3.7}$$

which is an expected result and (3.7,a) corresponds to surface mode.

4. CONCLUSIONS

By the comparison of (2.9) and (3.1) we write $U_j = A_j e^{ikx}$ $(j = 0,1,2)$ and substitute this form into the equation (2.11) to obtain

$$A_1 = - (\kappa e^{ika} + e^{2kai})A_2, \quad A_0 = \{\kappa(\kappa + e^{ika}) - 1\}e^{2kai}A_2 \tag{4.1}$$

$$(\kappa e^{ikx} + 1)A_0 + A_1 e^{ika} e^{ikx} = 0 ; \quad x\varepsilon I \tag{4.2}$$

Here $\kappa = (\omega a/c_1)^2 - 2$. Now considering the continuity condition at $x = a$ which is $u_0(a) = u_1(a)$, we find

$$\kappa + 2 \cos ka = 0 \tag{4.3}$$

which tells us that the surface effects cease in an atomic distance a. Because (4.3) is exactly the same dispersion relation obtained by the equation (2.14). Or the comparison of this result with the corresponding problem as the monoatomic model, we understant that the surface mode passes into a wave-like mode. Also the corresponding solutions of the continuous model to the results (3.6) in lattice dynamics, give similar solutions. $\omega = 0$ corresponds to translation and $\omega = \sigma$ may represent surface mode. We find sim ilar results for $x = 0$. Writting $x = 0$ in the equation (4.2) we obtain

$$(\kappa+1)\left[(\kappa^2 - 1) + \kappa e^{ika}\right] = \kappa + e^{ika} \tag{4.4}$$

Here again, we find translation for $\omega = 0$ and $ka = \pi$ for cut-off frequency.

REFERENCES

|1| E. İnan and C. Eringen "Nonlocal Theory of Wave Propagation in Thermoelastic Plates" Submitted to Acoustical Society of America Journal.

|2| C. Eringen and B.S. Kim "Relation Between Non-local Elasticity and Lattice Dynamics" Crystal Lattice Deffects, 1977, Vol.7. pp.51 - 57.

|3| S. Anderson, Surface Science, Vol.1, 1975, Vienna.

CYLINDRICAL SHELLS UNDER DYNAMIC AXIAL IMPACT

IORDANKA IVANOVA

The variational principle of Hamilton is used to predict the quasi-static behaviour of the axialy impact loaded thin cylindrical shells.The full energetic functional is determined on some axisymmetric deformed middle shell surface,which gives the extremal value for the functional.This surface is given by the parametric equations and approximate well enough the middle surface of a deformed shell obtained as a result of experiments.The shell material is strain rate and curvature rate sensitive. The full energetic functional W=K-U+A ,where K is the kinetic energy,U is the deformation energy and A is the work performed by the external force is calculated on the approximated deformed middle surface and is averaged with respect to time.Some theoretical predictions are used to define the wave number and the time for forming one semilob(wave).Looking for the critical crushing force we obtain an algebraic equation which minimal real rooth defines exactly the geometry of the deformed middle surface.Using this method and some simple mechanical prediction the theoretical values for the crushing loading ,the final axial reduction of the length of shell and the number of axisymmetric waves are obtained.The comparisson between the theoretical results and experimental ones,obtained in the Department of Mechanical Engineering ,The University of Liverpool gives a good agreement.This method can also be used in the case of progressive asymetric buckling of a shell.In the dynamic case the main problem is to find a good approximation for the deformed shell surface in order to obtain analytical formulas for the dynamic crushing load and the final axial reduction of the shell length .

1.Introduction

In the recent years many articles and books have been published on the problems of static and dynamic crushing of cylindrical shells [1-6] .In [2] it is mentioned that dynamic crushing could be regarded as progressive or dynamic buckling of a shell.In this paper the progressive buckling of plastic cylindrical shells is considered.The theoretical results for the crushing force and crushing axial final reduction of the shell length are compared with the experiments,obtained in the Department of Mechanical Engineering,The University of Liverpool (see [2] ,Tabl.1) .

2.Theoretical predictions

We consider the middle length cylindrical shell ,simply supported,with a radius R ,length L ,thickness h under axial impact P(t) , initiated by a drop hammer rig,having a tup mass M with an initial velocity v_0 .We suppose that the form of the deformed middle shell surface is axisymmetric.The shell material is strain rate and curvature rate sensitive.The empirical Cowper-Symonds uniaxial constitutive equations have the form:

$$\frac{\sigma^d}{\sigma_0} = 1 + \left(\frac{\varepsilon}{\varepsilon_0}\right)^{1/p} \quad , \quad \frac{M^d}{M_0} = 1 + \left(\frac{\kappa}{\kappa_0}\right)^{1/p} \tag{1}$$

where $\varepsilon_0, \kappa_0, p$ are the material constants, σ_0 is the uniaxial yield stress, $M_0 = \frac{\sigma_0 h^2}{\sqrt{3}\ 2}$, σ^d and M^d are the dynamic stress and the bending moment,respectively.

We assume that the geometry of the deformed middle surface could be approximated by means of some surface of revolution.For convinience we shall consider only one semilob (semiwave) of this surface of revolution,described by the following parametric equations:

I part $\qquad x= r \sin \phi \quad , \quad y= R + r(1 - \cos \phi) \quad , \qquad 0 < \phi < 5\pi/6 \quad , \tag{2}$

II part $\quad x=r - r \sin (\phi - 2\pi/3), y=r(1+\sqrt{3}) + R - r \cos (\phi - 2\pi/3) , 5\ \pi/6\ < \phi < 5\pi/3 , \tag{3}$
where r is a function of the time t .
(i) Let t_0 be the time moment for forming one semilob.With 1 we denote the unknowen parameter $r(t_0)$.The other unknowen parameter is k -the number of semilobs (semiwaves),which could be determined from the entire part of the ratio $[t_1/t_0]$. Here t_1 is the total time from the velocity-time history experimental curve of a test spesimen,striken by a tup mass.
(ii) The approximation of the velocity -time history experimental curve,obtained using a laser doppler velocimeter system [2] could be regarded as
$$v(t) = v_0 (1 - t/t_1) \tag{4}$$
The lenght of a semilob could be obtained from (2) and (3) and it is given by:
$$s(t) = 5\pi r(t)/3 \tag{5}$$
where the final value of s(t) could be reached in the moment t_0. Now,integrating $r(t) = 3v_0(1 - t/t_1)$ from 0 to t_0 we obtain,that $r(t_0) = 1 = 3v_0 t_0(1 - 1/2k)/5\pi$.

Supposing that k is sufficiently large, we have:
$$1 = 3v_0 t_0/5\pi \tag{6}$$

2.Theoretical method and results

In [2] the problem of the dynamic plastic crushing of cylindrical shells under axial impact is investigated.Here both the axisymmetric (concertina mode) and asymmetric (dyamond mode) are considered using a good engineering approximation for them.In the case of progressive crushing the authors ignored the kinetic energy from the full energetic functional,based on the modified Alexander, s method [3,4] .They obtained the mean crushing loads and the effective crushing distances and got a good comparison with their experimental values.In this paper becides the abovementioned characteristics the other characteristics like the exact geometric form and the number of waves are obtained.

In order to obtain all characteristics,depending on the unknowen parameter 1 we use the variational principle of Hamilton-Ostrogradski.The exact geometric form ,determined by the parameter 1,minimizes the full energetic functional:

$$W = \int_0^{2t_0} (K - U - A) \, dt \qquad (7)$$

where K is the kinetic energy,U is the deformation energy,A is the work performed by the external force.K, U and A are calculated over the deformed axisymmetric surface of revolution.For the external force $P(t_0)$,obtained from the condition that the first variation of the functional W is zero,we look

for its critical value $\dfrac{dP(t_0)}{dl} = 0$, and obtain the algebraic equation in order to find the unknowen

parameter l.Everywhere, when integrating with respect to t,we use the theorem of the average value of the integrals.It holds only in the case of progressive (quasi-static) crushingSubstituting the parametric geometric equations (2) and (3) ,the constitutive equations (1) in the full energetic functional W we obtain:

$$U = 2k[\, 2\pi \int_{0 \, \text{I} \, \text{II}}^{2t_0} \int y \sqrt{x'^2 + y'^2} \{ \int_0^{\kappa_c} M^d \, d\kappa \} d\phi dt + 2\pi \int_{0 \, \text{I} \, \text{II}}^{2t_0} \int y \sqrt{x'^2 + y'^2} \{ \int_0^{c_c} \sigma^d \, dc \} d\phi dt \,] \qquad (8)$$

$$K = 2\pi \int_{0 \, \text{I} \, \text{II}}^{2t_0} \rho v^2(t) \int \sqrt{x'^2 + y'^2} \, d\phi dt \qquad (9)$$

$$A = 2\pi \int_{0 \, \text{I} \, \text{II}}^{2t_0} P^d(t)[-RL + 2k \int y \sqrt{x'^2 + y'^2} d\phi] dt \qquad (10)$$

The values of the κ_c and c_c could be obtained from the following considerations.
(j) While the curvatures of the two parts equations of one semilob of the deformed shell are equal

$$\kappa_I = \kappa_{II} = \frac{1}{r(t)} \quad \text{then} \quad \kappa_c = \kappa_{I,II}(t_0) = -\frac{r(t)}{r^2(t)} \quad . \text{For } \kappa_c \text{ we have:} \quad \kappa_c = \kappa_{I,II}(t_0) = -\frac{3v_0}{5\pi l^2} \qquad (11)$$

(jj) While the deformation c in the radial direction Oy is $c = \dfrac{y}{R} - 1$,then $c = \dfrac{y}{R}$,and   .

$$c_c = c_{max}(t_0) = \frac{3v_0(2 + \sqrt[2]{3})}{5\pi R} \qquad (12)$$

Substituting (11) and (12) into (8) to (10),integrating with respect to c , κ and t ,we obtain:
(We note here,that to obtain the energy functional ,calculated over the hole surface it is enough to multiply the result by 2k)

$$U = 4\pi t_0 2k \, \{ \, \frac{3v_0\sigma_0(2 + \sqrt[2]{3})}{5\pi R} [\frac{p}{1+p}(\frac{-3v_0(2 + \sqrt[2]{3})}{5\pi Rc_0})^{1/p} + 1](\frac{5\pi Rl}{3} + \frac{5\pi(\sqrt[2]{3}l^2)}{6}) - \frac{3Mv_0}{5\pi}[\frac{1}{l} + \frac{p}{1+p}$$

$$(\frac{-3v_0}{5\pi\kappa_0})^{1/p}\frac{1}{l}^{1+\frac{2}{p}} \,](\frac{5\pi R}{3} + \frac{5\pi(2 + \sqrt[2]{3})l^2}{6}) \qquad (13)$$

$$K = 4\pi t_0 2k\rho \frac{v_0^2}{2} [\frac{5\pi Rl}{3} + \frac{5\pi(2 + \sqrt[2]{3})l^2}{6} \,] \qquad (14)$$

$$A = 4\pi t_0 2k [\frac{5\pi Rl}{3} + (2 + \sqrt[2]{3}) \frac{5\pi l^2}{6} - \frac{RL}{2k}] P(t_0) \, . \qquad (15)$$

In order to obtain the value of the parameter l we derive the functional W with respect to l. After that we look for the critical value of $P(t_0)$ and we obtain the following algebraic equation:

$$\left\{\frac{-3v_0}{5\pi\kappa_0}\right\}^{1/p}(2+\sqrt[2]{3})^2\frac{2}{p}1^2+3(2+\sqrt[2]{3})R1^{1+2/p}+\frac{(p+2)(2p+3)}{p(1+p)}\left\{\frac{-3v_0}{5\pi\kappa_0}\right\}^{1/p}(2+\sqrt[2]{3})1+$$

$$2R^21^{2/p}+\frac{(p+2)^2}{p(1+p)}\left\{\frac{-3v_0}{5\pi\kappa_0}\right\}^{1/p}R^2=0 \tag{16}$$

Due to the mechanical reason the value of 1 is the minimal one among the real roots of (16). Detrmining 1, we can determine the critical value of the crushing force $P(t_0)$. Finally we obtain the following asympthotic formula for $P(t_0)$:

$$P(t_0)=\frac{Mv_0^2}{2\delta_f} \tag{17}$$

Here δ_f is the final axial reduction of the shell length L .When we take equations (2) and (3) ,we ignore the thickness h of the shell. In fact the effective crushing distance in the axial direction Ox is $\delta_e=2k(1+h)$,so taking into account the thickness h, we obtain more correct value $\delta_f=L-\delta_e$.

4. Comparison between theoretical and experimental results.

The experimental results are given in Table 1 of [2].The shell material is a mild steel with $c_0=40.4s^{-1}$,$p=5$,$\kappa_0=2c_0/h\sqrt[2]{3}$. The radius of the cylindrical shells is R=28.032 mm, the thickness h is 1.2 mm, the length L varies from 100 to 178 mm. The following comparison between theoretical and experimental values for δ_f and $P(t_0)$ are obtained:

N	Experiment	Theory
T4	δ_f= 90 mm	δ_f= 91 mm
	P^d= 44KN	P^d= 43.47 KN
T5	δ_f= 69.1 mm	δ_f= 71.4 mm
	P^d= 44.6 KN	P^d= 43.14 KN
T45	δ_f= 29.3 mm	δ_f= 28 mm
	P^d= 47.1 KN	P^d= 46 KN
T46	δ_f= 42.6 mm	δ_f= 39.8 mm
	P^d= 44.8 KN	P^d= 47.9 KN

5.Conclusions.

The variational method used in this paper for determining the crushing load and the final crushing axial reduction of the length of the cylindrical shell under dynamic impact, gives a good agreement with the experimental results.This method can also be effectively used for asymmetric (diamond) shape of the deformed cylindrical shell when progressive or dynamic buckli-ng is considered.

Acknowledgements - The author wishes to acknoledge the Bulgarian Committee of Sciences for the financial support of this study through grant number 1022.The author thanks also Professor N.Jones from the Liverpool University for his valuable comments on this paper

REFENCES
1.T.Werzbicki,N.Jones,Structural failure,J.Wiley,New York(1989)
2.W.Abramowicz and N.Jones,Dynamic axial crushing of circular tubes,Int.J.Impact Engng.,Vol.2,3,263-281 (1984)
3.J.M.Alexander,An approximate analysis of the collapse of thin cylindrical shells under axial loading,Q.J.<ech.Appl.Math.13,10-15(1960)
4.W.Johnson,Impact Strenght of Materials,Edward Arnold,London,Crane Russak,New York(1972)
5.T.Wierzbicki and W.Abramowicz,Crushing of thin-walled strain rate sensitive structures,Rozprawi Inzynierskie Polska Akademia Nauk,29,153-163(1981)
6.K.R.F.Andrews,J,L,England and E.Ghani,Classification of the axial collapse of cylindrical tubes under quasi-static loading,Int,J,Mech,Sci.,25,687-696(1983) .

STABILITY OF DOUBLE-STEEL BEAMS ELEMENTS BEYOND THE ELASTICITY LIMIT

V. Kartopoltsev

Bridges and Structures Department
The Tomsk Institute of Civil Engineering
Solyanaya Square, 2
634003 Tomsk-3
USSR

In the process of work of double-steel beams beyond the elasticity limit the total and local stability calculation has some pecularities.

The total stability calculation is characterized by the introduction of the coefficient φ in the formula

$$M_{min}/\varphi \cdot n \cdot J_{nt} \leq \beta \cdot R_s,$$

where n - is a reduced coefficient.

In testing the local stability vertical walls are considered as totally homogeneous plastics. The loss of local stability is expressed in thin sheets bulging under the influence of normal and tangential stresses.

In case of one-sided flow in the walls of double-steel beams Kirchhoff-Levy's hypothesis concerning the material incompressibility is adopted. The moment of bifurcation of plastics balance is taken for a critical state. Depending on the plate flexibility, stressed state, limited conditions bifurcated loading may be considerably excelled due to the membrane stresses development. The decision is made on the basis of lasting loading conception at the moment of bifurcation with the invariability of forces is the middle plane by power way. In its turn, bifurcated state of plate is characterized by the zero-variations of forces beyond the elasticity limit.

To solve the problem a number of assumptions is adopted, including that the moment of bifurcation may be compared with stressed state of slightly curved plate with fixed loading.

Fig. 1

Test arrangement and the failure mode of a web--flange system.

COMPARISON OF THEORETICAL RESULTS WITH EXPERIMENTAL DATA

CHARACTERISTICS OR COMPUTATIONAL FORM	TEST GIRDERS						
	S-AS1	S-AS2	S-AS4	S-AS5	I-T1	I-T1	I-T2
	EXPERIMENTAL DATA						
Web depth h (mm)	815	1145	825	825	785	785	785
Aspect ratio μ	5,5	5,5	5,5	5,5	4,24	4,25	4,25
Web yield stress σ_T^e (kp/cm^2)	2360	2160	2360	2360	3050	3050	3050
Maximum axial stress σ_P^e in flanges (kp/cm^2)	4900	3950	3660	3050	6250	6250	6250
Average shear stress τ^e (kp/cm^2)	900	850	1230	1180	-	1230	-
Reason of failure	Shear				O.K.	Shear	O.K.
	THEORETICAL RESULTS						
	a) correction coefficient						
$\nu_1 = \sigma_T/\sigma_P = E_s/E$	0,482	0,547	0,645	0,774	0,488	0,488	0,504
$\Delta = \sigma_T/\tau$	0,482	0,547	0,645	0,774	-	0,503	-
$\nu_2 = c \cdot \nu_1$	2,622	2,541	1,919	2,000	-	2,480	-
$\nu^* = \Delta_2 + 27\nu_2/\Delta^2 + 27$	0,587	0,634	0,688	0,803	-	0,595	-
$\nu^* = \nu_1$	-	-	-	-	0,488	0,488	0,504
$1/\nu^*$	1,704	1,577	1,453	1,245	2,049	2,049	1,984
	b) computations according to SNip Code						
$\sigma_o = 7(100\delta/h_{sT})^2$	6743	3416	6581	6581	7270	7270	7270
τ_o	1234	6253	1105	1205	-	1331	-
$\sqrt{(\sigma_T/\sigma_o)^2 + (\tau/\tau_o)^2}$	0,809	0,647	1,081	1,042	0,418	0,945	0,418
	c) buckling characteristics with a correction coefficient						
$\frac{1}{\nu^*}\sqrt{(\sigma_T/\sigma_o)^2 + (\tau/\tau_o)^2}$	1,380	1,020	1,571	1,298	0,848	1,588	0,833

EXPERIMENTAL DATA ON LOCAL BUCKLING OF FLANGES OF HYBRID BEAMS

Test beam	Slenderness ratio h/t	Critical stress in flanges (MPa)	$\frac{M_{exp}}{M_{pl}}$	Type of failure
M-AC1	212	7000	1,0	Combined lateral and vertical buckling
M-AC2	300	6760	0,96	Vertical buckling
M-AA1	193	6900	0,98	Local later. buckling
M-AA2	252	6450	0,92	Dtto.

Fig. 2: Test girders after failure

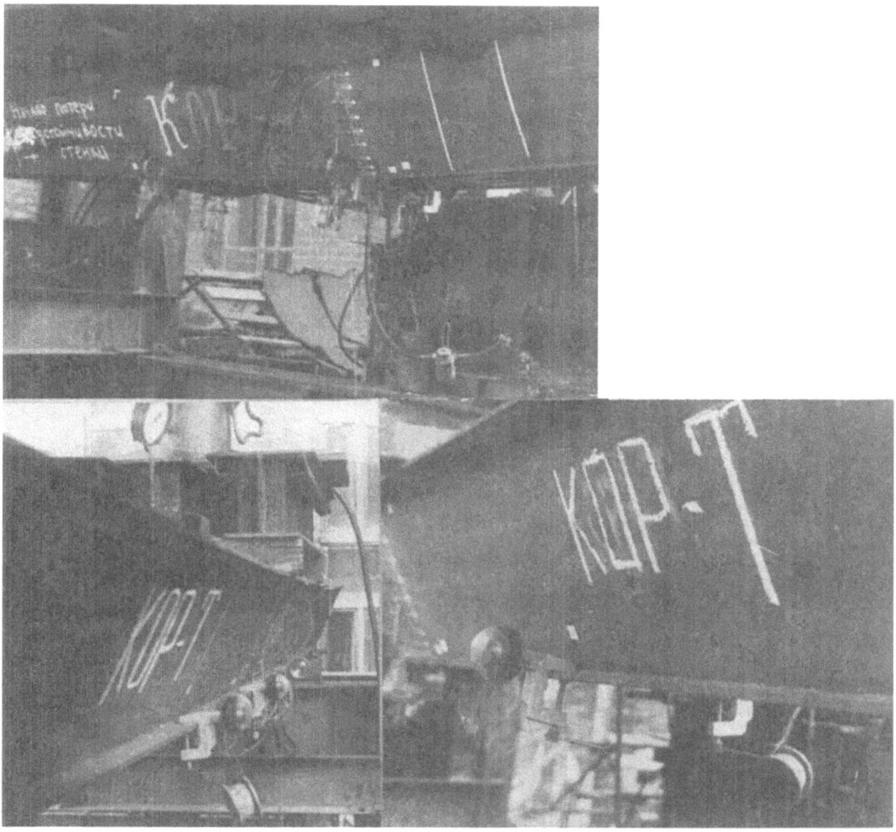

LOCAL BUCKLING EFFECTS ON THE SUPPORTS OF CONTINUOUS
TRAPEZOIDAL PROFILED STEEL SHEETS

GRAZIELLA MATEESCU, VICTOR GIONCU
BUILDING RESEARCH INSTITUTE TIMISOARA
STR. TRAIAN LALESCU 2
1900 TIMISOARA, ROMANIA

Corrugated sheet roof decking panels are more often continuous than simply supported. They are generally designed in the elastic range, to resist the maximum bending moments at the intermediate supports. A tentative plastic design must consider the differences in the behaviour of these corrugated sheet at the supports and that in the central zone of the span. While the postcritical behaviour under bending is stable (fig.1a), it becomes unstable under bending and shear (fig.1b).

In the case of a continuous flexural member, the decrease in the moments at the supports ΔM_B due to local buckling results in an increase in the central zone moments ΔM_1. If $\Delta M_B = \alpha M$, it follows that $\Delta M_1 = \beta M_1$.

Laboratory tests were carried out on 21 panels modelling the behaviour of continuous panels at the intermediate supports (fig.3 and 4), and 6 panels modelling the behaviour in the central zones of continuous elements (fig.3 and 5). It can be noticed that in the first case the behaviour is strongly unstable, while in the second case a horizontal plateau is present. For practical design purposes, the actual curve caracterizing the behaviour in the zones of intermediate supports was replaced by a trilinear diagram (fig.2b). The following values were obtained, based on test results:

$$\alpha_m = 0.481 \quad ; \quad \overline{s} = 0.077 \quad ; \quad \alpha = \alpha_m + 2\overline{s} = 0.635$$

For a two-bay continuous panel, one obtaines $\beta = (6 + \alpha)\alpha/9$ and for the value of determined experimentally, $\beta = 0.47$. This rise is very substantial and the strength reserves in the central zones are considerably reduced.

A two-bay panel was tested in order to check up the validity of the theoretical results. The panel failed immediately after the appearance of local buckling, thus the element did not exhibit any capacity of plastic adaptation.

Conclusion: The local effect of the shear forces at the intermediate supports is to reduce the bending moments in these regions (due to local buckling) and consequently to increase the moments in the central zones.

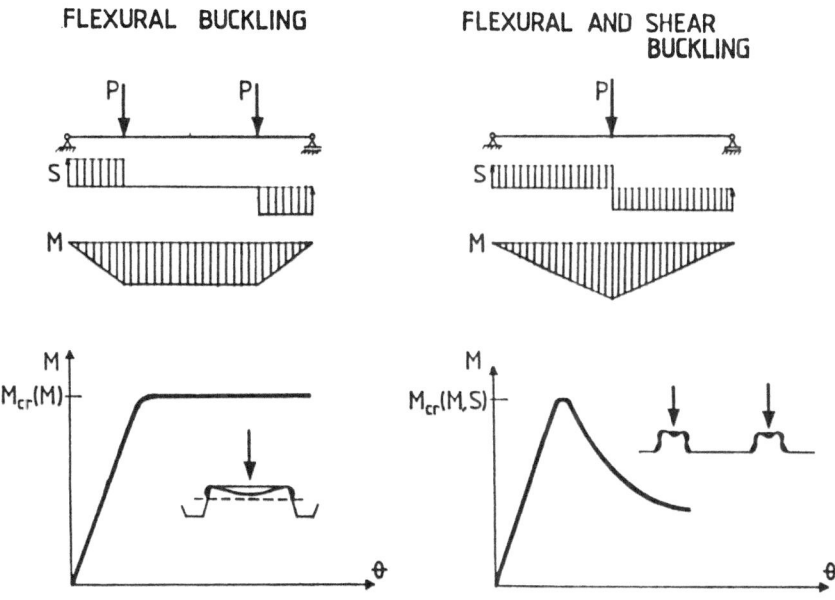

FLEXURAL BUCKLING

FLEXURAL AND SHEAR BUCKLING

FIG. 1

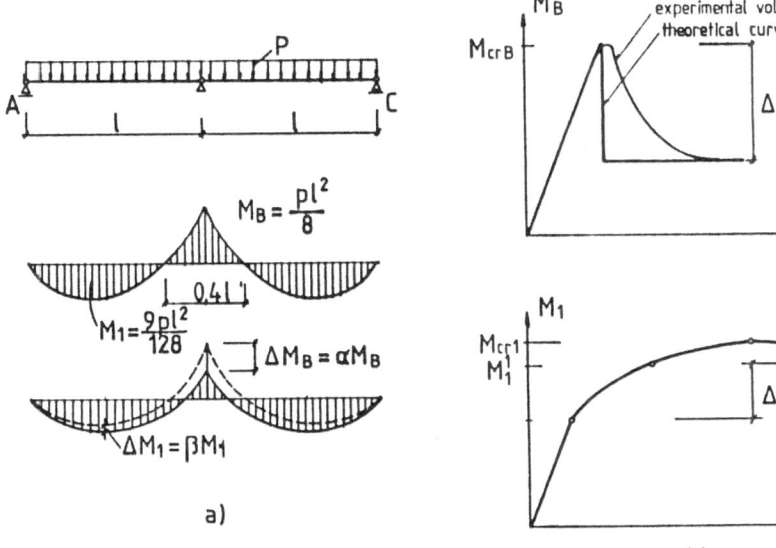

$$M_B = \frac{pl^2}{8}$$

$$M_1 = \frac{9pl^2}{128}$$

$\Delta M_B = \alpha M_B$

$\Delta M_1 = \beta M_1$

a)

experimental values
theoretical curve

$\Delta M_{crB} = \alpha M_{crB}$

$\Delta M_1 = \beta M_1$

b)

FIG. 2

EXPERIMENTS

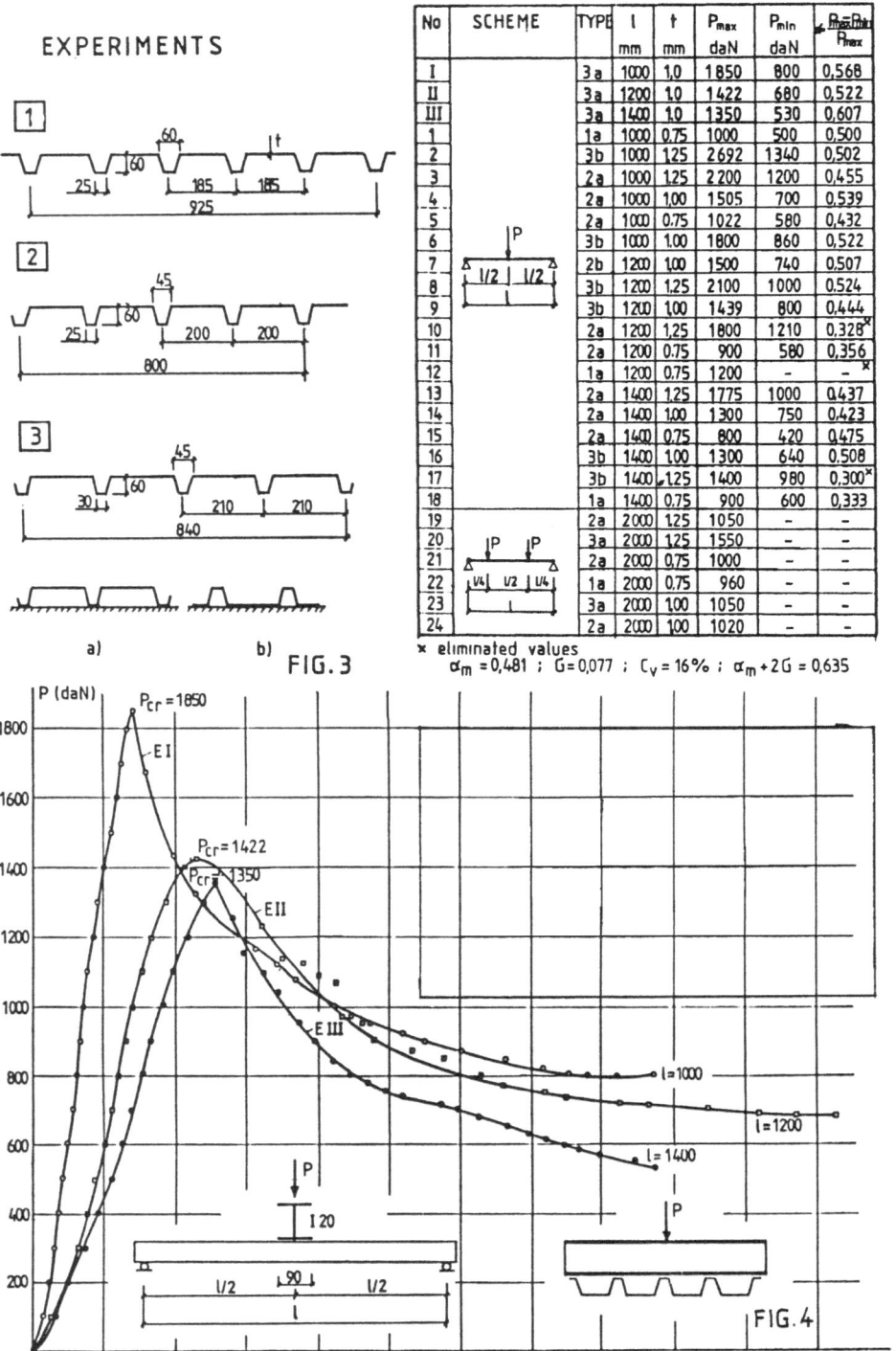

No	SCHEME	TYPE	l mm	t mm	P_{max} daN	P_{min} daN	$\frac{P_{max}-P_{min}}{P_{max}}$
I		3a	1000	1,0	1850	800	0,568
II		3a	1200	1,0	1422	680	0,522
III		3a	1400	1,0	1350	530	0,607
1		1a	1000	0,75	1000	500	0,500
2		3b	1000	1,25	2692	1340	0,502
3		2a	1000	1,25	2200	1200	0,455
4		2a	1000	1,00	1505	700	0,539
5		2a	1000	0,75	1022	580	0,432
6		3b	1000	1,00	1800	860	0,522
7		2b	1200	1,00	1500	740	0,507
8		3b	1200	1,25	2100	1000	0,524
9		3b	1200	1,00	1439	800	0,444
10		2a	1200	1,25	1800	1210	0,328 ×
11		2a	1200	0,75	900	580	0,356
12		1a	1200	0,75	1200	-	- ×
13		2a	1400	1,25	1775	1000	0,437
14		2a	1400	1,00	1300	750	0,423
15		2a	1400	0,75	800	420	0,475
16		3b	1400	1,00	1300	640	0,508
17		3b	1400	1,25	1400	980	0,300 ×
18		1a	1400	0,75	900	600	0,333
19		2a	2000	1,25	1050	-	-
20		3a	2000	1,25	1550	-	-
21		2a	2000	0,75	1000	-	-
22		1a	2000	0,75	960	-	-
23		3a	2000	1,00	1050	-	-
24		2a	2000	1,00	1020	-	-

a)   b)   FIG.3

× eliminated values
$\alpha_m = 0,481$; $G = 0,077$; $C_v = 16\%$; $\alpha_m + 2G = 0,635$

FIG.4

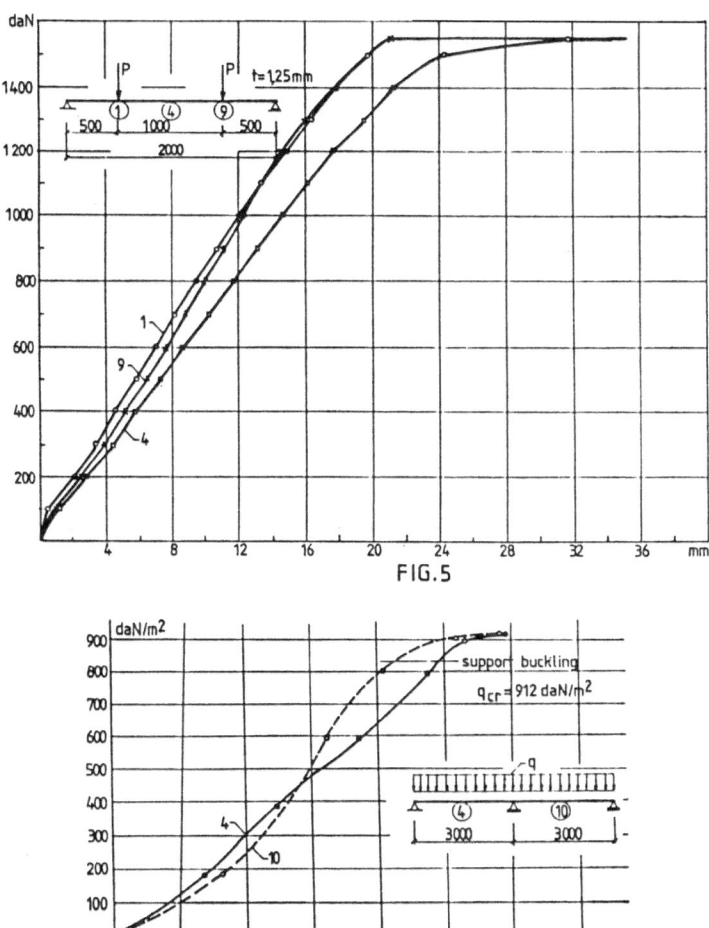

FIG.5

FIG.6

NONSTATIONARY VIBRATIONS OF RECTANGULAR PLATE EXCITED BY CONCENTRATED FORCE WITH LINEARLY VARIABLE FREQUENCY

J. Náprstek, O. Fischer

Dept. of Dynamics of Structures and Continua
Institute of Theoretical and Applied Mechanics
Vyšehradská 49, 128 49 Praha 2
Czechoslovakia

Many physical and technical problems can be modelled as a rectangular plate with certain edge-supports, loaded by forces with variable frequency (e.g. machines in industrial buildings, electrodynamic systems). Such systems can be solved numerically using finite element method, i. e. solving the system of ordinary differnetial equations in time. In some practical cases of frequency-variation this problem can be solved analytically; the advantage of such a procedure is the objectivity and the possibility of parametric analyses of the problem.

Let us examine a rectangular plate of constant thickness h according to Fig. 1, simply supported on all sides. A concentrated mass in the position ξ ,η is loaded by a force q(t) (1). As this is the only external loading, the equation of motion can be written in the form (2) - see [2]. Considering zero initial condition, Laplace transformation can be applied (3) and expressed by means of the influence function of the deflection at the point x, y, viz. $G(x,y,\xi,\eta,p)$, in which the frequency parameter p is included. This influence-function for the case in question is the solution to the equ. (7) and has the form of (8). The resulting motion of the plate w(x,y,t) will be obtained as the convolution (9); the function r(x,y,t) can be evaluated from its Laplace-image R(x,y,p) (6), which has been expressed as a sum of partial fractions (12) or (13) using the complex poles (11) - the radices of (10). For r(x,y,t) we thus obtain the sum (14) and for the final solution (15).

For linearly increasing frequencies of a rotating machine (16) the given general equations and formulae can be expressed in the form of probabilistic integrals like (17). Some examples of obtained results in graphical form are given in Fig. 2 - 4. It can be seen that the maximum amplitude will be reached with higher rotations than those corresponding to the natural frequency of the structure. This delay decreases with higher natural frequencies.

REFERENCES

/1/ Fischer O., Náprstek J.: Analytical solution to the response of a structure excited by a force with linearly increasing frequency (in Czech). UTAM Intern. report, 1990
/2/ Timoschenko S.: Vibration problems in engineering. D. V. Nostrand Co, N. Y. 1955
/3/ Ditkin V. A., Prudnikov A. P.: Handbook of operator calculus. (in Russian), Moskva 1965

$$\varrho = \varrho(t) \; ; \; \varrho(t) = 0 \; for \; t \leq 0 \tag{1}$$

$$D.\Delta\Delta w + 2\omega_b \varrho h \frac{\partial w}{\partial t} + \varrho h \frac{\partial^2 w}{\partial t^2} = \left(-M \frac{\partial^2 w}{\partial t^2} + \varrho(t)\right)\cdot Dir \, (x-\xi) \, Dir \, (y-\eta) \tag{2}$$

$$D = \frac{1}{12}\cdot\frac{Eh^3}{1-\nu^2} \; ; \; \omega_b - \text{ circular frequency of damping}$$

$$\Delta\Delta W - \varphi^4 W = \left(-\frac{Mp^2}{D} W + \frac{\varrho(p)}{D}\right)\cdot Dir \, (x-\xi)\cdot Dir \, (y-\eta) \tag{3}$$

$$\varphi^4 = \frac{\varrho h p^2}{D}\left(1 + 2\,\frac{\omega_b}{p}\right)$$

$$W(x,y,p) = \left(\frac{\varrho(p)}{D} - \frac{Mp^2}{D}\cdot W(\xi,\eta,p)\right) G(x,y,\xi,\eta,p) \tag{4}$$

$$W(x,y,p) = Q(p)\,\frac{G(x,y,\xi,\eta,p)}{D+p^2 MG(\xi,\eta,\xi,\eta,p)} = Q(p).R(x,y,p) \tag{5}$$

$$R(x,y,p) = \frac{G(x,y,\xi,\eta,p)}{D+p^2 MG(\xi,\eta,\xi,\eta,p)} \tag{6}$$

$$\Delta\Delta G - \varphi^4 G = \frac{1}{D}\, Dir \, (x-\xi)\cdot Dir \, (y-\eta) \tag{7}$$

$$G(x,y,\xi,\eta,p) = \frac{4}{Dab}\sum_{k=1}^{n}\sum_{l=1}^{n}\frac{\sin\frac{k\pi\xi}{a}\cdot\sin\frac{l\pi\eta}{b}}{\pi^4\left(\frac{k^2}{a^2}+\frac{l^2}{b^2}\right)^2 - \varphi^4}\cdot\sin\frac{k\pi x}{a}\cdot\sin\frac{l\pi y}{b} \tag{8}$$

$$w(x,y,t) = \int_0^t g(t) \cdot r(x,y,t-\tau)\, d\tau \tag{9}$$
$$r(x,y,t) = \mathcal{L}^{-1}(R(x,y,p))$$

inverse transformation:

$$D + Mp^2 G(\xi,\eta,\xi,\eta,p) = 0 \tag{10}$$

poles

$$p_j = -\alpha_j + i\beta_j \quad , \quad \bar{p}_j = -\alpha_j - i\beta_j \quad \Rightarrow \tag{11}$$

$$\Rightarrow R(x,y,p) = \sum_{j=1}^{\infty} \frac{A_j}{p-p_j} + \sum \frac{B_j}{p-\bar{p}_j} = \frac{P(p)}{U(p)} \tag{12}$$
$$A_j = \frac{P(p_j)}{U'(p_j)} \quad ; \quad B_j = \frac{P(\bar{p}_j)}{U'(\bar{p}_j)}$$

$$R(x,y,p) = \sum_{j=1}^{\infty} \frac{\psi_j\, p - \lambda_j}{p^2 + 2\alpha_j p + (\alpha_j^2 + \beta_j^2)} \tag{13}$$
$$\psi_j = 2 Re\,(A_j) \quad ; \quad \lambda_j = 2\left(-\alpha_j\, Re\,(A_j) + \beta_j\, Im\,(A_j)\right)$$

$$r(x,y,t) = \sum_{j=1}^{\infty} S_j e^{-\alpha_j(t-\tau)} \cdot \sin\left(\beta_j(t-\tau) + \varphi_j\right) = \sum\left(A_j e^{p_j(t-\tau)} + \bar{A}_j e^{\bar{p}_j(t-\tau)}\right) \tag{14}$$

resulting movement:

$$w(x,y,t) = \sum\left[A_j(x,y)\, T_j(t) + \bar{A}_j(x,y)\, T_j^+(t)\right] \tag{15}$$
$$T_j(t) = \int_0^t g(\tau) e^{p_j(t-\tau)}\, d\tau \; ; \; T_j^+(t) = \int_0^t g(\tau) e^{\bar{p}_j(t-\tau)}\, d\tau$$

excitation:

$$g(\tau) = g_0 \cdot (\varepsilon^2 t^2 + i\varepsilon) e^{-i\varepsilon t^2/2} \tag{16}$$

$$\Phi(z) = e^{-z^2} \int_0^z e^{\xi^2} \cdot d\xi \tag{17}$$

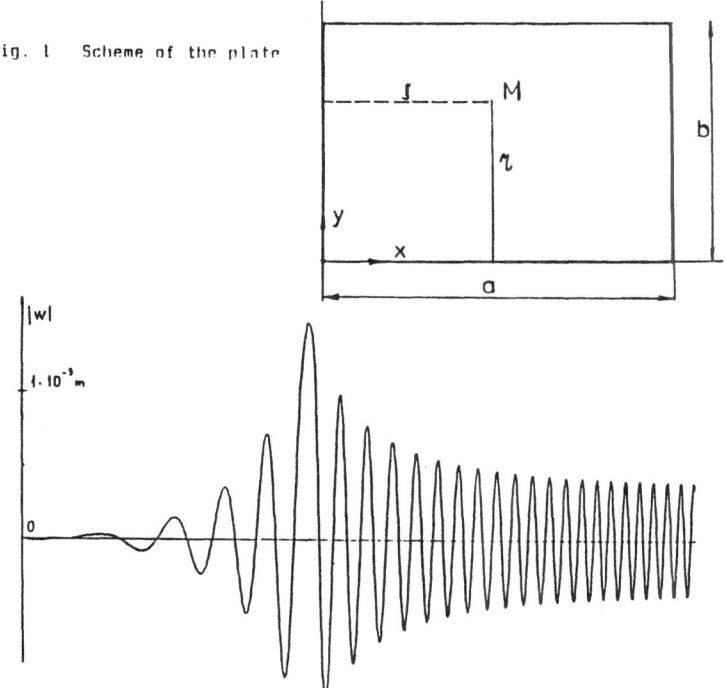

Fig. 1   Scheme of the plate

Fig. 2   Deflection of one point of the plate - the component
corresponding to the 1st natural mode

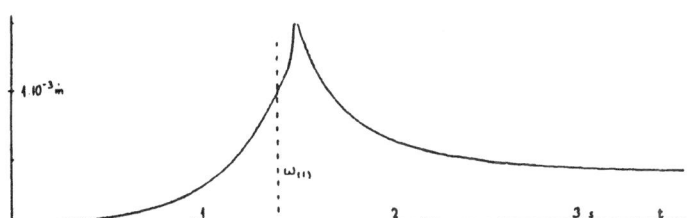

Fig. 3   Envelope of the amplitude - maxima in 1st natural mode

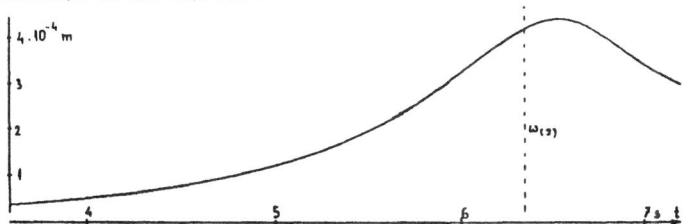

Fig. 4   Envelope of the amplitude - maxima in 2nd natural mode

295

GENERELIZED NONLINEAR THEORY OF SHELLS AND PLATES
APPLICABLE
FOR CONTACT LOADS PROBLEMS

B.L. Pelekh, M.V. Marchuk

Department for Mechanics of Thin-Walled Elements of Structures,
Institute of Applied Problems of Mechanics and Mathematics,
Naukova 3 "b",
290601 Lvov,
U.S.S.R.

Maximal simplicity of resolving equations and the completeness of accounting of physical mechanical characteristics of the used materials and load conditions are those natural requirements which must be satisfied by different variants of the theory of thin-walled elements of structures. The first condition, particularly for nonlinear statement of the problem of investigating stress-strained state of shells and plates is of importance for constructing solutions of corresponding boundary or initialy-boundary problems. The second condition represents the calculated model choice adequacy. Thus, for example, while investigating thin-walled constructions of most composites it is necessary to account in the calculated model, together with stiffness characteristics anisotropy, the mechanical pliability to transversal shear deformation and the reducing in thicness. Contact loading of shells and plates has its specifity, it is carried out by the interaction between thin-walled elements and with rigid and elastic bodies as well. In most cases interaction has a local nature with the unknown region of contact. In the linear statement theories of shells and plates, taking into account given factors, have been developed sufficiently enough. In nonlinear statement the majority of investigations rests upon the classical theory or on the technical theory based on the generalized shear model [1,2].

It should be noted that local effects, caused by contact loading of shells and plates can be detected using linear defined models of thin-walled element deformation. Though, for investigation of nonlinear effects occuring during such loading it is necessary to use geometricaly and in some cases physicaly nonlinear theories of shells and plates.

In the presented paper the previously developed approach to the construction of linear defined equatios of shells and plates applicable for contact problems [3,4,5] has been generalized for the case of geometrical nonlinearity. The essence of such approach consists in using the method of expansion of characteristics of the stress-strained state of thin-walled element in Legen polynomials from the normal coordinate to the median surface [7] with simultaneous satisfaction of boundary conditions in stresses on outward surfaces [5].

Equilibrium equations with reference to the component asymmetrical stress tensor S^{ij} [1], nonlinear deformation relation [1,2], the elasticity relationship and static boundary conditions on outward surfaces $\alpha_3 = \pm 1$ ($\alpha_3 = x^3/h$, fig. 1) were assumed to be initial ones. The approximation expressions along the coordinate

Fig. 1.

$$S^{ij} = S_0^{ij} P_0(\alpha_3) + S_1^{ij} P_1(\alpha_3), \ (i,j = 1,2)$$
$$S^{i3} = \sum_{\kappa=0}^{4} S_\kappa^{i3} P_\kappa(\alpha_3), \quad (i = 1,2; \ i \neq 3)$$
$$S^{33} = S_0^{33} P_0(\alpha_3) + S_1^{33} P_1(\alpha_3) + S_2^{33} P_2(\alpha_3) \quad\quad (1)$$

Fig. 2.

$$e_{ij} = e_{ij}^o P_0(\alpha_3) + e_{ij}^1 P_1(\alpha_3), \ (i = 1,2; j = 1,\overline{3})$$
$$e_{33} = e_{33}^o P_0(\alpha_3) \quad\quad (2)$$

$$u_i = u_i^o P_0(\alpha_3) + u_i^1 P_1(\alpha_3), \ (i = 1,2,3), \quad (3)$$

where $P_\kappa(\alpha_3)$ - Legendre polynomials from α_3 variable [5], were taken for stress tensor S^{ij} and strain tensor e_{ij} components and for vectors of elastic displacement u_i. Approximational expressions for stress symmetric tensor components σ^{ij} follows from (1) with allowance for the fulfilment of relations [1]

$$S^{ij} = 6^{i\kappa}(\delta_\kappa^j + \nabla_\kappa u^j) \quad\quad (4)$$

by the criterion of the method of least squares.

From the general variational principle of mechanics of deformable media [4] with the allowance for condition in forces on outward surfaces follows the complete set of equations of flexible shells taking into account the transversal shear deformation and the reducih in thickness:

a) equilibrium equations

$$\frac{\partial}{\partial x^1}(A_2 S_0^{11}) + \frac{\partial}{\partial x^2}(A_1 S_0^{12}) - S_0^{22}\frac{\partial A_2}{\partial x^1} + S_0^{12}\frac{\partial A_1}{\partial x^2} + A_1 A_2(k_1 S_0^{13} + X_1^-) = 0, (1 \rightleftarrows 2)$$

$$\frac{\partial}{\partial x^1}(A_2 S_1^{11}) + \frac{\partial}{\partial x^2}(A_1 S_1^{12}) - S_1^{22}\frac{\partial A_2}{\partial x^1} + S_1^{12}\frac{\partial A_1}{\partial x^2} + A_1 A_2(k_1 S_0^{13} - S_0^{31}/h + X_1^+) = 0, \quad (5)$$
$$(1 \rightleftarrows 2)$$

$$\frac{\partial}{\partial x^1}(A_2 S_0^{13}) + \frac{\partial}{\partial x^2}(A_1 S_0^{23}) + A_1 A_2(X_3^- - k_1 S_0^{11} - k_2 S_0^{22}) = 0 ,$$

$$\frac{\partial}{\partial x^1}(A_2 S_1^{13}) + \frac{\partial}{\partial x^2}(A_1 S_1^{23}) + A_1 A_2(X_3^+ - S_0^{33}/h - k_1 S_1^{11} - k_2 S_1^{22}) = 0,$$

$$X_i^\pm = [X_i(h) \pm X_i(-h)]/h , \ X_i(\pm h) = S^{3i}(\pm h)$$

b) deformation relations

$$e_{ij}^o = \frac{1}{2}(\mathcal{E}_{ji}^o + \mathcal{E}_{ij}^o) + \sum_{\kappa=1}^{3}(\frac{1}{2}\mathcal{E}_{\kappa j}^o \mathcal{E}_{\kappa i}^o + \frac{1}{6}\mathcal{E}_{\kappa j}^1 \mathcal{E}_{\kappa i}^1)$$
$$e_{ij}^1 = \frac{1}{2}(\mathcal{E}_{ji}^1 + \mathcal{E}_{ij}^1) + \frac{1}{2}\sum_{\kappa=1}^{3}(\mathcal{E}_{\kappa j}^1 \mathcal{E}_{\kappa i}^o + \mathcal{E}_{\kappa j}^o \mathcal{E}_{\kappa i}^1) \quad\quad (i,j = 1,2)$$

$$e_{i3}^o = \frac{1}{2}(\mathcal{E}_{3i}^o + \mathcal{E}_{i3}^o) + \frac{1}{2}\sum_{\kappa=1}^{3}\mathcal{E}_{\kappa 3}^o \mathcal{E}_{\kappa i}^o , \ e_{i3}^1 = \frac{1}{2}(\mathcal{E}_{3i}^1 + \sum_{\kappa=1}^{3}\mathcal{E}_{\kappa 3}^o \mathcal{E}_{\kappa i}^1), (i=1,2)_{(6)}$$

$$e_{33}^o = \mathcal{E}_{33}^o + \frac{1}{2}\sum_{\kappa=1}^{3}(\mathcal{E}_{\kappa 3}^o)^2 ,$$

$$\mathcal{E}_{11}^{\kappa} = \frac{1}{A_1}\frac{\partial u_1^{\kappa}}{\partial x^1} + \frac{1}{A_1 A_2}\frac{\partial A_1}{\partial x^2}u_2^{\kappa} + k_1 u_3^{\kappa}; \quad \mathcal{E}_{21}^{\kappa} = \frac{1}{A_1}\frac{\partial u_2^{\kappa}}{\partial x^1} - \frac{1}{A_1 A_2}\frac{\partial A_1}{\partial x^2}u_1^{\kappa},$$

$$\mathcal{E}_{31}^{\kappa} = \frac{1}{A_1}\frac{\partial u_3^{\kappa}}{\partial x^1} - k_1 u_1^{\kappa}, \quad (\kappa = 0, 1; \; 1 \rightleftarrows 2); \quad \mathcal{E}_{i3}^{0} = u_i^{1}/h, \quad (i = 1,2,3)$$

c) elasticity relationship

$$S_0^{11} = \mathcal{G}_0^{11}(1+\mathcal{E}_{11}^0) + \frac{1}{3}\mathcal{G}_1^{11}\mathcal{E}_{11}^1 + \mathcal{G}_0^{12}\mathcal{E}_{21}^0 + \frac{1}{3}\mathcal{G}_1^{12}\mathcal{E}_{21}^1 + \frac{1}{A_1}\mathcal{G}_0^{13}\mathcal{E}_{13}^0$$

$$S_0^{12} = \mathcal{G}_0^{11}\mathcal{E}_{21}^0 + \frac{1}{3}\mathcal{G}_1^{11}\mathcal{E}_{21}^1 + \mathcal{G}_0^{12}(1+\mathcal{E}_{22}^0) + \frac{1}{3}\mathcal{G}_1^{12}\mathcal{E}_{22}^1 + \frac{1}{A_2}\mathcal{G}_0^{12}\mathcal{E}_{23}^0$$

$$S_0^{13} = A_1(\mathcal{G}_0^{11}\mathcal{E}_{31}^0 + \frac{1}{3}\mathcal{G}_1^{11}\mathcal{E}_{31}^1) + A_2(\mathcal{G}_0^{1}\mathcal{E}_{32}^0 + \frac{1}{3}\mathcal{G}_1^{12}\mathcal{E}_{32}^1) + \mathcal{G}_0^{13}(1+\mathcal{E}_{33}^0)$$

$$S_1^{11} = \mathcal{G}_1^{11}(1+\mathcal{E}_{11}^0) + \mathcal{G}_0^{12}\mathcal{E}_{11}^1 + \mathcal{G}_0^{12}\mathcal{E}_{21}^1 + \mathcal{G}_1^{12}\mathcal{E}_{21}^0 + \frac{1}{A_1}\mathcal{G}_1^{13}\mathcal{E}_{13}^0 \qquad (7)$$

$$S_1^{12} = \mathcal{G}_0^{11}\mathcal{E}_{21}^1 + \mathcal{G}_1^{11}\mathcal{E}_{21}^0 + \mathcal{G}_1^{12}(1+\mathcal{E}_{22}^0) + \mathcal{G}_0^{12}\mathcal{E}_{22}^1 + \frac{1}{A_2}\mathcal{G}_1^{13}\mathcal{E}_{23}^0$$

$$S_1^{13} = A_1(\mathcal{G}_0^{11}\mathcal{E}_{31}^1 + \mathcal{G}_1^{11}\mathcal{E}_{31}^0) + A_2(\mathcal{G}_0^{12}\mathcal{E}_{32}^1 + \mathcal{G}_1^{12}\mathcal{E}_{32}^0) + \mathcal{G}_1^{13}(1+\mathcal{E}_{33}^0), \quad (1 \rightleftarrows 2)$$

$$S_0^{33} = A_1(\mathcal{G}_0^{13}\mathcal{E}_{31}^1 + \frac{1}{3}\mathcal{G}_1^{13}\mathcal{E}_{31}^1) + A_2(\mathcal{G}_0^{23}\mathcal{E}_{32}^0 + \frac{1}{3}\mathcal{G}_1^{23}\mathcal{E}_{32}^1) + \mathcal{G}_0^{33}(1+\mathcal{E}_{33}^1)$$

$$S_0^{31} = \mathcal{G}_1^{13}(1+\mathcal{E}_{11}^0) + \frac{1}{3}\mathcal{G}_1^{13}\mathcal{E}_{11}^1 + \frac{A_2}{A_1}(\mathcal{G}_0^{23}\mathcal{E}_{12}^0 + \frac{1}{3}\mathcal{G}_1^{23}\mathcal{E}_{12}^1) + \frac{1}{A_1}\mathcal{G}_0^{33}\mathcal{E}_{23}^0, \quad (1 \rightleftarrows 2)$$

$$\mathcal{G}_0^{ij} = c^{ijem}\mathcal{E}_{em}^{\kappa}$$

At $k_1 = k_2 = 0$ from (5)-(7) the equations for plates follow.

The received nonlinear equations are the equations of minimum order in which the faculty of shells and plates to transversal shear strains and the reducing in thicness is taken into account. Equations given in 6 follow from (5)-(7) as a particular case at certain limitations.

Rigidly hold isotropic plate under the influence of rigid parabolic stamp is considered as an example. In this case the resolving equations are of the form:

a) contact region $\qquad |x^1| \leq a$

$$\frac{d}{dx^1}S_0^{11} = 0; \quad \frac{d}{dx^1}S_1^{11} - S_0^{31}/h = 0$$

$$\frac{d}{dx^1}S_0^{13} + S_+^{33}/h = 0; \quad \frac{d}{dx^1}S_1^{13} - S_0^{33}/h + S_+^{33}/h = 0 \qquad (8)$$

$$u_3(x^1, h) = -\delta + \alpha x^2; \quad 2\int_0^a S_+^{33}(x_1)dx^1 = P$$

b) free region $\qquad a \leq |x_1| \leq \ell$

$$\frac{d}{dx^1}S_0^{11} = 0; \quad \frac{d}{dx^1}S_1^{11} - S_0^{31}/h = 0$$

$$\frac{d}{dx^1}S_0^{13} = 0; \quad \frac{d}{dx^1}S_1^{13} - S_0^{33}/h = 0. \qquad (9)$$

In fig. 3,a the dependence is given of the value of contact region α on pressing forse $\bar{p} = P/E$. The curve 1 corresponds to the results obtained from the above relations (8)-(9); curve 2 - without referense of compressibility of the plate in lateral direction. In fig. 3,b the dependence of the upsetting stamp δ on $\bar{p}$ in the given above cases is given. The plate has follow characteristics: $E = 2 \cdot 10^5$ MPa; $\nu = 0.25$; $h/\ell = 0.1$.

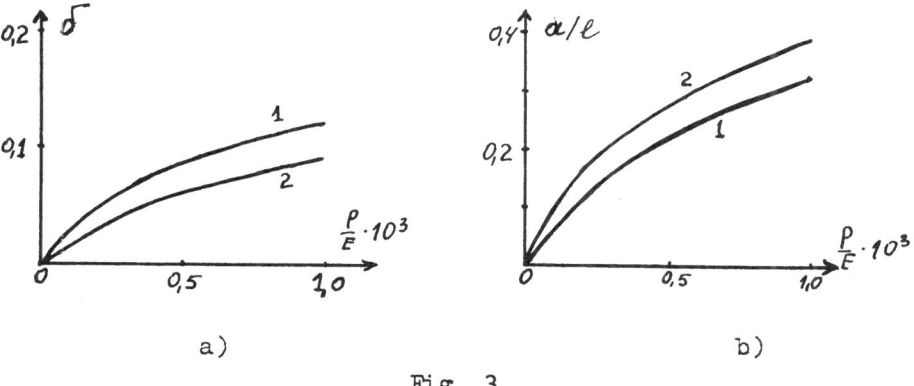

a)   b)

Fig. 3.

REFERENCES

1. Galimov K.Z. Theory of shells with taking into consideration of transverse shear. Izdatel'stvo Kazan University (in Russian), 1977.

2. Krys'ko V.A. Nonlinear static and dynamics of inhomogeneous shells. Izdatel'stvo Saratov University (in Russian), 1976.

3. Lazko V.A. Stress-strain state in anisotropic multilayer shells with zones of nonideal contact between layers: Part 2 - Generalised equations of orthotropic laminar shells subjected to discontinuous displacements at the interface. Mekhanika Kompozitnykh Materialov 18(1), Jan-Fab 1982, 77-84.

4. Pelekh B.L., Lazko V.A. Layered anisotropic plates and shells with stress concentration. Izdatel'stvo Naukova Dumka, Kiev (in Russian), 1982.

5. Pelekh B.L., Sukhorolski M.A. Contact problems of the theory of elastic anisotropic shells. Izdatel'stvo Naukova Dumka, Kiev (in Russian), 1980.

6. Vasilev V.V. Mechanics of structures of composite materials, Izdatel'stvo Mashinostroenie, Moscow (in Russian), 1988.

7. Vekua I.N. Shell theory: general methods of construction. Izdatel'stvo Nauka, Moscow (in Russian), 1982.

INFLUENCE OF THE WELD SIZE ON THE STRENGTH OF
RECTANGULAR HOLLOW SECTION JOINTS

J. Szlendak

Institute of Civil Engineering
Technical University of Bialystok
ul. Wiejska 45 E
15-351 Bialystok
Poland

ABSTRACT

Rectangular hollow section /RHS/ members could be very interesting structural elements for the frameworks, industrial buildings and the light loaded structures without siding. Design of such members is not the problem nowadays, however the accurate estimation of the safe load which could be carry be the joints between them is still in discussion. The behaviour of these connections could be described as the problem of the elasto-plastic local demage of the rectangular shell due to the local load applied to one or more of the shell walls.

So, the theoretical study are usually complicated and very often useless for the practical applications. Therefore the semiempirical or empirical formulae described the strength of such connections are very popular. Up to now, for the joints loaded by the static or predominant static load the presented design formulae do not take into account the size of the filled welds between the main and minor members. In this paper the problem of weld size in regard to the RHS joints strength is discussed. It is shown in which area of the connections geometry the size of welds could be taken into account and what are the diffrences compare with the standard formulae, if this effect is neglected.

GEOMETRICAL PARAMETERS OF JOINTS

The most common failure mode for T,Y and X joints is idealized by the yield line mechanism:
- for in plane moment load of branch /Fig. 1/
- for axial load of branch /Fig. 2/.

Generally, the main geometrical parameters of joints as:

$\beta = b_n/b_o$, $\eta = h_n/b_o$, $\lambda_o = b_o/t_o$

are internationally [1] calculated to take

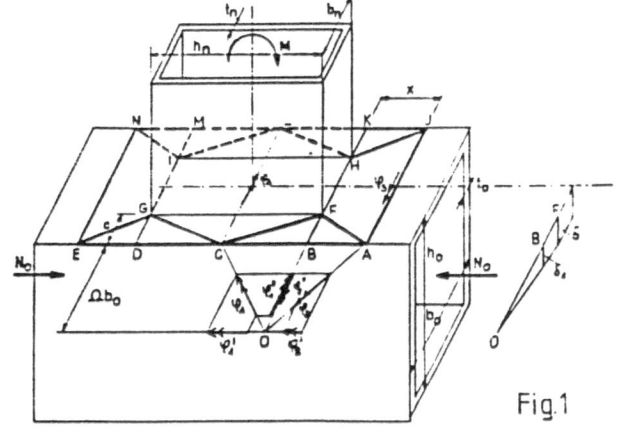

Fig.1

into account the noninal geometry (outside dimensions) of chord and branch menbers.

However, in the real connections the welds size could have the significant influence on the parameters β and η Specially, the change of parameter β (see Fig. 3) may be important. At least the following different definitions of parameter β could be studied:

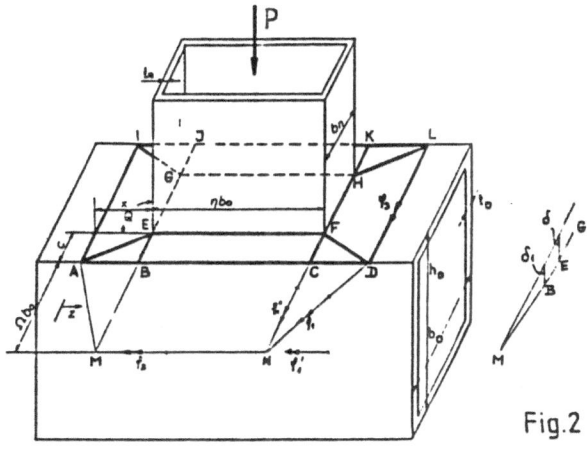

Fig.2

$$\beta^{i0} = b_n/b_0, \quad \beta^{i1} = /b_n + \sqrt{2}\ t_n//b_0 = \beta + \sqrt{2}\ k_t/\lambda_0,$$

$$\beta^{i2} = /b_n + 2\sqrt{2}\ t_n//b_0 = \beta + 2\sqrt{2}\ k_t/\lambda_0, \quad \text{where:}\ k_t = t_n/t_0.$$

DESIGN FORMULAE

Moment load
The base design formula is given as bellow [2]:

$$M = m_p h_n\ (4\ B_0\ +\ \eta B_0^2\ +\ \frac{2}{\eta}\ ) \tag{1}$$

where: $m_p = \sigma_{eo} t_0^2/4$ - plastic moment of resistance per unit length of hinge

$\eta = h_n/b_0$

B_0 - coefficient (see Fig. 4).

CROSS SECTION OF JOINT

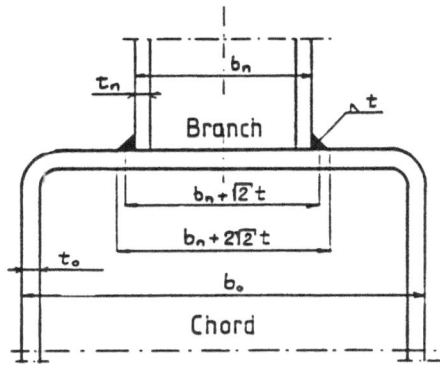

Fig. 3

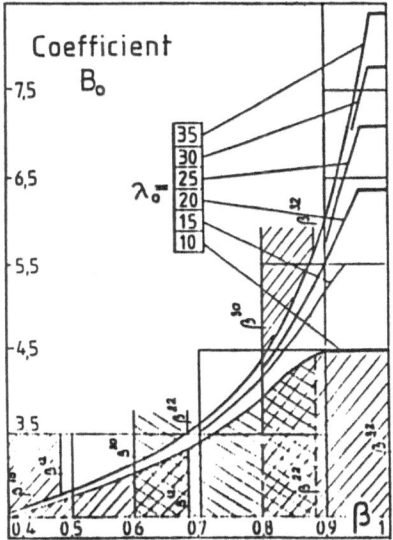

Fig.4

Axial load
The base design formula is given as bellow [3]:

$$P = 4\,m_p\,(\,\eta\,C + 2\sqrt{2}\,D\,)\tag{2}$$

where: C,D - coefficients (see Fig. 5 and 6).

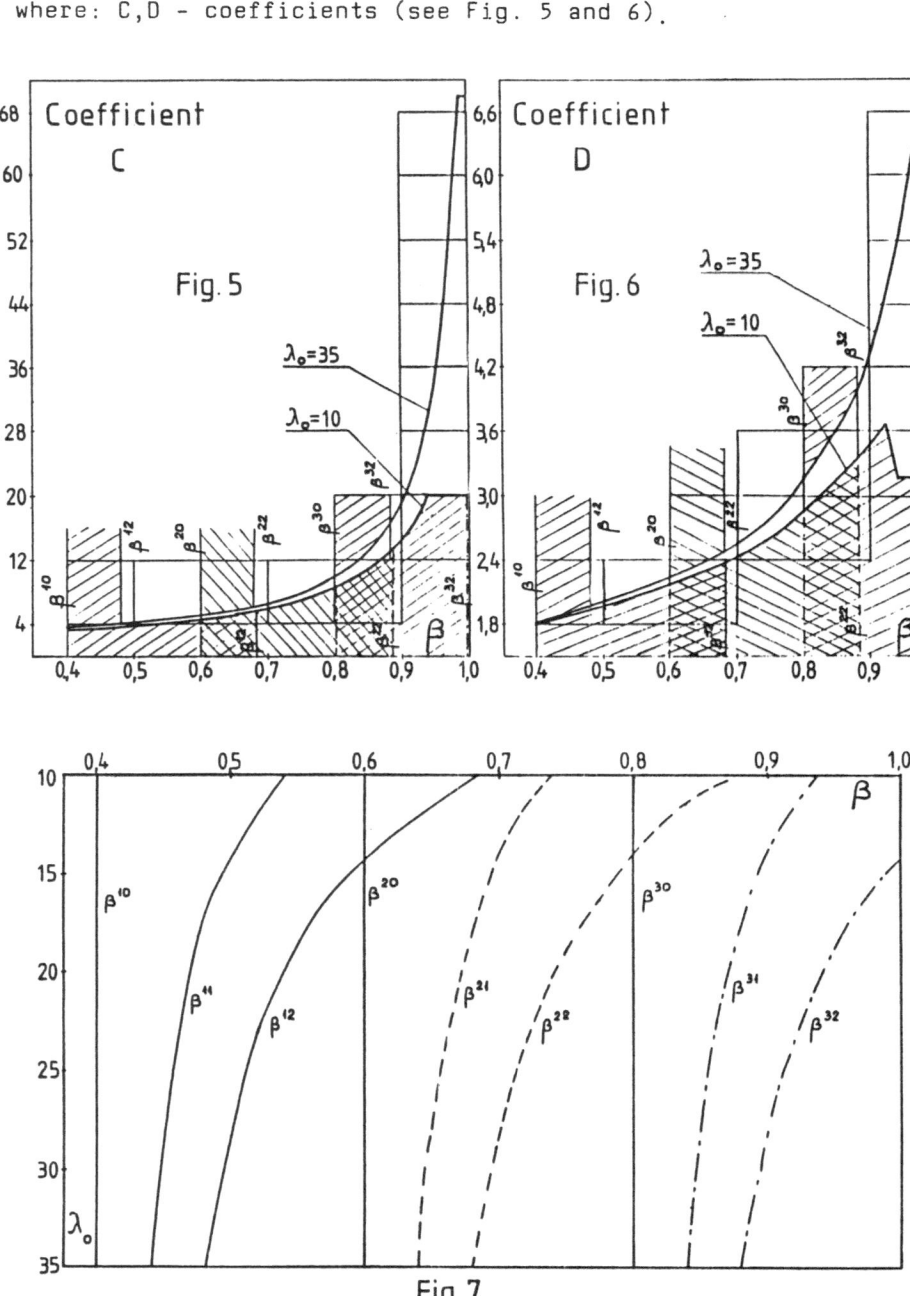

Coefficient C

Fig. 5

Coefficient D

Fig. 6

Fig. 7

INFLUENCE OF WELD SIZE

The influence of the weld size on the parameter β is shown in Fig. 7, where the base parameters $\beta^{10}_i = 0.4$, 0.6 and 0.8 are compared with the parameters β^{11} and β^{12}. In Table 2 the β^{11}/β^{10} and β^{12}/β^{10} ratio is presented

Table 1:

β / λ_o	$\beta^{10} = 0.4$		$\beta^{20} = 0.6$		$\beta^{30} = 0.8$	
	β^{11}/β^{10}	β^{12}/β^{10}	β^{21}/β^{20}	β^{22}/β^{20}	β^{31}/β^{30}	β^{32}/β^{30}
10	1.35	1.71	1.24	1.47	1.18	1.35
35	1.10	1.20	1.07	1.14	1.05	1.10

In Table 2 the dimensionless change of coefficient B_o, C and D and relative increase of joints strength with regard to $^{12}/^{10}$ ratio is shown /is assumed that $= /$.

Table 2:

β / λ_o	β^{12}/β^{10}					β^{22}/β^{20}					β^{32}/β^{30}				
10	1.34	1.70	1.32	1.37	1.64	1.47	2.72	1.53	1.79	2.33	1.14	2.43	1.10	1.36	2.08
35	1.09	1.21	1.09	1.06	1.16	1.11	1.24	1.11	1.16	1.21	1.28	1.65	1.09	1.48	1.37
	$\bar{B}_o$	$\bar{C}$	$\bar{D}$	$\bar{M}$	$\bar{N}$	$\bar{B}_o$	$\bar{C}$	$\bar{D}$	$\bar{M}$	$\bar{N}$	$\bar{B}_o$	$\bar{C}$	$\bar{D}$	$\bar{M}$	$\bar{N}$

CONCLUSIONS

/i/ Weld thickness could have the significant influence on the strength of RHS joints

/ii/ The reason of this influence is rather not the dimension of the weld itself but the significant change of parameter β and in the less degree η

/iii/ For the small slenderness of chord walls / $\lambda_o \cong 10$/ this influence seems to be the highest for $0.6 \le \beta^0 \le 0.8$ but for the high slenderness / $\lambda_o \cong 35$/ for $0.8 \le \beta \le 1.0$

REFERENCES

1. Wardenier J.: Hollow section joints. Delft University Press, 1982.
2. Szlendak J. and Bródka J.: Investigation into the static strength of welded T-moment unreinforced joints in rectangular hollow sections. IIW Doc. XV-538-83.
3. Szlendak J. and Brédka J.: Yield and buckling strength of T, Y and X joints in rectangular hollow section trusses. Proc. Inst. Civ. Engrs., Part 2, 1985, 79, Mar., 167-180.

VIBRATIONS OF COMPOSITE PLATE CONTAINING CIRCLE
DELAMINATION
ON BOUNDARY OF LAYERS

D.D. Zakharov and I.V. Simonov
Wave dynamics laboratory
Institute for Problems in Mechancs
Vernadskogo 101
117 526 Moscow
USSR

Two - dimensional equations for two-layered elastic pla-
te are derived by asymptotic method ($\varepsilon = H/L$ is a small parameter, L - is a characteristic length in longitudinal direction, characteristic time of process to $\sim \varepsilon^{-1}$). A coordinate
system allowing separate consideration of dynamic equation
for normal displacement and quasistatic one for tangential
displacements is given. The accuracy of equations is $O(\varepsilon^2)$.
All components of stress field, moments and forces are obtained. All the results are generalized for n-layered plate with
essential difference of elastic and geometric constants.

A problem for low-frequent vibrations of two-layered
plate with circle delamination on the interface is considered
under $H/R \ll 1$ condition. Effective amplitude of stress intencity factors is determined from the invariant integrale calculating.

After a closer parametrical analysis the following effects are discovered. When $\alpha = H_1/H_2$ is a small value the area
under the delamination behaves like a circle plate with fixed
edge. With α - increasing the first quasiresonance looses
its quality factor and when $\alpha > \alpha_*$ it bifurcates (Lower - and
Upper - quasiresonances respectively with intermediate anti-
resonance point). L - frequency becomes much lower and
 L - mode reminds synchronised motions of two included rigi-
de disks. After its amplitude diminuating U-quasiresonance
becomes antiphase and non-limited. High-order quasiresonances
are also investigated. It's turned out there is no such bifurcation phenomena in plane problem that we explain at the ex -
pense of considerable flux of energie. Less influent parame-
ters E_1/E_2 , ρ_1 / ρ_2 and then Poisson's ratio are also varied. In
the space of parameters a special surface of real spectrum is
found (so as $H_1/H_2 = 1$ in homogenious plate).

A comparaison of effective amplitudes of stress intencity factors in homogenious plate is effectuated for precious
and approximate statements.

Two-dimensional dynamic equations for two-layered plate

H_j - thickness $\quad\rho_j$ - mass density
E_j - Yang's modul $\quad\nu_j$ - Poisson's ratio
$H = (H_1 + H_2)/2, \; H_3 = 2H, \; \varepsilon = H L^{-1} \ll 1, \; t_0 \sim \varepsilon^{-1}$

Coordinate of interface:

$z_1 = (\varkappa H_2^2 - H_1^2)/2(\varkappa H_2 + H_1) \quad$ where $\quad \varkappa = E_2(1-\nu_1^2)/E_1(1-\nu_2^2)$

Normal load:

$\mathcal{D}_3 \Delta\Delta u_z + \rho_3 H_3 \partial^2 u_z/\partial t^2 = \sigma_z^{(2)} - \sigma_z^{(3)}, \quad \Delta = \Delta_{xy}$ - Laplace operator
$u_\beta = -z \partial u_z/\partial\beta \;, \quad \beta = x, y$
$u_z = O(\varepsilon^{-3}) \qquad u_\beta, \sigma_\beta, \sigma_{xy} = O(\varepsilon^{-2}) \qquad \sigma_{\beta z} = O(\varepsilon^{-1}) \qquad \sigma_z = O(1)$

Tangential load:

$\Delta \vec{u}_0 + p \cdot \operatorname{grad} \operatorname{div} \vec{u}_0 = 2q_+/E_1 \, (\vec\tau^{(2)} - \vec\tau^{(3)}), \quad \vec{u}_0 = (u_{ox}, u_{oy})$
$\mathcal{D}_3 \Delta\Delta u_z + \rho_3 H_3 \partial^2 u_z/\partial t^2 = H(p_1 \operatorname{div} \vec\tau^{(2)} - p_2 \operatorname{div} \vec\tau^{(3)})$
$u_\beta = u_{o\beta} - z \partial u_z/\partial\beta, \; p_j = p_0 q_0 + (-1)^{j+1} H_j/H, \; p = q_+/q_-$
$p_0 = \{H_1^2/(1-\nu_1^2) - H_2^2 E_2/E_1(1-\nu_2^2)\}/2H^2$
$q_0 = H\{H_1/(1-\nu_1^2) + H_2 E_2/E_1(1-\nu_2^2)\}^{-1}$
$q_\pm = \{H_1/(1 \pm \nu_1) + H_2 E_2/E_1(1 \pm \nu_2)\}^{-1}$
$u_z = O(\varepsilon^{-2}) \qquad u_\beta, \sigma_\beta, \sigma_{xy} = O(\varepsilon^{-1}) \qquad \sigma_{\beta z} = O(1) \qquad \sigma_z = O(\varepsilon)$

Effective parameters of plate:

$\mathcal{D}_3 = 2\{\mathcal{D}_1(2z_2 + z_1)/H_1 - \mathcal{D}_2(2z_3 + z_1)/H_2\}$
$\nu_3 = 2/\mathcal{D}_3\{(2 + 3z_1/H_1)\nu_1 \mathcal{D}_1 + (2 - 3z_1/H_2)\nu_2 \mathcal{D}_2\}$
$\mathcal{D}_j = E_j H_j^3/12(1-\nu_j^2)$

Vibrations of circle delamination on the interface

$$\tau \leqslant R, \quad z = z_1 : \quad \sigma_{zj} = (-1)^{j+1} q e^{i\omega t}, \quad q = const \quad \sigma_{z\beta j} = 0$$

$$\tau > R \qquad : \quad [\sigma_{z\beta}] = [\sigma_z] = 0, \quad [\vec{u}] = 0$$

$$\sigma_{\beta\gamma} = \frac{K}{\sqrt{2\pi\rho}} \Sigma_{\beta\gamma} + O(1), \quad \Sigma_{\beta\gamma} = O(1) \quad \rho \to +0 \text{ is a distance}$$
$$\text{from the front of delamination}$$

$$P_z[\vec{u}] = -\frac{1}{T_0} \int dt \int_S \operatorname{Re}\sigma_{z\beta} \cdot \operatorname{Re} u_\beta \, dS \geqslant 0, \quad T = 2\pi/\omega, \quad 0 \in S, diam.S \to +\infty$$

Asymptotic statement $(\delta = H_3/R \ll 1, \; w_j = u_{zj}/R)$:

$$(\Delta - k_j^2)(\Delta + k_j^2) w_j = a_j, \quad \Delta \equiv \Delta_\tau, \quad j = 1, 2, 3$$

$$\text{where } k_j = (\rho_j H_j \omega^2 R^4 / \mathcal{D}_j)^{1/4}, \quad a_j = (-1)^{j+1} q R^3/\mathcal{D}_j, \quad a_3 \equiv 0$$

in areas $\Omega_j = \{\tau < R, z_{j+1} \geqslant z \geqslant z_1\}, \quad \Omega_3 \; \{\tau > R, z_3 \leqslant z \leqslant z_2\}$

$$\tau > R : \quad P_z[w_3] \geqslant 0$$

$$\tau = R : \quad w_1 = w_2 = w_3 \qquad\qquad y_1 = y_2 = y_3$$
$$M_1 + M_2 = M_3 \qquad\qquad Q_1 + Q_2 = Q_3$$

Stress intencity factor and Γ-integrale:

$$\Gamma(t) = \frac{1}{2}\left(\frac{1-\nu_1^2}{E_1} + \frac{1-\nu_2^2}{E_2}\right) \cdot \frac{|K|^2}{ch^2 \pi \alpha_0}, \quad \alpha_0 = \frac{1}{2\pi} \ln \frac{(1+\nu_2)E_1 + (3-\nu_1)E_2}{(3-\nu_2)E_1 + (1+\nu_1)E_2}$$

$$2\Gamma(t) \approx \mathcal{D}_1 \xi_1^2 + \mathcal{D}_2 \xi_2^2 - \mathcal{D}_3 \xi_3^2, \quad \xi_j \text{ is a curvature in area } \Omega_j$$

Lineair system:

$$(1),(2) \Rightarrow w_j = A_j J_0(k_j \tau) + B_j I_0(k_j \tau), \quad w_3 = A_3 H_0^{(2)}(k_3 \tau) + B_3 K_0(k_3 \tau)$$

where $J_0, I_0, H_0^{(2)}, K_0$ are Bessel, Hankel and
$$\text{Macdonald functions}$$

$$(3) \Rightarrow \|m_{en}\|, \quad m = det\|m_{en}\| = m_1 + i m_2$$

Equation for resonance frequencies:

$$m = 0 \quad (m \approx m_2 \text{ under } \alpha = H_1/H_2 \ll 1)$$

$$m_2 = 0, \; dm_2/dk_1 = 0 \Rightarrow \alpha = \alpha_*(E_1/E_2, \rho_1/\rho_2, \nu_1, \nu_2, k_1)$$

$$(H_1/H_2)^2 = E_2 \rho_1(1-\nu_1^2)/E_1 \rho_2(1-\nu_2^2) \text{ is a surface equation for real spectrum}$$

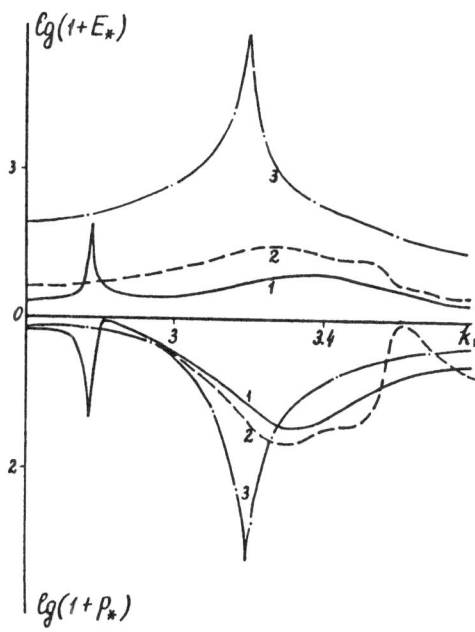

fig. 1 Energie and Power

fig. 2 Modes and Phases

fig. 3 Effective amplitude
of stress intensity factor

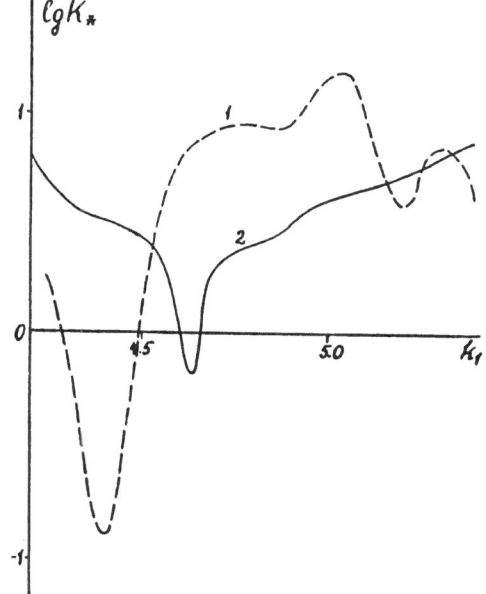

fig. 1 $\quad \alpha = 0.8, 0.4, 0.1$

fig. 2 $k_1 = k_1^{1L}, k_1^{1A}, k_1^{1U}, \alpha = 0.8$

$E_1/E_2 = \nu_1 = \nu_2 = 0.25, \rho_1/\rho_2 = 1$

$E_* = \{E(\Omega_1) + E(\Omega_2)\} \dfrac{E_1 \varepsilon^3}{q^2 R^3}$

$\rho_* = P_\tau \dfrac{\sqrt{E_1 \rho_1} \varepsilon^2}{(qR)^2}, w_* = w \dfrac{E_1 \varepsilon^3}{q}$

fig. 3 Homogeneous plate
1 Precious statement
2 Asymptotic
$H_3/R = 0.33, \nu = 0.3$
$H_1/H_2 = 0.2, K_* = K/q\sqrt{R}$

307